TRAUMA INFORMED PLACEMAKING

Trauma Informed Placemaking offers an introduction to understanding trauma and healing in place. It offers insights that researchers and practitioners can apply to their place-based practice, learning from a global cohort of place leaders and communities.

The book introduces the ethos and application of the trauma-informed approach to working in place, with references to historical and contemporary trauma, including trauma caused by placemakers. It introduces the potential of place and of place practitioners to heal. Offering 20 original frameworks, toolkits and learning exercises across 33 first- and third-person chapters, multi-disciplinary insights are presented throughout. These are organised into four sections that lead the reader to an awareness of how trauma and healing operate in place. The book offers a first gathering of the current praxis in the field – how we can move from trauma in place to healing in place – and concludes with calls to action for the trauma-informed placemaking approach to be adopted.

This book will be essential reading for students, researchers and practitioners interested in people and places, from artists and architects, policy makers and planners, community development workers and organisations, placemakers, to local and national governments. It will appeal to the disciplines of human geography, sociology, politics, cultural studies, psychology and to placemakers, planners and policymakers and those working in community development.

Cara Courage SFIPM, FRSA is a Culture and Place Consultant-Director and scholar, named in the top 10 of place-thinkers globally and a 'strategy angel' for the arts sector. She has authored and edited several books on the topic. Cara has a prodigious creative industries, cultural institutions and higher education professional career alongside a renowned academic one. She undertakes her own arts/place research and also for academy and sector-commissioned partners across the globe and UK, including for national arts networks, cultural institutions and local and central government.

Anita McKeown, FRSA, FIPM, is an award-winning artist, curator, educator and researcher working at the intersection of Inclusive Design, Creative Placemaking, Open Source Culture and Technology and STEAM education, across a range of inter- and transdisciplinary projects, processes and partnerships. Anita is the director of SMARTlab Skelligs, SMARTlab Academy and co-founder of Future Focus21c, Anita's work uses an adaptive change method reverse-engineered for 21st-century challenges and integrates design-thinking, circular economic principles, and the Sustainable Development Goals (SDGs) / Earth Charter to encourage a holistic growth mindset and develop strategies and tactics to build social, environmental and economic resilience and encourage systemic behavioural change.

TRAUMA INFORMED PLACEMAKING

Edited by Cara Courage and Anita McKeown

Routledge
Taylor & Francis Group

LONDON AND NEW YORK

Designed cover image: Artwork on a remnant of destruction from Supertyphoon Odette, Archie Oclos (2022), *Tulong Pangmatagalan Hindi Pansamantala*, which loosely translates as 'Long-Term and Not Temporary Support.' Image credit: Brian Jay De Lima Ambulo, 2022.

First published 2024
by Routledge
4 Park Square, Milton Park, Abingdon, Oxon OX14 4RN

and by Routledge
605 Third Avenue, New York, NY 10158

Routledge is an imprint of the Taylor & Francis Group, an informa business

British Library Cataloguing-in-Publication Data
A catalogue record for this book is available from the British Library

ISBN: 9781032443096 (hbk)
ISBN: 9781032443102 (pbk)
ISBN: 9781003371533 (ebk)

DOI: 10.4324/9781003371533

Typeset in Times New Roman
by Newgen Publishing UK

To join the trauma-informed placemaking conversation, please visit www.traumainformed.place

CONTENTS

ORIGINAL FRAMEWORKS, TOOLKITS AND LEARNING EXERCISES

Listing of the original contributions presented in this volume of frameworks, toolkits and learning exercises.

Introduction: A Pathway Through the Six Principles of a Trauma-Informed Approach and Placemaking
Cara Courage and Anita McKeown

Chapter 6: Slow Collaboration as a Sustainable Framework for Practices of Repair
Crawley Jackson, Krasniqi & Vojvoda
A model to develop a community of practice around placemaking in post-conflict contexts.

Chapter 7: Creative Placeproofing
Brian Jay De Lima Ambulo
An emergent pivotal approach in fostering community resilience and resistance, a union between creative placemaking and trauma-informed placemaking.

Chapter 8: Ethical Placemaking
Lisa Eckenweiler
An ideal and concept that can help support people concerned with migrant and refugee accommodation and health services.

Chapter 10: Focus Group Exercises, Questions and Sample of Questions from Oral History Guide
Carlton Turner, Mina Matlon, Erica Kohl-Arenas and Jean Greene
A series of prompts to aid an imaging of a/the future.

Chapter 15: Case Study Thematic Findings
Julie Goodman, Theresa Hyuna Hwang and Jason Schupbach
An effective framework of commonalities of the organisation-led regenerative
  trauma-informed practices.

Chapter 15: Advice for Practitioners
Julie Goodman, Theresa Hyuna Hwang and Jason Schupbach
How to center trauma-informed placemaking practice in organizational practice.

Chapter 16: Trauma-Informed Placemaking: Integrative Approach Framework
Joongsub Kim
A six-point social justice-informed approach to make placemaking more effective
  in addressing trauma in underserved communities:

Chapter 17: Thematic Framework for Nature Therapy Consideration
Lyubomira Peycheva
Across eight thematic topics of nature escape, freedom, sensory heaven, calming
  and relaxing nature, appreciation, change of perspective, memories, and life
  satisfaction.

Chapter 18: Thematic Framework for Placemaking and the Migrant Experience
John C. Arroyo and Iliana Lang Lundgren
Across four thematic topics of trauma caused by forced family separation, resulting
  from persecutive displacement, associated with cultural displacement and
  stemming from lack of visibility or misrepresentation.

Chapter 21: Valuing Creative Placemaking Toolkit
Cathy Smith, Josephine Vaughan, Justine Lloyd, and Michael Cohen
A range of value indicators of placemaking success across Social, Environmental
  and Economic impacts to evaluate creative placemaking initiatives.

Chapter 25: The Place-Healing Manifesto
Wolfe
A Manifesto to meet the challenge of place-healing and to create adaptation and
  transition pathways.

Chapter 26: MOTIF Framework
Theo Edmonds, Josh Miller and Hannah Drake
A Culture Futurist™ practice framework for Creative Placehealing, designed to
  guide practitioners in the process of transforming places and fostering cultural
  wellbeing

FIGURES

TABLES

CONTRIBUTORS

John C. Arroyo, PhD, AICP, University of California, San Diego, is an Assistant Professor in Engaging Diverse Communities and Director of the PNW Just Futures Institute for Racial and Climate Justice at the University of Oregon. At UO he is affiliated with the School of Planning, Public Policy and Management; Indigenous, Race, and Ethnic Studies department, Latinx Studies Minor, and the Historic Preservation program. Arroyo's research focuses on the political and cultural dimensions of immigrant-centered built environments in emerging gateways. He was an Andrew W. Mellon Foundation Fellow in Latino Studies at the School for Advanced Research in Santa Fe, New Mexico. He received a doctorate in Urban Planning, Policy, and Design from MIT. His civic service includes serving on the boards of UO's Center for Latino and Latin American Studies (CLLAS), the Public Humanities Network of the Consortium for Humanities Centers and Institutes (CHCI), and the School for Advanced Research (SAR).

Sharon Attipoe-Dorcoo, PhD, MPH, Principal of TERSHA LLC, is first and foremost grounded in her cultural identity as a Ghanaian-American and embraces her other intersectional facets of being a wife and mom in her work. She is a former board member of AcademyHealth, a member of the Education Council, author of a children's book, poet, and consultant. As a community scholar-activist, she found her path from engineering into public health and her work involves engaging national mobile clinic programs. The vision for her work is rooted in culturally responsive and equitable tools for co-designing research and evaluation initiatives with communities.

Sarah Barns, Vice Chancellor's Senior Research Fellow, RMIT School of Design & Social Context. Sarah Barns is a researcher and urban digital strategist in the area of city data strategy, smart cities and digital storytelling. Her work over the past decade has taken place at the intersection of urban transformation, placemaking and digital disruption. How technology innovation shapes our cities, our places, our infrastructures and our built and cultural fabric, has been a long-term preoccupation. From 2013 Sarah has held an Urban Studies Foundation Postdoctoral Research Fellowship at Western Sydney University where she developed an international research program addressing the rise of the data economy as a major game-changer for cities and policy makers. This work has informed a set of strategy and consultancy projects with organisations such as Data61/CSIRO, the NSW Department of Planning and Environment, the NYU Centre for Urban Science and Progress and the ABC to facilitate data discovery and governance strategies underpinning smart city policies and digital engagement opportunities. Sarah's work today builds on her early career in innovation policy and program development for organisations spanning the ABC, Arup, the Australia Council for the Arts, and the Creative Industries Innovation Centre.

Zohreh BayatRizi, Associate Professor, Department of Sociology, University of Alberta. Her main interests are history of sociology and the sociology of death and grief. She is the author of *Death Sentences: The Modern Ordering of Mortality*, as well as several recent articles that aim to offer a decolonized, transnational perspective on death and grief.

Katy Beinart is an Artist and Senior Lecturer in Architecture, University of Brighton. Her art practice, which emerged from a background in architecture and community work, includes installation, sculpture, film, performance and socially engaged projects. She has exhibited internationally and created temporary and permanent artworks on sites around the UK. Her current research project, Acts of Transfer, (with Dr Lizzie Lloyd) explores the spaces and subjectivities of socially engaged practice through making alternative documents and texts in collaboration with artists. Recent publications include: 'Don't Look Back: The Challenges of Public Art and Meanings of Authenticity' in *Heritage Contexts, Public Art Dialogue* (2002), 10(2) and 'Khlebosolny/Bread and Salt: a time-travelling journey to Eastern Europe (and back)', *Mobile Culture Studies* (2019), 4.

Katie Boone, PhD student at University of Minnesota, Qualitative Researcher and Community Engagement Coordinator, Department of Human Services, State of Minnesota. Katie is a social innovator who understands what it means to leverage potential, seeing the gaps between what is and what could be and building community through shared work to find and develop new ways forward. She believes in praxis, applying both theory and practice to the work she does. Her roots are in community building and designing collaborative and transformative systems change. Currently she is working on her PhD in Organizational Leadership and Policy Development Evaluation Studies at the University of Minnesota, working from her values and passion that are rooted in equity, story, place and community. She works as a Qualitative Researcher and Community Engagement Coordinator for the Minnesota Department of Human Services.

Tom Borrup, founder of Creative Community Builders. Tom is an international consultant, speaker, and lecturer addressing cultural planning, creative economy, and cultural district planning. As founder of Creative Community Builders, he consults with cities, foundations, and non-profits to develop synergy between arts, economic development, urban planning and design. His 2006 book, *The Creative Community Builders' Handbook*, remains a leading text in the field, and his *The Power of Culture in City Planning* was published in 2020 (Routledge). Tom serves as Director of Graduate Studies and Senior Lecturer for the University of Minnesota's Master of Professional Studies in Arts and Cultural Leadership. He is also a Visiting Professor at the University of International Business and Economics in Beijing, China and teaches for the University of Kentucky's PhD Program in Arts Administration and the Creative Placemaking Certificate Program at Purdue University.

Urmi Buragohain, Founder, PlaceMaking Foundation. Urmi is founder of PlaceMaking Foundation, a social enterprise that seeks to create dignified and healthy places, and alongside a number of voluntary and advocacy roles, is also co-founder of Neprise Consulting Private Limited, a service aggregator platform for multi-sector professional services. Urmi is an urbanist, a water champion and nowadays, an accidental entrepreneur. After having worked for over 20 years spanning Australia, South Asia and the Middle East with significant experience in public, private and not-for-profit sectors, Urmi chose to move back to India to pursue her vision of giving back and making a difference in her region of origin, North Eastern India. Urmi's key strength is her ability to work across disciplines and help people from diverse backgrounds find common ground.

Michael Cohen is the director of City People, an Australian organisation committed to strategising, planning and implementing arts and cultural programs that build stronger communities and better places. City People's projects include the Circular Quay redevelopment cultural framework, an arts strategy and arts and cultural accelerator for Western Sydney Parklands, and arts in health programs for diverse NSW Health infrastructure programs. Previously, Michael was Creative Producer, Sydney Harbour Foreshore Authority, Co-Artistic Director of Theatre Kantanka (1996–2006), Co-Director of Sydney's fringe festival, Live Bait (Bondi Pavilion 2004), Programme Director of Newcastle Live Sites (2004–8) and artist and convenor at Bundanon Trust, one of Australia's leading regional contemporary arts venues. Michael has a doctorate in Performance Studies (University of Sydney, 2002) and has published numerous articles about site-based and interpretive work, cultural representation and culture-led placemaking.

Amanda Crawley Jackson is Associate Dean for Knowledge Exchange at the London College of Communication, University of the Arts London. Previously Faculty Director of Knowledge Exchange and Impact (Arts and Humanities) at the University of Sheffield and Senior Lecturer in French Studies, she curated the 2020 exhibition, 'Invisible Wounds: Landscape and Memory in Photography' at the Graves Gallery in Sheffield. She has written extensively on the ways in which artists from France, Algeria and Morocco have engaged critically with post-traumatic landscapes, their representation and repair. In collaboration with Museums Sheffield, she co-edited *Invisible Wounds: Negotiation Post-Traumatic Landscapes* (2020) and was the Sheffield lead for Landscapes of Repair.

Cara Courage SFIPM, FRSA is a creative industries worker, arts and place consultant and scholar, named in the top 10 of place-thinkers globally and a 'strategy angel' for the arts sector; and author of *Arts in Place: The Arts, the Urban and Social Practice* (Routledge, 2017); co-editor, with Dr Anita McKeown, of *Creative Placemaking and Beyond* (Routledge, 2018); and editor of *The Routledge Handbook of Placemaking* (Routledge, 2021). Cara has a prodigious creative industries, cultural institutions and higher education professional career alongside a renowned academic one. Cara undertakes her own arts/place research and the same for academy and sector-commissioned partners in the USA, Australia, and across the UK, including for national arts networks, cultural institutions and local and central government. Cara speaks internationally on topics across culture, museums and heritage and place, placemaking and socially engaged art. Cara's first encounter with the power of place came from growing up as rural working class in 1980s UK, seeing first-hand the precarity of farming economies and the increasing polarisation of wealth and poverty in the extremes; and then, experiencing homophobic house attacks on the family home during the HIV/AIDS crisis in the 1980s.

Daria Dorosh, PhD, Artist. Daria Dorosh is an artist exploring the implications of art, fashion, and technology converging on the body. She is a co-founder and active member of A.I.R. Gallery, NY, a community of self-identified women artists founded in 1972 as an alternative to mainstream art institutions that excluded women. Fashion Lab in Process LLC, is her platform for testing new economies for artists, which have included presentations and exhibitions: 'Re-made in America', American Folk Art Museum, NY; 'Take Back Your Body' at VSMM2017; and 'Artists Policing Data', John Jay College of Criminal Justice, NY.

Gordon C. C. Douglas is an associate professor of urban and regional planning at San José State University, where he also directs the Institute for Metropolitan Studies. Gordon's research, teaching, and community work focus on questions of access, equity, and local cultural identity in urban planning and design. He is the author of *The Help-Yourself City: Legitimacy and Inequality in DIY Urbanism*, and his writing and photography have been published in *City and Community*, *Urban Studies*, *Journal of Urban Design* and other journals, magazines, newspapers, and blogs. His research has also appeared in the *Washington Post*, *Los Angeles Times*, *Fast Company*, *National Post* (Toronto), *La Presse* (Montreal), and multiple Bay Area news outlets. Born in London and raised in Northern California, Gordon received his doctorate in sociology from the University of Chicago and also holds degrees from the University of Southern California and the London School of Economics.

Hannah Drake, Chief Creative Officer, IDEAS xLab. Hannah Drake (she/her) is a blogger, activist, public speaker, poet, author of 11 books, and the Chief Creative Officer at IDEAS xLab. She writes commentary on politics, feminism, and race and her work has been featured in *Cosmopolitan Magazine*. Her work has been recognized by Colin Kaepernick, Ava DuVernay and *The New York Times*, and she was selected as a Muhammad Ali Daughter of Greatness.

Lisa A. Eckenwiler, PhD, is Professor and Chair of the Department of Philosophy, George Mason University, where she teaches bioethics and global health ethics. Current projects focus on the ethical closure of humanitarian projects and a new theoretical framework for humanitarian health ethics. She is at work on a book tentatively entitled *Placemaking for Health Justice*, with Routledge, in addition to an edited collection, *Forced Migration and Health Justice*, with Oxford University Press. Her previous books include *Long-term Care, Globalization and Justice: Migrant Care Workers, Aging, and Families* (2012) and *The Ethics of Bioethics: Mapping the Moral Landscape*, co edited with Felicia Cohn (2007). Lisa is Vice President of the International Association of Bioethics, Fellow of the Hastings Center, co-founder and current chair of the Resisting Borders/Migrant Health and Ethics Network (within the International Association of Bioethics); a founding member of the Independent Resource Group for Global Health Justice; and a member of the Humanitarian Health Ethics Network. She is a Collaborator with the Ethics and Health research axis at le Centre de Recherche en Éthique at the University of Montréal.

Theo Edmonds, Culture Futurist™ and Co-Founder, IDEAS xLab. Theo Edmonds is a Culture Futurist with a non-traditional skillset working globally across sectors to inspire, plan, design, and deliver sustainable cultural transformation projects, demonstrate their impact, and identify opportunities to evolve and grow. Theo is an experienced builder of industry-university collaborations, social entrepreneur, cultural analytics inventor, developer of next-generation initiatives in the creative economy, artist and poet; and is culture lead for several international work groups operating at the intersection of arts/media, creativity sciences, technology, neuroscience, and economics. Theo grew up in a nine-generation Appalachian family in south-eastern Kentucky.

Rezvaneh Erfani is a SSHRC Canada Vanier Scholar, Killam Scholar and a PhD candidate in the Department of Sociology, University of Alberta, Canada. She is also a visiting student in the Department of Political Science and International Relations at the Boğaziçi University, Turkey. Her research interests include environmental justice and activism, postcolonial theory, ecofeminism, gender and politics in the Middle East.

Moéna Fujimoto-Verdier, Erasmus University Rotterdam, The Netherlands. Moéna Verdier-Fujimoto is bi-cultural French/Japanese, growing up in France, and with Masters in Tourism, Culture and Society from Erasmus University of Rotterdam. With her interdisciplinary background and as a freelancer, she combines experiences in various domains (cultural, media and business). She is currently working for Yuzu Kyodai, a brand consultancy company based in Tokyo, with the ambition to bridge cultural gaps between Europe and Japan.

Pablo Gershanik is an actor, director, theatre pedagogue, and an Argentinian Mexican living in France. Graduating from the Ecole Internationale de Théâtre Jacques Lecoq, Pablo is Director of the Diploma in Performance and Interpretation with Masks and professor of the Bachelor of Performing Arts at the National University of San Martín, Argentina, a master's student of dramatherapy at the University of Paris, an actor in Cirque Eloize, Compagnie Philippe Genty, Compagnia Finzi Pasca, an artist, Cent Quatre Paris and the Cité Internationale des Arts, Paris, and creator of the Intimate Model Laboratory.

Rebecca Gordon is a researcher, lecturer, and writer in modern and contemporary art, specialising in the history and theory of contemporary art conservation. She has taught in the History of Art Departments at University College London and University of Glasgow, and has guest lectured at University of Amsterdam and New York University. Her current research focusses on collective care and emotional labour of social practice artists, the ethics of posthumanist care for museum futures, and the conservation of contemporary art as a counter-extinction activity. Rebecca works as a freelancer and is based in Glasgow, Scotland.

Jean Greene is the co-director of the Utica Institute Museum and co-project director for the IMLS Museum Grant for African American History & Culture Utica Roots: Building Community Through Shared Oral Histories. She served as director of the William H. Holtzclaw Library and Archives at Hinds Community College-Utica Campus from 2000 until her retirement in 2020. She was co-director of a National Endowment for the Humanities grant William Holtzclaw & the Black Man's Burden, a two-year research project to create a humanities course taught across disciplines to give students a broader understanding of William H. Holtzclaw and his pioneering efforts in African American education in rural Mississippi.

Julie Goodman is the Department Head for Arts Enterprise, joined Drexel University in the fall of 2011, and served as the graduate arts administration program's director from 2012–17. She was previously Executive Vice President for the Greater Philadelphia Cultural Alliance, where for twelve years she led advocacy, field research, community engagement, and grant-making efforts. She also works on consulting projects with arts organizations and artists and is a board member of the Association of Arts Administration Educators (AAAE), Citizens for the Arts in Pennsylvania, Tribe of Fools, and the Stockton Rush Bartol Foundation. She currently serves as Chair of the Mayor's Cultural Advisory Council for the Philadelphia Office of Arts, Culture, and the Creative Economy. Her research examines the strategies and impact of individual artists, cultural organizations, and communities. Recent publications include articles in Journal of Arts Management, Law and Society, American Journal of Arts Management, and Metropolitics

Juan David Guevara-Salamanca, PhD Candidate, Sociology, University of Alberta. Juan is a political scientist from Universidad del Rosario, Colombia, with an Interdisciplinary Master of Science from the University of Alberta, Canada. He has experience in social research, communitarian work and the private and public sectors. His research interests intersect socioeconomic development, participatory democracy, the social production of spaces, human geography and the socio-political Colombian armed conflict. He is currently working on his doctoral dissertation that aims to understand the socio-material culture of urban informality through the mobilities and rhythms of stairs, pathways, materials and informal transportation.

Sophie Hope is a senior lecturer and practice-based researcher at Birkbeck, University of London. Her work is often developed with others through the format of devised workshops exploring subjects such as art and politics in the year 1984, physical and emotional experiences of work and histories and politics of cultural democracy and socially engaged art. Current projects include: *Manual Labours* with Jenny Richards, *Meanwhile in an Abandoned Warehouse* with Owen Kelly, *1984 Dinners* and *Cards on the Table*.

Michal Huss is a Postdoctoral Fellow, at The Minerva Center for Human Rights, where her research deals with the analysis of public planning and construction struggles in Jaffa and Tel Aviv. Her ESRC-sponsored doctorate in the Architecture Department at the University of Cambridge focused on the urban resistance and activism of displaced populations within transcultural landscapes. She has set up numerous workshops on radical cartography in and around gentrifying spaces and migration detention centres.

Theresa Hyuna Hwang, Architect, Educator, Mother, Department of Beloved Places. She is a community-engaged architect, educator, and facilitator and founder of Department of Beloved Places. She has spent over 15 years supporting equitable cultural community development across the US with multiple organizations and campaigns. She received her Master of Architecture from Harvard Graduate School of Design (2007) and a Bachelor of Science in Civil Engineering from the Johns Hopkins University (2001). She is a licensed architect in California.

Gina Jiménez is a researcher from Colombia with a master's degree in development management and a bachelor's in social sciences. She works as a social professional in the Land Restitution Unit. Her professional and academic experience has focused on formulating, executing, monitoring and evaluating programs and projects that seek to benefit vulnerable communities and peasant-communities victims of the armed conflict in Colombia.

Molly Rose Kaufman is a Co-Founder and Co-Director of the University of Orange, a free school of restoration urbanism in Orange, NJ, that builds collective capacity for friends, neighbors, and partners to cultivate a just and equitable city. Molly co-founded the organization in 2008 alongside family members, local activists and community leaders. She serves as a member of the board of the Jersey City Public Library and as Chair of the Board of the World Fellowship Center. She teaches place-based organizing and popular education at the Eugene Lang College of the New School University.

Wilfred Keeble is a full-blooded Dakota Sioux native from the Crow Creek Agency in South Dakota, grounded in the traditions, culture and spirituality he was raised in. He has an extensive background in farming and ranching. He is well known throughout Indian Country as a strong advocate for all Native American Treaty Rights. He experienced the Bureau of Indian Affairs boarding school system, and later went on to earn his college degree in Business Management. He served as Tribal Chairperson for Crow Creek Agency. He has been part of the Dakota 38 + 2 Wokiksuye Horse Ride for 17 years, serving as Staff Carrier for the last eight years, providing leadership for the families and horses on the ride. He has a passion for educating youth and building the future horse nation, supporting rides throughout Indian Country to honor the ancestors, warriors and historically significant sites throughout the year.

Angela Kennedy, Director, Innovating for Wellbeing, and Consultant Clinical Psychologist. Angela was senior professional, leader and innovator in NHS mental health services for some decades after qualifying as a Clinical Psychologist in 1994. She has been at the forefront of trauma-informed and compassionate leadership developments and set up a UK wide network for change. She delivers programmes of wellbeing change-oriented work and embeds co-production and lived experience in all that she does. She has been on the board of Trauma charities and set up a regional staff support service during COVID-19, which commissioned lots of creative or collective healing opportunities. As part of that suicide prevention research, a documentary film *Life Well Lived* illustrated the stories of healthcare staff who survived attempts to end their lives. Angela is a feminist visual artist and delivers arts-for-wellbeing events.

Joongsub Kim, PhD, RA, AIA, AICP, is a professor at Lawrence Technological University, Michigan, and directs its architecture college's urban design program and Detroit Studio. A city planning and science in architecture studies MIT masters graduate and PhD from University of Michigan, he engages in applied research studies and projects in socially responsive design, urban design, community development, and environmental psychology. He served as a reviewer for grant competitions for the Social Sciences and Humanities Research Council of Canada and Kresge Foundation Innovative Projects. He has received numerous grants and other awards from National Endowment for the Arts, National Science Foundation, American Institute of Architects, Boston Society of Architects, National Council of Architectural Registration Boards, Association of Collegiate Schools of Architecture, Kellogg Foundation, and Graham Foundation. His work has been published in journals including *Journal of Urban Design, Urban Design International, The PLAN Journal, Places Journal, Environment & Behavior, Local Development and Society, International Journal of Community Well-Being,* and *Architectural Record*. His book *What Do Design Reviewers Really Do?* was published by Springer Press.

Erica Kohl-Arenas is an Associate Professor in American Studies at the University of California, Davis and the Faculty Director of Imagining America: Artists and Scholars in Public Life. She is a scholar of social movements, freedom struggles, and the politics of institutionalization, professionalization, and private philanthropy. Kohl-Arenas' research has focused primarily on the radical imaginations and deferred dreams of social movements that become entangled with the politics of institutionalization and funding. This work is captured in her book *The Self-Help Myth: How Philanthropy Fails to Alleviate Poverty* (2016) and in a diversity of publications including *Antipode: A Radical Journal of Geography, Social Movement Studies, Journal of Poverty, Geography Compass, Public: A Journal of Imagining America,* chapters in books with international scholars of philanthropy and poverty studies, and public scholarship platforms including HistPhil and Transformation (at Open Democracy).

Korab Krasniqi, forumZFD. Korab Krasniqi is a researcher, civil society, and memory activist, focused on dealing with the past, transitional justice, and memorialization. He is completing an MA in Political Sciences at the University of Prishtina, with a bachelor's degree in psychology. Korab has attended the fellowship program at Columbia University, NYC, on Historical Dialogue and Accountability and was part of numerous dealing with the past educational programs, locally and internationally. For the past nine years, Korab has been working with forumZFD, an international organization working on Peace Education and Dealing with the Past. Korab explores photography and art as a medium to foster a broader conversation on memory and the difficult past.

Teri Kwant, DreamLab founder. Teri Kwant, experience design consultant, designer, and public artist, founded and leads DreamLab, a human-centered experience design collaborative, with clients from healthcare to corporate to community-based experiences. She has designed public engagement at the scale of cities to inform urban design and has created placemaking and public art projects at the intimate scale of human interaction. Teri has led award-winning design teams for the Sandy Hook memorial, a children's hospice, and an eco-retreat and wellbeing center for BIPOC and LGBTQ communities. Teri has been a professional lecturer in the design of experience, environments and communications at the University of Minnesota School of Design, and currently teaches Human Centered Design for Well-Being with a trauma-informed, social justice lens as an architectural studio.

Sally Labern, Artist Director, the drawing shed [sic]. Sally Labern is a multi-disciplinary artist whose collaborative practice includes the work she directs and facilitates with the drawing shed, a social practice artist-led organisation hosted by communities of two linked social housing estates in east London, UK. Sally's practice led PhD (UEL 2021) looked at the radical imagination and foregrounding of co-authored practices through the lens of works made with others in Istanbul, Republic of North Macedonia, Charlottesville, US, and across the UK.

Friederike Landau-Donnelly, Assistant Professor for Cultural Geography, Radboud Universiteit, Nijmegen, The Netherlands. She is a political theorist, urban sociologist and cultural geographer interested in intersections between politics and space. In her dissertation *Agonistic Articulations in the 'Creative' City – On New Actors and Activism in Berlin's Urban Cultural Politics* (Routledge, 2019), Friederike conceptualized different modalities of political organization and representation amongst Berlin-based independent artists and cultural workers. In addition, she co-edited *[Un]Grounding – Post-Foundational Geographies* (2021), which discusses ontologies of space through a lens of conflict and contingency.

Iliana Lang Lundgren, MNM, Connected Lane County. Iliana Lang Lundgren is a graduate student research assistant at the University of Oregon's Just Futures Institute. She holds a master's degree in Non-profit Management from the University of Oregon and a Bachelor of Science from Georgetown University in Environmental Biology. She is committed to increasing the impact of non-profit organizations in the fields of environmental justice, reproductive health, and gender equality.

Brian Jay de Lima Ambulo is currently a PhD student in Culture Studies, The Lisbon Consortium, Catholic University of Portugal. He is an award-winning researcher and operations officer who worked for the Philippine Department of Trade and Industry, managing portfolios on start-ups and the creative and cultural industries, and drafting cultural policy recommendations. He is also an Artist Fellow of Singapore International Foundation's Arts for Good Fellowship. He received his BA in History and Political Science from the University of the Philippines (2012); and his International Master's in Dance Knowledge, Practice, and Heritage from Norway, France, Hungary, and the UK (2016) through the Erasmus+ programme. His PhD dissertation, *The Future is Culture: Imaging, Managing, and Imagining Creative Placemaking in Post-Disaster Recovery*, continuously navigates the intersections of creativities, cultures, grassroots, governance, development, and policymaking.

Justine Lloyd is a cultural sociologist in the Discipline of Sociology at Macquarie University. Her research focuses on the relationship between spatial and social change, particularly investigated through cultural histories of media and urban space. She has also published in the area of transnationalism and border theory.

Ryan Lugalia-Hollon serves as the CEO for UP Partnership, an education and youth development backbone that seeks to ensure all young people in San Antonio are ready for their future. He has previously served as an executive director for Excel Beyond the Bell San Antonio and the YMCA of Metropolitan Chicago's Youth Safe and Violence Prevention department. Ryan's book, *The War on Neighborhoods*, was published by Beacon Press and tracks the devastating impact of mass incarceration on one Chicago community area. He is an active leader in the national StriveTogether network and serves as the Board Chair for the Children's Funding Project.

Anna Marazuela Kim, London Metropolitan University. Anna Marazuela Kim is an international consultant for culture with over a decade of experience in ground-breaking institutes of advanced of study and transdisciplinary research groups in the UK, US and Europe, leading projects, publications, and critical debate at the intersection of culture, conflict, urban futures and human thriving. Since 2013, she has been an Associate Fellow, Thriving Cities Project and Lab, University of Virginia and is currently member of a new Research Centre in Creative Arts, Cultures and Engagement at London Metropolitan University. Dr Kim brings academic expertise to broader publics and practice and writes for a wide variety of platforms, from NATO to Frieze and is often tapped as a keynote to speak on topics of urgent contemporary concern, from the public role of museums in the wake of humanitarian crisis, to iconoclasm and the culture wars, and is currently speaking on culture as a human right.

Annaclaudia Martini is Assistant Professor in Human Geography at the University of Bologna, Italy. She obtained her PhD in human geography from the University of Groningen, The Netherlands. Her research focuses on the coastal communities in Tohoku, Japan, hit by the 2011 earthquake, tsunami, and nuclear meltdown at the Fukushima nuclear power plant. She developed a keen interest and commitment in methods and frameworks for social participation, academic outreach and engagement outside academia.

Ayako Maruyama, M.C.P., Assistant Professor, Rhode Island School of Design, University of Orange Cities Research Group, and Design Studio for Social Intervention. Ayako Maruyama is a Filipina-Japanese designer, educator and illustrator, and Professor in the Industrial Design department at Rhode Island School of Design. She has co-created numerous public engagement strategies often around transportation, housing and community development and co-authored and co-illustrated, *Ideas Arrangements Effects: Systems Design and Social Justice.* Ayako has taught Sense of Place at Boston University's City Planning and Urban Affairs program since 2013, and the Experience of Public Engagement studio at RISD, and with the community at University of Orange as an Urbanist in Residence and in the Collective Recovery Team.

Mina Matlon is an arts organizer, researcher, attorney, artist, and cultural equity advocate. Her research and organizing interests lie in the intersecting areas between arts and community development, with her practice particularly focused on community memory, identity, and imagination, Black and Indigenous self-determination, and the protection and leveraging of traditional and Indigenous knowledge and cultural assets. This work weaves together more than twenty years in the cultural sector and has included positions within higher education, corporate law, domestic and international legal aid and policy organizations, and small to large arts, media, and cultural institutions and collectives. Matlon holds a J.D. from Harvard Law School and an MA in Arts Administration and Policy from the School of the Art Institute of Chicago. She is a Robert Wood Johnson Foundation Interdisciplinary Research Leaders Fellow.

Lynne McCabe, LMSW, CCTP, MFA, Certified Clinical Trauma Professional, Social Worker, Director, She Works Flexible, Houston, TX. Lynne McCabe is a multi-disciplinary socially engaged artist engaged in the vocabulary of relational aesthetics, feminist and durational performance practices, philosophy, and pedagogical theory. A recent recipient of The Hogg Foundation for Mental Health award, and research grant from The Texas Higher Education Coordinating Board's Minority Health Research and Education Program, McCabe has held a residency at The Cynthia Woods Mitchell Center for the Arts and was awarded the Artadia Houston award in 2008. McCabe has exhibited nationally and internationally, most notably as part of the 53rd Venice Biennale. Her writings have been included in several collections, including SFMOMA's Open Space, 2010 and 'Becoming Love, More Love: Art, Politics, and Sharing since the 1990s', Ackland Art Museum, North Carolina, 2013.

Anita McKeown, FRSA, FIPM, is an award-winning artist, curator, educator and researcher working at the intersection of inclusive design, creative placemaking, open-source culture and technology and STEAM education, across a range of inter- and transdisciplinary projects, processes and partnerships. Anita is the director of SMARTlab Skelligs, a satellite of the global research network SMARTlab Research Institute, UCD, the SMARTlab Academy and co-founder of Future Focus 21c, a research-driven SME working in STEAM Education and co-lead, Muinín Catalyst Sustainable STEAM, a Science Foundation Ireland / Dept. of Education two-year funded education project (2022–24) in the Munster Region. Anita's work uses an adaptive change method reverse-engineered for 21st-century challenges and integrates design-thinking, circular economic principles, and the SDGs / Earth Charter to encourage a holistic growth mindset and develop strategies and tactics to build social, environmental and economic resilience and encourage systemic behavioural change. Anita is co-editor, with Dr Cara Courage, of *Creative Placemaking and Beyond* (Routledge, 2018) and was one of seven sub-editors with Cara on *The Routledge Handbook of Placemaking* (Routledge, 2012). Anita grew up in Northern Ireland in the 1970s developing an acute awareness of colonialism and place construction and exclusive practices. This has driven Anita's placemaking work beyond the realms of spatial planning into developing inclusive resilient processes that see placemaking as an existential process of being, belonging and becoming.

Jacque Micieli-Voutsinas, PhD, Assistant Professor & Co-Director, Graduate Program of Museum Studies, Affiliated Faculty, Department of Geography, University of Florida, Gainesville. Jacque is a critical museum and heritage studies scholar with research and teaching expertise on 9/11 memory and landscapes of terrorism, broadly defined. Her research program explores the evocative power of places of difficult heritage to cultivate public emotion (such as fear, empathy, and hope) and generate a collective sense of community in the wake of traumatizing events. She is particularly interested in trauma-informed museum praxis and the pedagogical power of heritage landscapes to advance or impede social change. Drawing on anti-racist, queer, and feminist theories of intersectionality, affect, and emotion, her work on heritage landscapes critically interrogates dominant narratives of cultural memory and questions of historical justice.

Josh Miller, MBA, (he/him) is the co-founder + CEO of IDEAS xLab, and owner of Josh Miller Ventures. He is a queer artist, explorer, international public speaker and entrepreneur whose work has been featured by *The New York Times*, and who was selected as a Nonprofit Visionary Leader in 2022 by Louisville Business First and for Business Equality Magazine's Forty LGBTQ+ Leaders under 40 in 2020.

Ellen Pearlman is a new media artist, curator, critic and educator. A Fulbright Scholar and Fulbright Specialist in Art, New Media and Technology she is a Research Fellow at MIT. She is also a Senior Research Assistant Professor at RISEBA University in Riga, Latvia. She created Noor, an interactive immersive brainwave opera in a 360-degree theater and AIBO, an emotionally intelligent artificial intelligent brainwave opera. She is also Director and Founder of ThoughtWorks Arts a global technology research lab.

Lyubomira Peycheva, graduated with a MSc degree in Environmental Science from Trinity College, Dublin and a BA in Geography and Chinese (Mandarin) language from Dublin City University. Her interests range from environmental, social, and corporate governance, sustainability strategy, and consultancy to urban planning and the implementation of a holistic way of life for nature and society, where nature is present in everyday life. Her interest shows through her master's thesis research, 'Abandoned landscapes as places of potential for Nature Therapy: Glendalough, Ireland', where she explores the inherent connection between people and nature and the effects of this connection on mental health.

Elena Quintana, PhD, is a Clinical and Community Psychologist who has dedicated her professional life to supporting healing where it is critically needed. Elena has worked nationally and internationally to inventively contribute to the fields of violence prevention, justice reform, and trauma-responsivity. She focusses on creating healing spaces in systems and communities and dedicates herself to expanding restorative justice alternatives to detention, promoting healing and education for incarcerated people, and promoting healing opportunities between people who have been harmed and those who have harmed them. Elena Quintana is the Executive Director of the Institute on Public Safety and Social Justice at Adler University.

Aisling Rusk, architect and Director of Belfast-based Studio Idir. Aisling's research explores the practices of locating 'in-between': pushing the boundaries of architectural practice to explore other ways of knowing and practicing in the leftover space of the margins. She tutors part time at Queen's University Belfast and is the founding chair of RSUA's Women in Architecture group, supporting and promoting women architects in Northern Ireland. In 2020, Aisling was named one of the Architect's Journal's '40 Architects under 40', a celebration of the UK's most exciting emerging architectural talent.

Jacek Ludwig Scarso is Reader in Art and Performance at the School of Art, Architecture and Design, London Metropolitan University. He is a University Teaching Fellow and Course Leader on the Public Art and Performance MA. He is Deputy Director of CREATURE, The Centre for Creative Arts, Culture and Engagement. Jacek is Trustee of The Line, London's first dedicated Public Art Walk, and Senior Advisor in Public Space for City Space Architecture. He is also Curator at Fondazione Marta Czok in Rome and Venice.

Jason Schupbach is the Dean of the Westphal College of Media Arts and Design at Drexel University. He was formerly the Director of the Design School at Arizona State University. In this position, he started the ambitious ReDesign. School project to reinvent design education for the 21st century and is a key advisor to ASU on diverse projects such as the Center for Creativity and Place, Roden Crater, the Creative Futures Lab, and ASU's Los Angeles downtown home. Previous to this position he was Director of Design and Creative Placemaking Programs for the National Endowment for the Arts, where he oversaw all design and creative placemaking grant-making and partnerships, including Our Town and Design Art Works grants, the Mayor's Institute on City Design, the Citizens' Institute on Rural Design, and the NEA's Federal agency collaborations. He has written extensively on the role of arts and design in making better communities, and his writing has been featured as a Best Idea of the Day by the Aspen Institute.

Rita Sinorita Fierro, CEO, Fierro, Consulting LLC and President, Home for Good Coalition. Dr Fierro has spent almost three decades studying systemic racism and draws upon her experiences as a trauma survivor, activist, artist and researcher to transform herself and our world. She founded Fierro Consulting in 2008, a firm that supports organizations to improve their external impact by having a more cohesive, creative, and equitable internal culture; and the Home for Good Coalition; and the Collective Power Podcast. Her PhD in African American studies, and 20+ years of listening to parents who lost custody of their children in Philadelphia is rendered in her unpublished manuscript, *In the Lion's Mouth: How the USA System Kidnaps Children.* Her latest book, *Digging up the Seeds of white Supremacy*, exposes the historical racist mechanisms embedded in our systems and shines a light on how we can transform the culture of fear at its foundation.

Cathy Smith is an Australian architect, interior designer, and Senior Lecturer in Interior Architecture at UNSW where she has also been the inaugural Turnbull Foundation Women in Built Environment scholar. She was a Richard Rogers Fellow (Harvard University GSD, Fall 2018) and Visiting Professor at Carleton University (Winter 2019, 2020). Both her design practice and interdisciplinary scholarly research concentrates on issues of social equity, property tenure, placemaking and urban renewal, focusing particularly on the DIY procurement methodology of meanwhile interiors and architecture, and has been published in a mixture of industry and scholarly journals.

Mindy Thompson Fullilove, MD, is a Professor of Urban Space and Health at The New School in New York City. She has studied the psychology of place for many years and written the highly regarded urban restoration trilogy, *Root Shock*, *Urban Alchemy* and *Main Street*.

Karen E. Till, Department of Geography, Maynooth University. Karen is an interdisciplinary feminist researcher, curator, creative writer and activist; Professor of Cultural Geography; Council Member of the Royal Irish Academy; and Co-convener of the Feminist Counter-Topographies reading group. She regularly partners with community groups, artists and activists to advance practices of place-based memory-work and an ethics of care in cities wounded by the legacies of state-perpetrated violence, including in Berlin, Cape Town, Dublin and Minneapolis. Most recently, Karen was awarded an Irish Research Council New Foundations grant, with Pavee Point and TravAct. Karen is author of The New Berlin: Memory, Politics, Place, numerous edited volumes, scholarly articles and creative essays. She is author of the forthcoming book, *Wounded Cities*, and co-editor of *Earth Writings*, Cork University Press. Karen is also Director of the Space&Place Research collaborative and co-convener of the international network Mapping Spectral Traces.

Samira Torabi is a PhD student in the Anthropology Department at University of Alberta. In her MA research she has examined the intersection of space, place, and gender. For her PhD project, she is looking at how certain sexual expressions and everyday life politics in the Middle East are challenging the dichotomies of modern discourse.

Carlton Turner is an artist, agriculturalist, researcher, and co-founder of the Mississippi Center for Cultural Production (Sipp Culture). He currently serves on the board of First People's Fund, Imagining America, Project South and the National Black Food and Justice Alliance. Carlton is a member of the We Shall Overcome Fund Advisory Committee at the Highlander Center for Research and Education and is the former Executive Director of Alternate ROOTS and is a founding partner of the Intercultural Leadership Institute.

Josephine Vaughan is a lecturer in Construction Management at the University of Newcastle, Australia. With a background that includes architecture and design, Josephine has been involved in community-based urban redesign and place activation projects over the last decade, Josephine is a member of an NSW Government collaborative project to develop a toolkit for determining the value of creative placemaking. Continuing her interest in the relationship of the natural environment with constructed landscapes and buildings, Josephine researches, lectures, and supervises students in the areas of placemaking, environment, and community.

Laura Van Vleet, MS, University of Orange Cities Research Group. Laura Van Vleet is a recently retired Middle School Physical and Earth Science teacher. Her focus in recent years, following a sabbatical study of local geology, has been to teach principles of geology using a local perspective and place-based model. Prior to teaching in a public-school setting, she worked for Cooperative Extension, bringing educational and occupational opportunities to young people, living in at-risk communities. She has her bachelor's degree from Cornell University in Ecology Systematics and a Master of Science from Elmira College in 7–12 Biology Education. She is now an organic farmer in Lodi, New York and operates a farm stand with her husband and family. She has developed an interest in Korean dramas and was part of the University of Orange 2021 TV club.

Liam Van Vleet, Blue MBA Candidate, University of Rhode Island and University of Orange Cities Research Group. Liam Van Vleet is a designer and builder based in Providence, Rhode Island working on objects, spaces, and experiences. Alongside work that has spanned product development, user interface and exhibit design, he has also been involved in organizing Feast Mass Boston, VentilatorProject.org, and Design by Rhode Island's design resiliency challenge.

Alexander Vojvoda, forumZFD. Alexander Vojvoda is community media activist and sociologist. He holds a MSc in Sociology and a MA in Political Communications and worked in the area of community media, community building and conflict-sensitive journalism in Austria, Germany, Cameroon and Kosovo. Currently, he is working with forumZFD Kosovo and is project lead of the Landscapes of Repair project (www.landscapesofrepair.org), focusing on public space, violence and trauma.

Charles R. (Chuck) Wolfe is founder and principal advisor of Seeing Better Cities Group, has practiced at several law firms, and has served as a long-time affiliate associate professor in the College of the Built Environments at the University of Washington in Seattle. He has been a frequent radio and podcast guest, and written regularly for many publications, including *The Atlantic, The Atlantic Cities/CityLab, Governing, CityMetric, Planetizen, The Huffington Post, Grist,* and *Crosscut.* He blogs at www.sustainingplace.com and has authored *Seeing the Better City* (2017) (finalist for a 2018 UK National Urban Design Award) and *Urbanism Without Effort* (rev. ed. 2019), both from Island Press.

Marwa N. Zohdy Hassan is a researcher, archivist, and storyteller, and BA English Literature, Linguistics, and Translation from Ain Shams University, Cairo.

Dedicated to trauma theory study in an attempt to capture an inclusive, therapeutic understanding of life and the human experience, Marwa's research interests include sci-fi theory, performance theory, memory studies, and anthropocentrism and posthumanism in a post-pandemic world. Previously, Marwa worked at SARD: Shubra's Archive for Research and Development, Cairo, mapping residents' oral narratives.

PRESENTATION OF ABSTRACTS BY CURATED SECTION

Foreword: What it Means to be Trauma-informed

Angela Kennedy

This Foreword moves from a consideration of Western Modernist ideals that have turned us inward looking, fuelling an illusion of an independent sense of self divorced from context, to calling attention to the need for such a contextual understanding and practice attuned to trauma and healing, in place. Traumatic reactions play out in our biology and emotions but also in our beliefs about the world, the way we relate to each other, how integrated we experience our sense of self, somatic disturbances in our body, our ability to function and the continuity of our memory. As such, traumatisation can pass from one generation to the next, its origins getting lost in time but transmitted via the way others are able to relate to us and through the biases we create in our communities. It asks us, how would we see ourselves and mental wellbeing differently?, and importantly, would that lead us to more interesting ways forward as a society? What role does 'place' have in the system that creates our felt sense of wellness and connection to others and purpose?

Section 1 – Understanding and Developing Our Trauma-in-Place Sensitivity

1 – Towards Trauma-Informed Placemaking: Mutual Aid, Collective Resistance and an Ethics of Care

Lynne McCabe

This chapter presents a first-person account of resiliency in the Gulf Coast region, presenting the concept of Mutual Aid as a voluntary, reciprocal, sharing of resources, skills, and ideas, and as something well understood within the Gulf Coast region. While many look at the Gulf Coast, and other regions similarly impacted by climate change, as places in need of rescue, the chapter presents the community practices developed within the Gulf Coast's impacted communities as a resource, as the effects of climate disasters begins to spread globally. This chapter shares examples of disaster resistance and recovery as both a resource for future disasters and in opposition to the top-down governmental response rooted in heteropatriarchal white supremacy.

2 – (em)Placing Trauma: The Wounds Among Us

Jacque Micieli-Voutsinas

This chapters revisits a growing spatial turn within trauma studies, as well as the introduction of trauma studies to the field of human geography. Human geographers have been foundational in theorizing the spatial processes of traumatic events as they impact afflicted communities and places. But trauma is by no means place-bound, traveling with and through the bodies of the traumatized across spaces and time, trauma is both in-situ and ex-situ, as human geography has aptly documented. Offering a re-evaluation of the interdisciplinary spatial fields of trauma studies, this essay also considers the proliferating discourse of trauma in contemporary daily lives amidst the compounding nature of novel COVID-19, as well as the global movement for Black Lives.

3 – The DMZ and the Laundry: Lessons From K-drama for Trauma-Informed Placemaking

Ayako Maruyama, Laura Van Vleet, Molly Rose Kaufman, Liam Van Vleet and Mindy Thompson Fullilove

The South Korean drama, *Crash Landing on You*, is an international sensation, depicting the difficulties of a North Korean army officer and a South Korean entrepreneur who accidentally meet in the DMZ, fall in love and struggle with the barrier of partition which threatens to separate them forever. The drama threads together several placemaking stories, which contribute to our understanding of the manner in which trauma can inform our work. The first of these storylines is that of the wealthy South Korean who is suddenly in a North Korean military village. To her eyes, the village is profoundly deprived, yet as she joins with the women, she experiences a healing community. The second storyline involves Switzerland, which is accessible from both Koreas, and offers the lovers a neutral haven and space of encounter. The third storyline is that of the 'forces of evil' on both sides of the border, who are able to use criminal channels to attempt to kill our heroes. The series of contrasts helps us to get past the trappings of wealth to see what a healing place really is: accepting, inclusive, and truly international.

4 – Flying, Fleeing, and Hanging on: Trauma Baggage at Airports

Rezvaneh Erfani, Zohreh BayatRizi and Samira Torabi

Airports are known as 'emotion-laden environments' (Malvini Redden, 2013, p. 122). They are places of joy and pain, reunion and separation, gateways to dream vacations and departing points toward disaster and doom. For immigrants and refugees, the story of airports is slightly different from others. This chapter opens a conversation about how diasporic populations experience airports as traumatic places both in normal times – as a place of separation and uprooting – and in extraordinary times – in the aftermath of air disasters and during war and conflicts. We specifically focus on Tehran Imam Khomeini International Airport (IKA) and the Ukrainian Flight 752 tragedy in 2020, while also making brief references to the Kabul's Hamed Karzai International Airport (KBL) in the aftermath of the American military withdrawal in 2021, to discuss airports as disaster areas where collective traumatic stories and experiences are repeatedly (re)shaped.

5 – Trauma: The Counterproductive Outcome of the Land Restitution Program

Juan David Guevara-Salamanca and Gina Jimenez

The Colombian armed conflict has produced the uprooting of peasants and farmers displaced from their lands that had to move to the cities where they survived in a hostile environment. These displacements ruptured spatialisations by leaving rural territories abandoned and later occupied by armed actors. These abandoned territories are places of violence and of collective trauma.

In 2011, the Colombian government created a program to recognize the victims of dispossession and abandonment of lands and their property rights on their farms – the Land Restitution program (LRp). This chapter discusses the LRp as a narrative/ discourse that creates an overlapping and saturation of trauma, conversing with critical approaches of collective and institutional trauma, and discussing the complex web that surround the victims, their subjectivities (re)produced, the bureaucratization of the LRp and the limited role of the land as a non-human actor of the program.

6 – Landscapes of Repair: Creating a Transnational Community of Practice with Sheffield- and Kosovo-based Researchers, Artists and Civil Society on Post-traumatic Landscapes

Amanda Crawley Jackson, Korab Krasniqi and Alexander Vojvoda

Landscapes of Repair (2021) was a knowledge exchange project between forumZFD Kosovo and researchers from the arts and humanities at the University of Sheffield, in cooperation with Kosovo artists, researchers, cultural workers and civil society activists. Through an online exhibition featuring the work of six Kosovo-based artists and architects, and a programme of international cross-sector talks and masterclasses, Landscapes of Repair gave space to questions on the role of informal networks of communities, cultural workers, artists and civil society activists in tangible, transformative interventions in dealing with traumatic pasts in public space. This chapter critically reflects on the methodologies and tools of knowledge exchange through a decolonial lens as a means of building a transnational forum for trauma-informed reflection and praxis with regard to the creation and re-appropriation of public space in post-conflict Kosovo.

7 – The Filipino Spirit is [Not] Waterproof: Creative Placeproofing in Post-Disaster Philippines

Brian Jay de Lima Ambulo

With a complicated confluence of geographical positioning, exposure, vulnerability, lack of coping and adaptive capacities, the Philippines is one of the world's most disaster-prone and climate vulnerable countries. Drawing a deeper inquiry from the world's strongest typhoons to ever be recorded, Super Typhoon Yolanda (2013) in Tacloban City, this paper examines the Philippines' vulnerability to natural disasters, focusing on the critical analysis of the prevailing resilience narratives, such as 'The Filipino spirit is waterproof', and the need for comprehensive and long-term disaster risk reduction strategies. It argues that these narratives oversimplify Filipinos' experiences, potentially obscuring the root causes of vulnerability, such as socio-economic inequalities and poor governance, and allowing powerful actors to evade accountability.

The concept of creative placeproofing is introduced as an alternative approach that bridges post-disaster recovery and placemaking by combining creative placemaking and trauma-informed placemaking.

Section 2 – Exploring the Dimensions of Trauma-Informed Placemaking

8 – Ethical Placemaking, Trauma, and Health Justice in Humanitarian Settings

Lisa A. Eckenwiler

This chapter reviews ethical critiques of refugee camps for their complicity in perpetuating injustice against traumatized populations and moves on to argue that humanitarians should embrace the ideal and practice of ethical placemaking (EPM) for its potential to respond to health injustice in this context. EPM is interpreted here as an obligation of transitional justice and is offered as a remedial responsibility – an 'ethic of the temporary' – for advancing health justice, for it can mitigate the harms linked to formal encampment and segregation in urban enclaves and moreover, create an environment that supports the capability to be healthy for traumatized peoples. Current efforts by UNHCR are reviewed, along with the ideas of architects concerned with displacement that do not focus on health and trauma but to some extent align with the ideal and practice of EPM, applauding them and suggesting how they might go farther to advance health equity in the short- and long-term. Examples of EPM, including an intervention aimed at women sufferers of gender-based violence in conflict zones, are also offered.

9 – Beyond Dark Tourism: Reimagining the Place of History at Australia's Convict Precincts

Sarah Barns

How Australians think about the histories of our places, and our connection to them, is in flux. For custodians of convict-era institutions, there is a need to confront histories of state-sanctioned dispossession, violence, trauma and abuse, to reveal how colonial impositions of power relied heavily on Benthamite forms of control to subordinate vulnerable populations, including most notably Australia's First Nations people. But what if Australia's convict precincts looked beyond the popular tactics of 'dark tourism' and reimagined these sites as places of restorative social justice and compassionate care? Such provocations are increasingly a part of contemporary dialogues around the value and purpose of Australia's convict heritage precincts, as programs of urban placemaking and revitalisation seek to define new First Nations and community programming outcomes in rapidly changing community contexts. How, then, might a deeper engagement with the troubled meanings etched into these places open up possibilities of wider dialogue – and care practices – with more diverse and perhaps vulnerable communities today?

10 – Equitable Food Futures: Activating Community Memory, Story, and Imagination in Rural Mississippi

Carlton Turner, Mina para Matlon, Erica Kohl-Arenas and Jean Greene

This chapter tells the story of the Mississippi Center for Cultural Production (Sipp Culture) in Utica, Mississippi. Sipp Culture uses art, agriculture, and collaborative action research as integrated approaches to comprehensive community and cultural transformation. The chapter introduces the regional context and story of Sipp Culture, and then focuses on a collaborative action research project Equitable Food Futures aimed at documenting and activating community agricultural knowledge and assets towards a more sustainable, equitable, and healthy food culture and economy. Integrating oral histories, arts-based, descriptive, and survey research to unearth and share Utica's agricultural and food stories and assets, this project seeks to restore community memory and make visible existing community knowledge around healthy ways of feeding the community, both physically and spiritually. Engaging with a history of slavery, sharecropping, and subsequent discriminatory agricultural policies and practices that have driven the drastic loss of Black farmland over the past century, Equitable Food Futures supports a communal process of healing from traumatic relationships with the land towards a renewed relationship that advances Black food sovereignty, leadership, and liberation. Ultimately, the project demonstrates how creative methodologies can catalyze historic and new knowledge in ways that inspire a more expansive imagination of healthy, locally owned, and equitable food futures.

11 – Language Is Leaving Me: *An AI Cinematic Opera of The Skin*

Ellen Pearlman

Language Is Leaving Me – An Opera Of the Skin is a work-in-progress immersive performance combining AI, computer vision, biometrics and epigenetic, or inherited traumatic memories of cultures of diaspora. It uses GPT-3, an artificial intelligence algorithmic database containing over six billion parameters using predictive verbal models to form sentences. The work will combine these models with computer vision transformation technologies like VQGan and CLIP that work with visual data sets by converting speech to images. The opera/work in progress interrogates the tensions between traditional forms of verbal, non-verbal, inherited and symbolic memories with their AI algorithmic manifestations. Using short films containing personal epigenetic stories and visuals emanating from the 1000-year-old geographic 'Pale of Settlement' of Eastern European Jews, the work illustrates how memory and meaning are erased by complex smoothing algorithms and transformation technologies that re-categorize and decontextualize cultural and individual identities. An AI music composition technology will create an accompanying soundtrack based on live human biometric indicators, thus linking the human to the machine, and memory to place.

12 – *Trauma and Healing in the Post-conflict Landscape of Belfast*

Aisling Rusk

This chapter explores ways in which trauma-informed placemaking has taken place within the existing built environment of the contested, post-conflict city of Belfast, enacted by the ordinary people of that place, as part of their collective healing. It considers two contrasting scales – a landmark building, the locally iconic, frequently bombed Europa Hotel, and the back alleys of houses – and temporalities – after two traumatic events during the Troubles, and the COVID-19 pandemic. Typically, places where atrocities occurred in Belfast city centre have been obliterated in an effort to forget these places of pain and shame. The chapter highlights how existing built fabric, reconceived, can contribute to healing after collective trauma, questioning what policymakers and place-shapers can learn from these bottom-up interventions and interpretations.

13 – Anticolonial Placemaking

Karen E. Till and Michal Huss

This chapter advocates anticolonial placemaking practice and argues that those who have inherited past and ongoing chronic place-based trauma should lead placemaking practices. Despite advances in participatory practice, 'experts' remain seen as the leading 'placemakers'. The authors draw upon anticolonial, Indigenous, feminist, and geographical theory to argue for an understanding of places as meaningful small worlds inhabited by multiple lives responsible to each other and the places in which they live(d). The chapter describes anticolonial tours and community mapping projects led by Palestinian and Traveller researchers in the wounded cities of Yaffa and Dublin, whereby participants and partners learn from emergent place-based stories of hope and pain to imagine more just futures.

14 – Placehealing in Minneapolis: Before and After the Murder of George Floyd

Teri Kwant and Tom Borrup

Social, physical, and psychological landscapes in Minneapolis were radically changed after the murder of George Floyd in 2020 during a time when a global pandemic already devastated the lives of so many. Located on traditional, ancestral, and contemporary lands of Indigenous Dakotah people, Minneapolis, like many cities, attempts to erase historical realities of those killed, driven from their homelands of thousands of years, and enslaved. Based in colonialism, practices designed to erase and forget people and events continue. Through the example of Minneapolis this chapter examines the nature of place trauma and how placemaking interventions can be employed across spectrums of cultural sensitivity, historic awareness, and acknowledgement of spiritual and sacred meanings. In the wake of significant recent unrest and destruction, contrasts in these approaches and outcomes are explored.

15 – Our Place, Our History, Our Future

Julie Goodman, Theresa Hyuna Hwang and Jason Schupbach

This chapter examines the regenerative practices occurring in the City of Philadelphia in the post-pandemic era. Philadelphia, the birthplace of the American project and America's sixth largest and poorest city, has been plagued by endemic poverty, racist policies and extreme violence in the years leading up to the pandemic. The pandemic has amplified the loss of black and brown business-owners produced a major uptick in violence. This chapter will summarize recent efforts to combat these negative forces with a focus on the asset-based, healing-engaged empowering and ground-breaking efforts of organizations Village for Arts and Humanities and Monument Lab. Through an honest assessment of these organizations work, a broader story of the history of our place, our history and the potential future will be told.

Section 3 – Crafting Spaces of Resilience and Restoration

16 – Trauma-Informed Placemaking: In Search of an Integrative Approach

Joongsub Kim

In the United States, we have long-standing policies and practices in the built environment that have caused trauma in underserved communities of color. This chapter focuses on an integrative methodology for trauma-informed planning and design practice, informed by concepts of social justice, kindness, and hope. Placemaking is often used to promote community engagement, but, this chapter argues, placemaking is not the most effective approach to addressing trauma. Placemaking and place-cultivating share many common characteristics: particularly, both promote community engagement. However, place-cultivating is more effective in addressing trauma because it advocates relevance, equity, and governance coupled with community engagement.

17 – Theorizing Disappearance in Narrative Ecologies as Trauma-Informed Placemaking

Marwa N. Zohdy Hassan

This chapter examines imaginary and fictionalized cities and the micro practices of resistance narrativized within, arguing that the pre-existent, fictionalized layer of reality can be accessed through Trauma theory, Science Fiction theory, Performance theory and Memory Studies. Real cities have always had, within their ontological fabric, pseudo-heritages and pseudo-memories, yet they are largely experienced through their pre-existing power discourses. This chapter proposes that marginalized narratives can take place in the mainstream canon despite the segregation performed by repressive and/or ideological state apparatus. If, within real cities, there are pre-existent, fictionalized cities which are revisited through the self-narratives of life performativities – because memories are always already textualized – then marginalized accounts can survive forced disappearances.

18 – Abandoned Landscapes as Places of Potential for Nature Therapy: Glendalough, Ireland

Lyubomira Peycheva

This chapter discusses an ambitious piece of research as it aims to classify an abandoned mining site/famous touristic site in Ireland – Glendalough – as an area of potential for nature therapy. The chapter considers that abandoned areas do not have to become permanent scars in the landscape; instead, they can be reimagined and re-purposed to serve for the benefit of humanity's mental health and ecological restoration. The research sought to uncover the links between natural spaces and human mental wellbeing and went on to highlight the importance of the presence of nature in our everyday lives and that people developed a greater appreciation for nature in times of uncertainty for the future during the COVID-19 pandemic. This indicates that in hardships people found their way back to nature, immersed into it, became part of it and found solace in it. The research concludes that we have inherent connection to nature, and when re-connected to it, we experience joy, inner peace, enter a state of relaxation, and ultimately appreciate life (ours and nature's) more.

19 – The Promise of Trauma-Informed Migrant Placemaking: Arts-based Strategies for Compassion and Resilience

John C. Arroyo and Iliana Lang Lundgren

This chapter broadly examines how fear and invisibility influence agentic immigrant strategies in new immigrant destinations across the US experiencing varying degrees of immigrant trauma. What concerns and/or opportunities do immigrants encounter when they evoke placemaking practices to reshape their built environment for safety and security? How do elements of tacit, intergenerational traditions influence how placemaking practices offer meaning for immigrants in a culture of anti-immigrant fear and tension? Where immigrants live in the US plays a critical role in how they adapt to their host society – and how their host society reacts to their presence in a policy context. In a 21st-century America defined by exponential diverse population growth and polarizing attitudes about immigration, ethnically oriented placemaking serves as both agent and canvas for expressions of cultural self and trauma-informed placemaking. We argue that despite an era of pervasive fear and heightened deportation tensions surrounding immigrant communities, new immigrant destinations provide immigrants flexible opportunities to assert their identity in suburban and exurban geographies unprepared for seismic population dynamics.

20 – Painting Back: Creative Placemaking in Vancouver's Hogan's Alley

Friederike Landau-Donnelly

The chapter discusses the complex relations between historical racial discrimination, paired with (predominantly white) settler colonialism, and contemporary efforts at reconciliation in the City of Vancouver, British Columbia, Canada. The chapter unpacks responses against some of the past and ongoing intersectional socio-spatial violences in Vancouver through public art and acts of creative placemaking in the neighborhood of Hogan's Alley, formerly populated by Black residents, as well as entertainment, culinary, religious and cultural facilities. Specifically, the chapter examines the mural, *Hope Through Ashes: A Requiem for Hogan's Alley*, by artist Anthony Joseph, commissioned by the Vancouver Mural Festival in the context of the unique curatorial Black Strathcona Resurgence Project in 2020, which spans 45 meters across the materialized instruments of Hogan's Alley's destruction, i.e., the Georgia Viaduct. In short, the chapter argues that places bear multiple wounds inflicted by colonialism and racism within themselves, and that public art can stimulate mo(ve)ments of resurgence through creative placemaking, that enable space for healing.

21 – Wanna Dance? Using Creative Placemaking Value Indicators to Identify COVID-lockdown-related Solastalgia in Sydney, Australia

Cathy Smith, Josephine Vaughan, Justine Lloyd and Michael Cohen

Using a Valuing Creative Placemaking (VCPM) methodology and associated toolkit, this chapter explores the traumatic experiences of the pandemic's impacts on public space through a case study of a creative placemaking-led intervention, Wanna Dance. Developed by City People (and led by one of the co-authors of this paper, Michael Cohen) in Sydney for the duration of June 2021, Wanna Dance invited the community to temporarily occupy public space by dancing in an inner-city laneway during a period in which dancing indoors was intermittently a banned activity. To consider the project and the relationship between creative placemaking and the reoccupation of public space in COVID-19 lockdown affected cities more broadly, the chapter turns to the notion of solastalgia: the disruption of place attachments due to their entanglements with significant traumatic events, whose sites are also referred to a traumascapes. The intersections between trauma-informed placemaking and creative placemaking are identified and evaluated, as an example of how placemakers might highlight, value and thus better respond to negative or traumatic experiences of place.

22 – Healing From Trauma in Post-disaster Places?: Placemaking, Machizukuri and the Role of Cultural Events in Post-Disaster Recovery

Moéna Fujimoto-Verdier and Annaclaudia Martini

Ishinomaki in Miyagi prefecture is one of the places hardest hit by the disaster that affected the North-eastern part of Japan in 2011. This chapter focuses on the Reborn-Art Festival, a biannual festival inaugurated in 2017, looking at how and in what ways the locals involved in the Festival, directly or indirectly, negotiate trauma and placemaking processes after the disaster. Cultural events can be considered as transient but essential placemaking tools, where displaced communities can meet, and through active or passive participation, form a new imaginary of the town. This new imaginary can greatly help interaction between people in the community, creating networks between communities and with visitors, allowing recovery from mental and emotional damages and also supporting economic activities. Our findings reveal that while the RAF can be considered as a successful event, which has potential to aid processes of trauma-healing, local's engagement is ambivalent, resulting in placemaking as well as healing processes that are unequally enacted.

23 – Placemaking, Performance and Infrastructures of Belonging: The Role of Ritual Healing and Mass Cultural Gatherings in the Wake of Trauma

Anna Marazuela Kim and Jacek Ludwig Scarso

Pandemic isolation, polarising politics and growing disparity in the right to spatial goods, whether through borders of national control or global capitalism, continue to exert a profoundly deleterious effect on the infrastructures of belonging at the foundation of a healthy, inclusive society, diminishing the psychological agency that would enable us to challenge these conditions. This article brings affect theory on depression as a public and political phenomenon into dialogue with ideas of placemaking as an infrastructure of belonging, to consider how mass cultural gatherings in the urban context might foster the conditions for re-envisioning and re-founding the potential for inclusive, civic life.

24 – Rethinking Placemaking in Urban Planning Through the Lens of Trauma

Gordon C. C. Douglas

This chapter develops trauma as a key conceptual frame for urban planning. In addition to the litany of traumatic actions and events that define the history of planning itself, the author posits that trauma and traumatic histories are key concepts for teaching and understanding some contemporary placemaking efforts. What's more, particular acts of placemaking, both formal and informal, can speak directly to traumatic contexts while expressing and empowering the voice and identity of those often excluded from urban citizenship. The chapter draws lessons from the author's experience leading a unique urban design studio course motivated by the crises of 2020 – from pandemic urbanisms to the Movement for Black Lives to extreme climate events. It also analytically examines three specific cases of trauma-informed placemaking in order to refine the meaning and relevance of the term. The author draws out the value of these cases for planning practice and pedagogy as well as their real-world significance as acts reclaiming place and community for marginalized groups. The author argues that this sort of placemaking, as a response to traumatic events, sometimes even as an act of resistance to them, poses important questions for mainstream planning and demands that we reconsider the real meaning and value of placemaking in urban development.

Section 4 – Our Call to Action: Nurturing Healing Through Action

25 – The Place Healing Manifesto

Charles R. (Chuck) Wolfe

Amid pandemic and protest, the need for urban mending has become abundantly clear. Place-healing is a mending-based call to action. Place-healing appeared in recent civic expression. Spontaneous clean up and damage repair followed violent demonstrations. Protesters crossed ideological lines to aid those injured. This type of place-healing is built on empathy, relatedness and potential to recover a sense of community. Other recent place-healing adaptations showed how the appearance and experiences of urban places can transform. Diversity and remnants from other eras allowed for a comforting continuity. Other examples included improvised entryways, alfresco dining, period furniture to frame restaurant take-out windows, and bifurcated tube (subway) cars. Businesses may also live on in a different form because they honour an ethic necessary to place-healing: respect for the people involved. Bobtail Fruit, once a fruit stall in Covent Garden, transitioned to five brick-and-mortar outlets around the city, and more recently to a web-based delivery business of quality baskets of fruit, and milk. A customer service imperative adapted to customer needs in a new form. The Place-Healing Manifesto is the catalyst for adaptation and transition.

26 – Leadership Horizons in Culture Futurism and Creative Placehealing

Theo Edmonds, Josh Miller and Hannah Drake

This chapter presents three types of futurist leadership horizons: exploratory, combinatory and transformational, as flexible frameworks and emerging innovation insights for creative placehealing aligned with the emerging future world of work, health, education and government, with case studies from arts, media and population health, and leadership practice tips required for working with six core strategies: Hope, Trust, Belonging, Creativity, Curiosity and Compassion. These are things most leaders think they understand but, in reality, few truly recognize their scientific underpinnings. Developing a high-level working knowledge of this can open a world of new possibilities: this chapter guides leaders through the process of aligning culture change with innovation efforts in order to unlock the innate creative potential of both organizations and communities of practice.

27 – Where Healing Happens: A Working Theory on Body, Relationship, and Intentional Structure for Restorative Placemaking

Elena Quintana and Ryan Lugalia-Hollon

Peace Rooms are a very specific place where props, room design, and human training can all prepare the way for healing, transformative relationships and, often, racial justice. Drawing from the authors' efforts to reduce the school-to-prison pipeline in two US cities, the chapter outlines key programmatic elements, processes, and policies needed to build and activate a Peace Room as intentional, trauma-informed places that support restorative, rather than punitive, responses to harm. They are inviting and accountable to all members of a community, with deep roots in Indigenous cultures and Tribal communities. Drawing from concrete examples from Chicago and San Antonio, the chapter illustrates the role Peace Rooms have played in advancing trauma-informed restorative justice at the level of both individual school campuses and big citywide systems. It explores how dull physical spaces can be transformed to help reground relationships in trust, reconciliation, and accountability, thereby enabling interconnectedness to ourselves, each other, and nature. While Peace Rooms themselves cannot eliminate harm, they can be key anchors in larger community efforts to reduce and address harm, vital touchstones that can help to sustain healing culture and social movements.

28 – The Art of Place

Daria Dorosh

This chapter presents a way to build a space of trust from which to take action. The chapter examines coping strategies coming from the authors art practice and that of other artists. Finding a place in the world means having space for one's personal narrative to unfold. To facilitate that, this text provides readers with space and engaging visual cues for their own story creation as they read the chapter. We have lost our cultural narratives and have not yet replaced them. We have the tools, but it may require freeing ourselves from the management of our personal narrative by commercial interests to find our stories. Trauma and stress from economic instability, habitat insecurity, political turmoil, confinement, or simply working in a mobile culture, may be with us indefinitely. The current pandemic is underscoring the importance of finding and adapting new tools of communication to keep our connection alive. If we combine material and digital tools that can cross-pollinate knowledge tribes, we can share our coping strategies in dimensions across time and space. Having options in our tool kit is key to addressing the many unpredictable factors in our lives today. Perhaps we can consider these difficult circumstances as a springboard for imagination and reflection, and one that directs us to self-compassion as a platform for the change we need to make to continue. The trauma of displacement and confinement is fueled by fear of scarcity, but the universe shows infinite generosity if only we dare to see and implement it.

29 – Unravelling Memories: The Metaphor as a Possibility of Resilience

Pablo Gershanik

Gershanik's Intimate Models Lab explores resilience through art. Working with a reconstruction process that combines art, psychology, and human rights, modelisation arose as a tool of transposition (creation of aesthetic mechanisms of intermediation with a painful experience) to work with cases of social trauma that a person, a family, or a community has suffered in recent years. This process explores the capacity of reconstruction and resilience of us as people through an artistic approach. Intimate Models Laboratory is a space to 'reconstruct' difficult events of your experience with a playful, panoramic and resilient gaze using aesthetic tools. Through case studies two main concepts will be deployed: poetic drift and poetic justice.

30 – Healing Place: Creative Place-remaking for Reconstructing Community Identity

Katy Beinart

This chapter investigates attempts to remediate, reclaim and transform community identity, by using spatial and artistic interventions to 'heal' place, examining definitions of 'place' and reasons for and issues around the exclusion and breakdown of communities, through the case study example of District Six Museum, Cape Town. Drawing on Gerd Junne's 'Architecture for Peace' (2006) and using metaphors of healing and medicine, the author establishes a set of criteria for interventions that 'heal' place.

31 – A Reconciliation Framework for Storytelling: A Trauma-Informed Placemaking Approach

Katie Boone, Wilfred Keeble, Rita Sinorita Fierro and Sharon Attipoe-Dorcoo

The fish in water doesn't even know it's wet – this applies to placemaking efforts around trauma and healing. The intergenerational trauma we carry is held and felt within the land we are on, and it impacts the processes we steward. In this chapter, the authors approach how we navigate our own social identities, personal, intergenerational, and collective trauma to support trauma-informed placemaking in the groups we work with, both individually, and as a collective. Weaving our various stories through the chapter, the authors explore truth telling and the collective healing that ensues when we tether our work in trauma to place.

32 – Allowing a Conversation to Go Nowhere to Get Somewhere: Intra-personal Spatial Care and Placemaking

Sally Labern, Sophie Hope and Rebecca Gordon

This chapter troubles the notion of placemaking and the role of the artist in that process, taking the form of a tripartite dialogue between socially engaged artist/ scholar practitioners, through which trauma-informed placemaking is being rethought as intra-personal spatial care. The form and function of this chapter is a slow curated conversation-to-text, intentionally to nowhere, presenting open-ended themes prescient to the endeavour of trauma-informed placemaking. The element of care that opens this up requires the opening of space in which to think/ process/do more slowly, together. It is a 'caring with' and a place of resistance to ongoing spatial violence. By beginning differently, agency can become active and shouldered laterally in solidarity. The eventual curation from dialogue to text presents key themes that emerged in order to share ideas with the readers that we believe are prescient. The culmination of such collaborations – and defiance against individualism of authorship and instrumentalisation of community – can become a cumulative act of reconfiguring place in defiance of the continual traumas of exclusion in place.

Closing Remarks: Being Accountable as Placemakers

Placemaking and the Manipur Conflict

Urmi Buragohain

This chapter is written from the live context of the ethnic conflict that started in May 2023, in Manipur, India, and asks three essential questions to placemakers from this purview, of when and how to work in places of trauma, and what practices are needed.

The author contends, that, if placemaking is about bringing people together and getting them to talk to each other, surely our duty-of-care extends to finding enabling pathways for traumatised people to heal and one day get talking to each other again? It discusses placemaking principles, concepts and approaches that operate at the nexus between placemaking, human psychology, mental health and trauma and their common thread assertion – that the knowledge and practices of these disciplines need to be imbued with a deeper understanding of the effect of trauma on cities and their contingent realities.

ACKNOWLEDGEMENTS

The Editors would like to thank all those who have contributed a chapter to this textbook. Their time, dedication to and enthusiasm for this project as a process undertaken most often outside of usual working hours or as an effective volunteer, and often during times of stress and duress in their personal and professional lives, is something we are ever thankful for. Thank you also to placemaking leader, Leo Vasquez, for his ongoing support, and to Archie Oclos for permitting depiction of their artwork, to Brian Jay De Lima Ambulo for the image, and to Kate Banjamin for image editing.

Our thanks extend to those that our contributors have worked with, the residents, the neighbourhood groups, the policy makers the volunteers, the archivists, the protesters, the colleagues, the dancers – 'it takes a community to create a book', as it were.

Thank you too to all those placemakers that work to hold the practice accountable still to its radical, progressive, people-first integrity. We want to express our gratitude to those who have embraced our project and helped us to pave a way forward, a collective journey into unexplored landscapes that call us to transcend boundaries, redefine perspectives, and contribute to a discourse that is as transformative as it is pioneering.

The textbook contains material of a highly sensitive nature that may be triggering for some individuals.

Both UK- and US-English appear in this textbook, recognising the voice of the contributors and direct quotes from other texts, artists and project and research participants.

This book is written by a global cohort of place practitioners, researchers, theorists, development workers and creatives, and as such, its use of terms across its pages are place- and sector-context dependent. We ask our readers to extend kindness and generosity to the writer if any offence is caused in the choice or use of words.

Citation as Courage and McKeown (eds.) (2024).

FOREWORD

What it Means to be Trauma-informed

Angela Kennedy

Mental health and wellbeing are significantly influenced by our environment: our social, relational, biological and earthly context. We are intricately entwined with the adversities and benefits that impinge on us from outside of our brains. The science of medicine has shown us how much of our mind emerges from the physicality of the brain. However, the primary purpose of the brain as an organ is to interpret, respond and adapt to the outside world and our experiences in it. Western Modernist ideals have turned us inward looking, fuelling an illusion of an independent sense of self divorced from context and understandable by a collection of implicit cognitions. We in turn have turned our attention to the psychology of the mind and pushed the responsibility for wellbeing to that of the individual, their coping, cognitions or self-care. The science of mental health has become more about defining the way an individual's beliefs have been distorted and the biology associated with that. But what if we turned our attention to the context in which those issues arise, and heal? How would we see ourselves and mental wellbeing differently? Importantly, would that lead us to more interesting ways forward as a society? What role does 'place' have in the system that creates our felt sense of wellness and connection to others and purpose?

Trauma is usually defined as an event or context that causes fear for one's life and overwhelms coping. However, as social animals we can be traumatised by so much more than potential death. Being shamed by breaking taboos can act as a trauma too. Being betrayed by someone we trust or rely on can distort our expectations of others and how we see ourselves. The absence of emotional or physical nurturing can be traumatic. Witnessing these in others is sometimes sufficient. Cultural issues such as racism and poverty play large too. Losing our homes in natural disaster illustrates how important a safe context is.

Traumatic reactions to such events do play out in our biology and emotions but also in our beliefs about the world, the way we relate to each other, how integrated we experience our sense of self, somatic disturbances in our body, our ability to function and the continuity of our memory. It can fragment families and influence our sense of meaning and purpose. It can disconnect us from others and the world around us. As such, traumatisation can pass from one generation to the next, its origins getting lost in time but transmitted via the way others are able to relate to us and through the biases we create in our communities.

We know that trauma is a huge causative factor in many psychiatric and social problems. Trauma is linked to suicides, psychosis, job problems, life expectancy, addictions. It wreaks havoc on our lives. Being trauma-informed requires facing some basic truths about our nature: that much trauma is inflicted on us by other people, often intentionally. It isn't accidental. Being trauma-informed requires us to face something that we would rather not acknowledge. It can be tautological in that it requires resilience to build resilience. It therefore needs a supportive context and attention to the welfare of the healthcare workers, parents, etc., who are at the frontline. Ironically, it is often those who have been traumatised who see the reality more clearly and have so much to teach us. The question is: can we hear them?

The term 'trauma-informed' is an attempt to redress the purely biological bias and put some of the context back into our experience of mental health. It seeks to actively heal the impact of the trauma and hopes to minimise any further harm. A trauma-informed approach would understand that we react to context in predictable ways when confronting extreme adversity that make sense as survival strategies. Traumatic events are moderated by having good social support, by safer spaces, by access to nature, by opportunities for physical activity and healthcare, by cultural references in the arts that bear witness to our lives, by a sense of belonging. The world around us can therefore very powerfully provide a context for wellbeing and growth. Healing or wellbeing takes place in the context of relationships with others and is enabled by the systems and processes and environments that enable such growth to occur.

As a term 'trauma-informed' is becoming ubiquitous, usually spearheaded by psychologists keen to showcase the abuses and injustices in the lives of people who present to psychiatric services, point out the phenomenology of being traumatised and develop the model of recovery via talking and the therapy room. Of course, I have seen first-hand the value that this has. However, being trauma-informed is so much more than being aware of a few things in childhood that created fear reactions and so much more than specialist trauma services. It is a responsive philosophy that puts research evidence, words and observable actions to wisdom about how to proactively co-create mental health and healthy functioning. That may be via biology but also the social, political and environmental context.

As social animals, we have the gift to regulate each other's neurobiology and ultimately evolve together societies that embed the trauma-informed principles of safeness, choice, justice, collaboration, cultural competence, solidarity with those traumatised, egalitarian support, relational repair, coordination across agencies. As animals that exist as part of an eco-system, we can also choose to tread more kindly on the planet that supports us, and avoid omnipotent arrogance in relation to our natural world. We can use design, planning, the arts and architecture to create coherent spaces of joy that enable such social contexts to flourish. We can talk about people in their context, kindly with less stigmatising language. We can pay attention to potential unintentional harm that we may cause people as a consequence of our actions, policies, cultures or physical environments and make efforts to avoid these or make reparations. A place-based trauma-informed philosophy can put lived experience of adversity at the core of our attempts to create healthy communities.

Such an approach involves acknowledging that whilst our species is flawed, causing deep pain and destruction to each other and our contexts, it also has a remarkable ability to evolve, grow and heal through collective action. Being trauma responsive acknowledges that attention to our context is not just what can harm us but is that which can also heal. It deeply motivates us towards a context for growth. We are still learning but need to be more courageous in our endeavours towards this if we are to avoid further loss of progress towards addressing these.

Key Texts

Bloom, S. (2013). *Creating Sanctuary: Toward the Evolution of Sane Societies*. Abingdon: Routledge.

Herman, J. (2023). *Truth and Repair: How Trauma Survivors Envision Justice*. New York: Basic Books.

Russo, J. & Sweeney, A. (eds.) (2016). Searching for a Rose Garden: challenging psychiatry, fostering mad studies. Wyastone Leys: PCCS Books.

PREFACE

The Trauma Informed Placemaking Endeavour

Cara Courage and Anita McKeown

Trauma Informed Placemaking is a praxis conceived by Dr Cara Courage and Dr Anita McKeown. It originated in their observations of the traumatic rupture in the people-place relation during the COVID-19 pandemic and the sites of Black Lives Matter protests and police brutality; people were either 'removed' from their public realm through lockdown/shelter-in-place strictures, or the streets they had peacefully inhabited became sites of violent unrest. This endeavour has been two lifetimes and two careers in the making, from formative childhood experiences, to working in sites of social planning, to looking at eco-anxiety with children in the classroom, and more. Indeed, each of the chapters in this textbook represent the post-traumatic growth in all the textbook's contributors, their communities in question, or both. This is not just an academic text – each chapter is a hard-won individual and collective engagement in trying to make a difference.

The Concern of the Trauma-Informed Placemaking Endeavour

Place is seen as ranging from being contested, where people and politics clash, or as convivial, as sites of serendipitous meeting and celebration, for example; but wherever placed on that spectrum, the potential of place to have been of a traumatic experience is not widely, openly or consistently acknowledged.

Broadly speaking, the endeavour has a mission to change placemaking's practice to be foregrounded by an understanding of trauma and the ruptures this can cause – and the healing processes and potentials of a trauma-informed approach. It holds an aim now, met through the publication of this textbook, to offer a primer for place-based practitioners from which to apply this knowledge to their own research and/or professional practice. This endeavour – as community of praxis and as book – brings together place academics, practitioners and thought-leaders, with those from psychology, the arts, behavioural and social sciences, in a transdisciplinary investigation of trauma-informed placemaking and calls for this approach as foundational for any understanding of, or working in, place.

Research-as-publication Form

This textbook is positioned as the novel *research-as-publication*, being the first time that many of the contributors have looked through their practice from a trauma-informed placemaking lens and being a generative research-through-writing process. The textbook has been described as *implicitly polemical*, in its attempt to redefine *place-based work through research, practice of trauma-informed knowledge.* It is curated, an approach established by the editors to progress academic publishing and to aid the readers navigation of texts in their previous publication (Courage, 2021; Courage & McKeown, 2019). It features: a wide global range of voices, from academic, professional, practitioner and community cohorts, and of those that work across place, arts, culture and the psychosocial; a wide variety of historic, contemporary and future-signalling contributions; and Indigenous and non-Western, non-Northern Hemisphere knowledges. It presents a wide number of case studies and examples, alongside learning exercises, frameworks and toolkits for direct or developed use in the field and spotlights the unique contributions to practice and to the placemaking lexicon. The editors and contributor cohort have been engaged in cooperative and co-creative activity though the creation of this textbook around the development of the project, and its theoretical underpinnings, terminology, values and practices – including through a World Café, kindly hosted with Leo Vasquez, then of Creative Placemaking Communities.

The form of a textbook permits us to explore an innovative approach that lies at the intersection of trauma, place, and practice, and that is concerned with the realm of human experiences. It stands at a crossroads, offering multiple trajectories and journeys that inevitably converge and diverge. These journeys are to be undertaken with curiosity and in preparedness for ambiguity – indeed an approach that is necessary for trauma-informed placemaking practices. The chapters challenge us as practitioners to think beyond disciplinary boundaries to reflect on the intersection of trauma and place, envisioning collective and collaborative approaches. We encourage a curiosity with the ideas, emotions and perspectives that contribute to how we perceive, interact and create the places that in turn have the power to shape us. In this sense, we perceive placemaking as an existential activity of being, belonging and becoming.

The Curatorial Form

Our curatorial conceit proposes the symbiosis of trauma-informed practice and place-healing as an urgent, critical and necessary foundation to placemaking practice, offering a field of belonging for all who are seeking holistic, generative and healing understandings of place, while offering an initial repository of histories, experiences and emotions. Such a proposition demands rigorous review of placemaking practices, necessary to present a measured review of the nuances and challenges within the field; and its sensitivity to trauma requires a critical examination of the ways in which diverse disciplines and practices can converge to foster understanding and healing and build resilience within the very creation of our environment.

With this initial idea, driven by our curiosity to make sense of these experiences, cognisant of a similar concern within our professional practices, we opened a call to collate a consideration of trauma-informed placemaking, to share novel, tested approaches to place-based teaching, research and practice that acted on the axis of disciplinary boundaries. We sought to highlight situated and embodied aspects of human experience within place-based practices, integrating trauma awareness with placemaking practices involving those who may not have previously seen their work as or within this context. This call garnered over 80 immediate responses; more have continually followed via the project website, www.traumainformed. place.

The chapters in this book present themes that are of profound importance to the sector, extending the possibilities for, and highlighting the concerns of, those who operate or chose to operate within an increasingly expanding inter/transdisciplinary landscape of placemaking. Exploration of migration and displacement, preservation and disruption of memory and heritage, recovery from disaster and conflict are all thematic intersections for contemporary placemaking and its evolution.

As creative practitioners (artists, curators and placemakers) we approached the book as we might an exhibition or curated conference panel: presenting information as well as setting up polemics within the textbook's architecture through the positioning of the chapters in a narrative arc. The themes discussed within the chapters, particularly in relation to migration/displaced communities, memorialisation/preservation and disaster/post-conflict recovery, have a growing significance for researchers and practitioners working within the expanded and interdisciplinary field of placemaking. Rather than presenting a consideration of trauma-informed placemaking and of place healing head-on, a more open, fluid, yet still structured, approach to the textbook's content was called for. What emerged through a curatorial process is a relational presentation and exploration of complex, multivariant and multifaceted situations and often emotionally heightened and subjective experiences of place.

In this way, the development of the textbook's form afforded an opportunity for it to inform and initiate new approaches to the teaching, research and practice of placemaking across disciplines such as urban planning, arts and heritage, architecture and design, and public health, amongst others. Further, unlike conventional textbooks, as an emergent field and dynamic practice, additional questions are posed through the relationships between the chapters. When placemaking is considered as dynamic and processual, provoking deeper questioning rather than giving fixed answers is a method of learning that becomes embedded in the form of the book – much like other pedagogical practices such as discovery or inquiry-based learning.

Through the book we embark on a journey to delve into the depths of our understanding of trauma-in-place sensitivity, an exploration that transcends the boundaries of conventional disciplines and invites us to engage with diverse perspectives. This journey is guided by a series of chapters, each shedding light on various facets of trauma's intricate relationship with the places and spaces we inhabit. This curatorial progression was also informed by the therapeutic trauma experience path – realisation, recognise, respond, resist re-traumatisation. The four main sections are bookended by the Foreword and the Closing Remarks.

Foreword

The Foreword is given by Dr Angela Kennedy, a clinical psychologist with a vast and significant public health and trauma expertise. It offers a foundational cornerstone of what it means to be trauma informed in the context of place and communities, a diverse yet concrete framing for the four sections following.

Section 1 – Understanding and Developing Our Trauma-in-place Sensitivity

The concept of trauma-informed placemaking calls for a profound shift in our approach to urban and rural design, urging us to consider the historical and present traumas that have shaped communities. This paradigm insists that placemaking is not only about physical structures but also about honouring the emotional, psychological, and social dimensions of spaces/places. In examining the ways communities respond to trauma, we encounter the principles of mutual aid, collective resistance, and an ethics of care, an intricate web of trauma imprinted upon our landscapes, is presented. This section draws attention to wounds, individual and collective, long-standing and contemporary, that persist and influence our lives.

Section 2 – Exploring the Dimensions of Trauma-Informed Placemaking

Section 2 is a deeply profound exploration into the praxis of trauma-informed placemaking, where theory meets practice and narratives intersect with spatial design. The chapters in this section invite us to recognise the intricate threads that are woven across and hold together the spaces we inhabit, the histories we carry, and the traumas we navigate. Each chapter presents another aspect of the multifaceted nature of trauma and consequently offers insights for developing trauma-informed placemaking, understanding through gaining an understanding of the complexities, challenges, and ultimately the transformative potential of a trauma-informed approach to placemaking. This section traverses the displacement of communities and dis-belonging due to conflict, disaster or placemaking practice that deliberately or inadvertently re-enforce dominant narratives and exclusive practices and begins to develop a bridge between the past, present and future practices of what trauma-informed practices could be and should be mindful off.

Section 3 – Crafting Spaces of Resilience and Restoration

The places we live, work and play, and the landscapes we inhabit are integral components of our emotional, psychological, and social narratives. Trauma, whether individual or collective, leaves an indelible mark on our bodies and our landscapes, shaping our experiences and perceptions of place. The awareness of, and growing interest in, the importance of situated and embodied experiences of place in relation to urban renewal, regeneration and other official placemaking practices form the North Star of this section. It traverses responsive trauma-informed placemaking, where the traces of trauma are catalysed into physical narratives of healing, transformation, and empowerment.

Section 4 – Our Call To Action: Nurturing Healing Through Action

As placemakers, we are entrusted with a profound responsibility – the power to shape the spaces and places that weave the fabric of human experience. The chapters in this section seek to highlight the intricate relationship between trauma, healing, and the environments we co-create. As we stand at this juncture, poised between insight and action, it becomes imperative to turn our awareness into accountability and our knowledge into transformative change. Beginning with a manifesto, a demand for an urgent response for place healing, the call to action, urging us to take concrete steps to acknowledge, address, and mend the traumas imprinted within our surroundings is grounded.

Closing Remarks

We close with a set of open questions, posed by Urmi Buragohain, founder of PlaceMaking Foundation, who gives us a reflexive rendering of placemaking practice, from her context of a live conflict situation in Manipur, North Eastern India.

Urmi's account reminds us all that placemaking can be brutal and that our species, wired for security, will often brutalise and dehumanise each other, and that each location, each community, will all require their own placemaking process, there is no formula. Yet these questions are as pertinent for any placemaking attempt, any traumatic situation or traumatised individual – as we discuss, events do not have to be rare or overtly traumatic to overwhelm us. In presenting this textbook, and the whole trauma-informed placemaking endeavour, we are not seeking to provide definitive answers, as there are none to give, and we intentionally close with questions to provoke your own reflexive interrogation.

Community of Praxis

A Community of Praxis is forming as an outcome of the trauma-informed placemaking endeavour. Taking a lead from Community of Practice (Wenger, 1998, 2006; Wenger & Snyder, 2000; Wenger et al., 2002) theory, approaching the endeavour as a Community of Praxis not only engenders the forming of a knowledge exchange community around the notion of trauma-informed placemaking, but also provides a step change in the practice of placemaking through the practical application of theory and learning from experience. This textbook offers an invitation and a welcome to embark on this journey with us as collaborators and contributors to an emerging and dynamic discourse on trauma-informed placemaking. The chapters offer pause points on this journey, guiding us all through the landscape of complexities, challenges and opportunities our present and futures may hold. Courage and McKeown are working to nurture this Community of Praxis through continued cohort exchanges, their collective divergent>convergent>lateral explorations, its aimed for outputs and outcomes, and in the application of learning-into-action to change sector practices.

References

Courage, C. (ed.) (2021). *The Routledge Handbook of Placemaking*. Abingdon: Routledge.

Courage, C. & McKeown, A. (eds.) (2019). *Creative Placemaking: Research, Theory and Practice*. Abingdon: Routledge.

Creative Community Builders (n.d.). Available: https://creativecommunitybuilders.com/. [Accessed: 21 August 2021.]

Wenger, E. (1998). Communities of Practice: Learning as a Social System. *Systems Thinker*. Available: https://thesystemsthinker.com/communities-of-practice-learning-as-a-social-system/. [Accessed: 8 April 2020.]

Wenger, E. (2006). *Communities of Practice: a brief introduction*. Available: https://scholarsbank.uoregon.edu/xmlui/bitstream/handle/1794/11736/A%20brief%20introduction%20to%20CoP.pdf?sequence=1&isAllowed=y. [Accessed: 8 April 2020].

Wenger, E. C. and Snyder, W. M. (2000). *Communities of Practice: The Organisational Frontier*. *Harvard Business Review*. Available: www.psycholosphere.com/Communities%20of%20Practice%20-%20the%20organizational%20frontier%20by%20Wenger.pdf. [Accessed: 8 April 2020].

Wenger, E., McDermott, R. and Snyder, W. M. (2002). *Cultivating Communities of Practice*. Boston: Harvard Business School Press.

NOTES TO THE TEXT

The textbook contains material of a highly sensitive nature that may be triggering for some individuals.

Both UK- and US-English appear in this textbook, recognising the voice of the contributors and direct quotes from other texts, artists and project and research participants.

This book is written by a global cohort of place practitioners, researchers, theorists, development workers and creatives, and as such, its use of terms across its pages are place- and sector-context dependent. We ask our readers to extend kindness and generosity to the writer if any offence is caused in the choice or use of words.

Citation as Courage and McKeown (eds.) (2024).

ABBREVIATIONS

AAA	Argentine Anti-Communist Alliance
AAAE	Association of Arts Administration Educators
ACE	Adverse Childhood Experiences
ACLU	American Civil Liberties Union
ADHD	Attention deficit hyperactivity disorder
AfVT	Association of Victims of Terrorism
AHRC	Arts and Humanities Research Council
AI	Artificial Intelligence
AIA	American Institute of Architects
AIBO	Artificial Intelligence Brainwave Opera
AIDS	Acquired immunodeficiency syndrome
AIM	American Indian Movement
AMK	Dr Anna Marazuela Kim
APA	American Psychological Association
ARCH	Art Remediating Campus Histories
ART	Attention Restoration Theory
ASU	Arizona State University
ATM	Automated Teller Machine
BFA	Bachelor of Fine Arts
BIPOC	B/black, I/indigenous, and other P/people of C/colour/color
BLM	Black Lives Matter
BSRP	Black Strathcona Resurgence Project
CAVE	Community Anti-Violence Education
CBD	Central Business District
CBO	Community Benefit Ordinance
CDC	Centers for Disease Control and Prevention

CDC	Community Development Corporation
CEO	Chief Executive Officer
CHCI	Consortium for Humanities Centers and Institutes
CLLAS	Center for Latino and Latin American Studies
CLOY	Crash Landing on You
COTT	Cards on the Table
COVID(-19)	Coronavirus disease (-19)
CPM	Creative placemaking
C-PTSD	Chronic Post-Traumatic Stress Disorder
CRXLAB	Creative Action Lab
D6	District Six
DACA	Deferred Action for Childhood Arrivals
DAPA	Deferred Action for Parents of Americans
DM	Discursive Materiality
DMZ	Demilitarized Zone
DNA	Deoxyribonucleic acid
EEG	Electroencephalogram
EFF	Equitable Food Futures
EU	European Union
FTC	Friends of Tri-County
FTE	Full Time Equivalent
GBV	Gender-based Violence
GPCA	Greater Philadelphia Cultural Alliance
HBCU	Historically Black College & University
HIV	Human Immunodeficiency Virus
IDP	Internally Displaced Palestinian
IKA	Imam Khomeini International Airport
IOM	International Organization for Migration
IRA	Irish Republican Army
IRGC	Islamic Revolutionary Guard Corps (
JEP	Jurisdicción Especial para la Paz
JFK	John F Kennedy [Airport]
JLS	Jacek Ludwig Scarso
JTDC	Juvenile Temporary Detention Center
KBL	Karzai International Airport
KIPP	Knowledge Is Power Program
km/h	Kilometre per hour
LGBTQAI+	Gay, Lesbian, Bisexual, Transgender, Queer, Intersex, and Asexual
LRJ	Land Restitution Judge
LRp	Land Restitution program
LRU	Land Restitution Unit
MCCP	Mississippi Center for Cultural Production
MfCA	The Museum for Contemporary Artists

MIT	Massachusetts Institute of Technology
MPA	Master Plan Approach
NAFTA	North Atlantic Free Trade Agreement
NATO	North Atlantic Treaty Organization
NDRRMC	National Disaster Risk Reduction and Management Council
NEAR	Neurobiology, Epigenetics, Adverse childhood experiences, and Resilience
NER	Northeastern Region (of India)
NGO(s)	Non-governmental Organisation(s)
NHS	National Health Service
NPO(s)	Non-Profit Organisation(s)
NSW	New South Wales
PAGASA	Philippine Atmospheric, Geophysical and Astronomical Services Administration
PAR	Philippine Area of Responsibility
PAR	Participant Action Research
PBMR	Precious Blood Ministry of Reconciliation
PPC	People's Paper Co-op
PTSD	Post-Traumatic Stress Disorder
RAF	Reborn Art Festival
rDNA	Ribosomal DNA
RJ Hubs	[Community] Restorative Justice Hubs
RSUA	Royal Society of Ulster Architects
SAMHSA	Substance Abuse and Mental Health Services Administration
SAR	School for Advanced Research
SARD	Shubra's Archive for Research and Development
SCM	Spanish Colonial Missions
SDG(s)	United Nations Sustainable Development Goal(s)
STACC	Sexual Trauma & Abuse Care Center
STR	Stress Restoration Theory
THR	Mehrabad Airport
TICIRC	Trauma-Informed Cre Implementation Resource Center
UN	United Nations
UNDRR	United Nations Office for Disaster Risk Reduction
UNEP	United Nations Environment Programme
UNHCR	United Nations High Commissioner for Refugees
UNSW	University of New South Wales
UO	University of Oregon
US	United States
USA	Unites States of America
USD	United States Dollars
USDA	United States Department of Agriculture
USICE	United States Immigration and Customs Enforcement

USICE	United States Immigration and Customs Enforcement
VCPM	Valuing Creative Placemaking [Toolkit]
VMF	Vancouver Mural Festival
VSP	Victoria Square Project
WASP	White Anglo-Saxon Protestant
WHO	World Health Organisation
WWII	World War II
YMCA	Young Men's Christian Association

INTRODUCTION: PATHWAYS TO A PRAXIS

Cara Courage and Anita McKeown

In a world of profound interconnectedness and shared experiences, the spaces we inhabit hold a complex interplay of memories, emotions, and histories that shape our individual and collective identities. From urban centres to rural landscapes, each place carries its own stories, often interwoven with both joy and pain, euphoria and tragedy. It is therefore essential to recognise that these places are not simply backdrops or local development plans; they are the very containers of human experience, laden with the imprints of trauma, healing and resilience.

We contend that, as placemakers, it is critical for us to understand that a trauma response can be triggered in *people-in-place* as a consequence of our actions as placemakers, of the place processes we are part of, and as a result of place as site of historical and contemporary events. We contend that we need to have this awareness to inform how we conduct ourselves in place; how we approach working in place and with the people of that place; and how we hold ourselves and all those around the placemaking table accountable to an ethical practice that is engaged in its own betterment and that works to the imperatives of intersectional and climate and ecological justice.

Trauma responses can come equally from the most mundane and seemingly harmless situation or action, as they can from the high-impact, high-visibility major event, or from centuries-old and ongoing systemic injustices: wherever on such a notional spectrum that trauma creates stress reactions and traumatic stress-related disorders that comprise a specific constellation of symptoms and criteria (American Psychological Association, n.d.). It is this situated and embodied traumatic experience of place that is the foundation of the trauma-informed placemaking praxis. As praxis, it gives placemaking, and indeed, anyone who works in and with place from any professional, applied or theoretical purview, opportunity to delve into previously unchartered territory, yet one that draws on

DOI: 10.4324/9781003371533-1

previous practice, names contemporary practice, and creates a transdisciplinary hybrid space. In this new thirdspace space, a *community of praxis* is forming (and, see Preface, this edition). A *communitas* of trauma- and healing-concerned place practitioners, interconnected through shared experience and intention, engaged in sharing knowledge and learning from others, and applying that knowledge to the live context of working in place. Throughout the textbook, practitioners define and redefine practice and language, affirming the need for fluid, responsive states and embodying the diversity of language, form and critical debates across those that intersect with people in place.

Priming Our Understanding of Trauma

Dr Angela Kennedy, Director, Innovating for Wellbeing, and Consultant Clinical Psychologist talks powerfully of trauma and its impacts in the *Foreword* to this textbook; and building on the collective call to be trauma-informed on our place-based practices that is this textbook, we now offer a primer to an understanding of trauma. We offer this with a caveat: the field of trauma studies is a wide one, it is complex and holds a multitude of views and experiences. Thus, what we offer here as a primer can only be just that, a primer, an entry point into trauma awareness to anchor the reader in their journey through this textbook and then further into their own trauma-informed inquiry.

Trauma emerges from profound, often disturbing, experiences our bodies are wired to protect us from, making concrete definitions and articulation difficult. As a spectrum of embodied responses, individual's experiences are unique, as are the possibilities and potential for both healing and re-traumatising practices. A composite understanding of trauma presents it as an emotional response to exposure to an event or series of events that can cause, or be perceived to cause, physical harm, emotional harm, and/or life-threatening harm. This event(s) could be to the person, or to the person's environment, or to their community, and may operate at the level of the systems that surround a person and/or their community. The event(s) do(es) not have to be rare to be traumatic; it is traumatic because it overwhelms usual coping mechanisms (APA, n.d.; TICIRC, n.d.; SAMHSA, n.d.; STACC, 2016), operating as an 'inner injury' that causes a 'a lasting rupture or split withing the self' (Maté, 2022, p. 20).

Both the brain and the body respond to trauma. The bodily response to trauma is a disruption to the limbic system of the brain – the part of the brain which stores emotional responses to experiences and controls fear reactions, the key structures being the hippocampus and the amygdala. Trauma activates the amygdala, which performs a primary role in emotional control and processes, memory and learning. The amygdala is the most ancient part of the brain and operates within a binary 'off' or 'on' status. It does not, nor cannot, discern between real or perceived threats and therefore, it is reactive rather than able to perform a nuanced response, with these functions performed by other parts of the brain.

An activated, or triggered, amygdala interferes with the hippocampus, itself involved with recall of memory, particularly long-term memory. Thus, when recalling trauma, memories may be lacking in detail, confused, fragmented and out of chronological order. When triggered, the prefrontal cortex begins to function less effectively. This is the decision-making/choice-taking part of the brain and responsible for, amongst others, rational thinking (e.g., planning effective responses, weighing up consequences of actions), discernment between perception and reality, and remembering important information (University of Northern Colorado, n.d.). This is what psychotherapist Peter Levine calls 'the tyranny of the past' (in Maté, 2022, p. 16) – shocks to the system 'can alter a person's biological, psychological, and social equilibrium to such a degree that the memory of one particular event comes to taint, and dominate, all other experiences, spoiling an appreciation of the present moment' (ibid.), ultimately inhibiting the ability to respond from/or to the present situation. It should also be noted that with our wiring for security, responses to situations can be disproportionate to the event or present situation.

Four responses are often mentioned when describing the neurobiological trauma response: *fight*, as in fighting back; *flight*, removing oneself emotionally, mentally or physically from the situation; *freeze*, an immobility to fight or flight responses; and *appease*, a tonic or collapsed immobility, a 'going along with' the traumatic experience or mitigating it through short-term decision making to allay its affects. All are part of the brains and body's survival mechanisms (STACC, 2016.; University of Northern Colorado, n.d.). Experiences that may be traumatic include, but are not limited to: physical, sexual, and emotional abuse; childhood neglect; living with a family member with mental health or substance use disorders; sudden, unexplained separation from a loved one; poverty; racism, discrimination, and oppression; violence in the community, from war, or from terrorism.

Trauma symptoms may include (and again, are not limited to), in the short-term, shock or denial, anger, not feeling safe, not being able to trust others or feeling betrayed; and in the longer term, unpredictable emotions, such as anger or aggression, a tendency toward shame, dissociation, avoidance, emotional numbing, isolation, flashbacks, strained relationships, and physical symptoms like headaches or nausea. It is common to develop often maladaptive and compound behavioural coping mechanisms in the face of trauma (from unhealthy eating habits to substance abuse, for example.) Prolonged exposure to traumatic experience(s) has the potential risk for chronic health conditions and health-risk behaviours, lasting adverse impact on mental, physical, social, emotional, and/or spiritual wellbeing and functioning. At the root of these symptoms are a 'a fracturing of the self and of one's relationship to the world' (Maté, 2022, p. 23), and a feeling of being neither witnessed or known by ones loved ones, one's community or the world at large (van der Kolk, in Maté, 2022, p. 23).

Trauma can happen to anyone at any age and from any demographic, but it is found to have an especially debilitating impact on children, BIPOC populations, those with low income, those who are unemployed or unable to work, and those who identify as LGBTQAI+. It is especially common in the lives of people with mental and substance use disorders. This again is where trauma operates at the level outside of the individual – the globally impacting system and ethic of capitalism is a cause of stress(ors) on human health, the human body as much as the body politic, shouldering an 'allostatic load' of trying to maintain an equilibrium in the face of pervasive and continual change, and a load carried more heavily by the politically disempowered and economically disenfranchised (McKeown, in Maté, 2022, p. 48).

Trauma is considered to be relational, impacting one-to-one, one-to-many, and many-to-many relationships, and to impact the relationship with and the functionality of the delivery systems that support individual, community and public need (APA, n.d.; TICIRC, n.d.; SAMHSA, n.d.; STACC, 2016). As such, the trauma is not solely defined by the event, but by the individual's or group's reaction to the event, and, in a group setting, is as much about the individual's own response as it is their response to other's responses (Zinner & Williams, 1999, p. xix). Additionally, at the group level, there may be a feeling of a collective threat to existence, an increase or decrease in social connection, a fracturing of community bonds (ibid., p. 7). How a community recovers from trauma is dependent on the community leaders' capacity to lead such efforts by both openly and candidly acknowledging the traumatic event and then turning to rebuilding community equilibrium (Zinner & Williams, 1999, pp. 8–9). Key examples of this at scale can be seen through peace and reconciliation initiatives such as oft-cited South Africa and Northern Ireland, which have had varying degrees of success.

When considering avoiding or proactively mitigating trauma, protective factors – such as supportive close relationships with friends, family and community members, for example – are often cited positively. Such relationships are thought to have protective measures and help build resilience. Trauma-informed care at the systemic level is thought to help process traumatic experiences and can include specific treatments and healing programmes (TICIRC, n.d.). A picture then is building of trauma being a behavioural and public health concern, with necessary healing and recovery processes (SAMHSA, n.d.). However, we face an inherent friction in this approach of treatment for systemic trauma at the level of the individual: 'In an atomised, materialistic culture people are induced to take everything personally, to see their own mental and physical distress as misfortunes or even failures belonging to them alone' (Maté, 2022, p. 276).

If trauma has a behavioural, cultural and public aspect, then it also has a place one – beyond the occasion of people inhabiting place, but as the site of trauma occurrences, live and memorial. The traumatic experiences of poverty, racism, discrimination, oppression, violence in the community, war, and terrorism all operate with a place component. And the memories of such events linger through the stories we tell, rituals we enact, memorials we create, the names we give our streets... a 'placemind' epigenetics spatiality (an *epi-spatiality*?), in the sense of place that we feel in the DNA of place, even if we can't articulate it, when we are in or recalling a place and its conditions.

Thus, we return to our contention that, as people who work in place with the people of place, we have to understand trauma and our role in it, both positive and negative, and understand the traumatic impact of the systems we work in and with. What then, is a trauma-informed placemaking approach? The following section offers a pathway into this thinking and to creating an eventual practice by moving though some of the common facets of placemaking thinking and putting this through the lens of a trauma-informed approach. We offer this fully cognisant that this is the beginning of a journey and necessarily, as a community of praxis, this thinking will and should be evolved through application, time, place, and experience.

The Pathway to Trauma-Informed Placemaking

When well applied, trauma-informed efforts at the individual level have been proven to significantly reduce symptoms of physical and mental illness, and to support interconnectedness and wellbeing. The general trauma-informed approach originated in the context of the organisation, of a sustained programme of activity, and at the systemic level. For these contexts to be trauma-informed they have to: *realise* the widespread impact of trauma and understands potential paths for recovery; *recognise* the signs and symptoms of trauma in clients, families, staff and others involved in the system; *respond* by fully integrating knowledge about trauma into policies, procedures and practices; and seek to actively resist *re-traumatisation*. What does that mean, if we extrapolate from this origin context to that of place, placemaking and placemakers? What does a trauma-informed approach tell us of those common facets of placemaking, of place attachment and place identity, for example?

Firstly, let's turn our attention to our own placemaking, as individual practitioners, and our own trauma-informed accountability. Placemaking can work to either recreate or dismantle allostatic strain and can either promote or inhibit a 'response flexibility' (the capacity to choose how we respond to our conditions of living, Maté, 2022, p. 29). As placemakers we have an obligation to be aware of this within our own practices. This does not mean that we must be social workers, therapists or clinicians, more that we acknowledge that we ourselves are vulnerable and should be aware of how our bio-chemical operating system is wired for security. Situations that may challenge or make us feel uncomfortable can also mean we may be in a condition of reaction rather than of response. Secondly, as placemakers, we may be part of, or work closely and regularly with, a placemaking organisation. A scalable step for trauma-informed placemaking would be for those placemaking organisations to adopt a trauma-informed approach to their internal culture and mode of operation, developing internal and external practices that are empathic, considerate and compassionate and mindful of the processes and impacts of our behaviour in the field. Although culture change is difficult, and what is being proposed goes against long-standing practices that are motivated and dominated by geo-political neoliberal agendas around power and money, this should not dissuade us from moving towards a trauma-informed approach. Furthermore, practitioners must be aware of dominant narratives and how this obfuscates what may not only be exclusive practices but also be harmful and retraumatising, such as with displacement or creating instances of dis-belonging (Bedoya, 2013). To this end the book offers an array of principles to be aware of and apply within a situated practice of placemaking, where the activities emerge out of the diverse context in which they occur.

Now, to turn our attention to the trauma and place interplay. Placemaking is first and foremost a cultural practice, concerned as it is with the people of place and the lives they live therein. Maté not only views health and wellbeing, and ergo, trauma, as issues for a cultural collective, but also that culture acts on our wellbeing being via various biopsychosocial pathways – from epigenetic causes to matters of public health such as access to healthy food and clean air (2022, p. 287). Cultures of care to address wellbeing and more are evolving in a number of contexts, from healthcare to university ethics in relation to fieldwork. Therefore, it follows that placemaking, which often happens in the field and with living organisms should also evolve its cultural practices to include radical care and compassion, key aspects when working with traumatised individuals and communities. This viewpoint is supported by Besel Van Der Kolk, Professor of Psychiatry at Boston School of Medicine and President of the Trauma Research Foundation, Brookline, Massachussets. Van Der Kolk's research affirms four fundamental truths, quintessential aspects of humanity with cultural dimensions and critical for trauma healing, two of which are critically pertinent to placemaking and our role as placemakers: the first, 'our capacity to destroy one another is matched by our capacity to heal one another. Restoring relationships and community is central to restoring well-being' and the fourth, 'we can change social conditions to create environments in which children and adults can feel safe and we can thrive' (2014, p. 38). Van der Kolk contends that neglecting these truths deprives people of healing and emphasises the pivotal role of 'safe connections [as] fundamental to meaningful and satisfying lives' (ibid., p. 354). These can and should inform placemaking principles of practice – we can either support or collude with an acculturation of place-based trauma, or we can challenge and dismantle it, and rebuild differently.

Van der Kolk (2014) also recognises the importance of the imagination in such work, stating it as critical to our lives and important for healing trauma, where we can imagine new possibilities and a launchpad for realising hopes and dreams. Placemaking is a critical, reflexive and innovative practice and it inherently holds the potential of the radical imagination, 'the ability to imagine the world, life and social institutions not as they are but how they might otherwise be' (Haiven & Khasnabish, 2014, p. 3). This is not an imagining for its own sake, but an imagining to make real in the present and inspire 'common cause' (ibid.) and continual action across those communities of place, impact and interest that placemakers work with and are part of. To do this effectively though, we must be cognisant of the histories and lives of people-in-place and hold ourselves accountable to such a solidarity: 'Education, training, and information can be instructive and guiding; but they are not necessarily prescriptive. Effective helpers allow those they help to instruct them about what or who was lost and the meaning of the loss' (Zinner & Williams, 1999, p. 6). A reflexive exercise for placemakers is to ask of ourselves and each other, when have we been instructive, let along guiding, and how can we be effective and accountable to that?

We are building a picture, then, of placemaking as an essential activity for wellbeing and security, and as a remedy to a sense of estrangement and displacement (Aravot, 2002) – and we can invoke here the concepts of place attachment and place identity. Maté, amongst many others, asserts that the two essential needs of the human are attachment and authenticity (2022, p. 10) – these have a place dimension. Place attachment is the person/people-to-place emotional, psychological, and cognitive bonding to spatial settings over time; place identity is the 'personal identification with places that are emotionally and socially significant' (Low, 2023, p. 152). Both resonate at the group level through a sense of shared place attachment and identity within a community of people, whether this be *communities of place* (think, those groups of people that form a neighbourhood, for example), to *communities of impact* (think, a group of walkability campaigners, for example) and *communities of interest* (think, a group of young people that have formed a skate community in a park, for example) (Courage, 2022).

Place, as a realm of meaning and not simply a 'framework of geometric relationships' (Hubbard et al., 2004, p. 5) is recognised as more than a geolocation (Malpas, 2006). Rather, place holds cultural and social construction, interwoven with ideas of emplacement, embodiment and localised knowledge (Casey, 1996). This situated knowledge, as physical and psychological experience, represents the 'fusion of self, space and time' (ibid., p. 9). If there is a rupture in the relationship that people have with a particular place, their place attachment and sense of place identity will be called into question, Bedoya's dis-belonging (2013), leading to detachment or dissociation from their surroundings. This is the people-in-place equivalent of the individual trauma-arousing event: 'The rupture of this relationship through disaster, immigration, forced relocation, and even a desired move often generates a sense of personal loss and dislocation of those affected' (Low, 2023 p. 152).

As biopsychosocietal beings (Maté, 2022, p. 275), our connection to place and ability to be in a public space is part of our psychosocial integration. When this is taken away, it is traumatic – there is a sense of social dislocation to others and alienation from public culture. It is one's own sense of place and the personal attachments in relationship with others that are important for creating a sense of belonging and purpose in our lives (Buttimer, 1980, Relph 1976; Tuan 1974, 1976).

This is what we term *place authenticity*, the capacity to act true to one's sense of place in tandem with holding a capacity to shape one's place, and that we see as holding potential in a trauma-informed placemaking practice. Trauma-in-place impacts our sense of place authenticity – we lose a sense of place-orientated purpose and agency, and our capacity for place authenticity will be called into question. The syndemic of 2020, of lockdowns (or however locally termed) of COVID-19 and the street protests of Black Lives Matter (BLM), are a case in point. During COVID, public space became a site of anxiety and literal and moral policing, and laid bare, if one had not known this before, the racial, class and economic stratification of, access to, and work of place. Public green and/or open spaces and the thirdspaces of community halls, churches, bars and cafés and the like were yearned for, for the 'co-presence' (Low, 2023, p. 6) they had once afforded us: one can walk the streets or sit in a café 'alone but together' with strangers in most usual times, but not during a lockdown. People cleaved on to place authenticity, expressed their identity and reached out for connection with others through window artworks, concerts on balconies, exercise classes on rooftops and the like. And then, a few months into the pandemic, public spaces in many parts of the world became full again as people took to the streets to protest the murder of George Floyd, on 25 May 2020, by a police officer in Minneapolis, Minnesota. Here was public space being reclaimed as formative for democratic purpose and dissent and racial equality. Together, COVID-19 lockdowns and BLM protests, 'The daily separation of us and them, or 'people like us' and 'others', increased spatial segregation and social prejudice' (Low, 2023, p. 3). Depression, anxiety and feeling socially isolated were matched with feelings of spatial dis-belonging and a loss of place authenticity. Our public culture making had to adapt, materially and symbolically, with place changed in the equation.

Recent research in neuroaesthetics and the built and natural environment (Weinberger et al, 2021) has identified the psychological impacts of design on the reward centres of the brain and the emergence of 'three psychological dimensions: *Fascination* (a scene's richness and interest), *Coherence* (analytic judgements about a scene's organisation and construction), and *Hominess* (feelings of warmth or coziness) (Weinberger et al., p1:2021). These findings are not only consistent with a growing body of research (Weinberger et al, 2021; Vartanian et al., 2021; Coburn et al., 2020) that evidences an intimate and physical relationship with our environment, but also add further weight to the importance of a trauma-informed approach to the built environment and therefore placemaking, with an added implication of the criticality of such an approach for our wellbeing. To negate this approach – particularly the Hominess dimension and the impact on the reward centres of the brain – would impact on our sense of security, thus alienating us from our environment and leading to what Bedoya, 2013 calls disbelonging. With our propensity to mitigate for feelings of insecurity this could lead to potentially harmful behaviour including short-term decision making and an inability to weigh up the consequences of our actions.

Additionally, the existential threat of climate and ecological emergency is now presenting as trauma, with many people already overwhelmed and operating within the flight/freeze/appease response. Young people are reporting anxiety around climate issues and that they are worried about their future as they perceive that they don't have one (McKeown et al., 2022) – and in this study, further investigation showed these young people felt that the adults around them were doing nothing about it, with this perceived lack of protection and lack of leadership increasingly the source of their anxiety. Author, cultural historian and award-winning educator, Maggie Favretti, acknowledges that 'the most dangerous position to be in is one where you have no power to change your situation (2023, p. 1). This is a traumatic and traumatising situation and one in which young people and adults are increasingly experiencing whether as eco-anxiety or futurephobia (ibid.; Brookins, 2021). Further, there are increasing numbers of Google searches on eco- and climate anxiety and calls for recommendations for group therapy for climate scientists (Haddaway & Duggan, forthcoming), as well as an increase in scholarly articles including the typology of eco-anxiety (Agostan et al., 2022) showing six categories of eco-anxiety. Collectively these independent incidents indicate trauma-informed practices will be necessary as we move forward.

A Pathway of Trauma-to-healing

The trauma-informed placemaking endeavour is about making a commitment to trauma awareness and trauma-informed practices, and with that comes a necessary commitment to place healing and context-responsive, community-located place healing practices. The two cannot be separated. Only by learning about trauma can we find our pathway to healing. Just as the form of healing may differ between individuals, at the group level, different types of trauma will require different practices and processes of healing – examples of which are presented in this textbook, all informed by personal experience and from learning from and with communities in place – and what form this takes will have an impact on how that group continues to experience its place and identity:

> 'Different types of traumatic events may lead to differing types of reactions and patterns of community healing. The social context of an event, the community perception and conceptualisation of the meaning of that event, and the method of community coping (cognitive or affective) can influence outcomes.'
>
> (Zinner & Williams, 1999, p. 12)

The trauma-to-healing process is one of meaning-making and storytelling, a mirror in practice to that of placemaking. This creative process is one of biopsychosocial growth that uses natural human creativity to 'integrate fragmentary experiences into cohesive narrative' (Winnicott, Menin & Samual, in Samuel, 2023, p. 114). When we work in place we are working with the meaning and stories of that place;

and we may also be creating new meaning and stories of places. Key to this aspect of practice though is those feelings of an agentive place authenticity at the group level – placemaking, and healing, are practices that are done for and with not to, the group in question, Haraway's 'sympoiesis – making with' (2016, p. 5), '...a word for worlding-with, in company' (ibid., p. 58). It follows then that what we do as placemakers, in the trauma-to-healing context, becomes part of this place narrative: 'the process of responding to any disaster becomes part of the story of the disaster' (Zinner & Williams, 1999, p. 239). As placemakers, we will impact community resiliency, either positively or negatively. We will impact how the community can first react and respond to the trauma-in-place in material, emotional and biopsychosocial ways, and how this will be passed through and on from the event and the group, as 'epi-place' and cultural narrative.

Haraway's phrase 'staying with the trouble' is apt to invoke here, calling on us as placemakers, in the singular settings of our sites of work as well as our communities of practice and praxis, to involve ourselves intentionally in the complexity of such work:

> *'Staying with the trouble does not require such a relationship to times called the future. In fact, staying with the trouble requires learning to be truly present, not as a vanishing pivot between awful or Edenic pasts and apocalyptic or salvific futures, but as mortal critters entwined in myriad unfinished configurations of places, times, matters and meanings.'*
>
> *(Haraway, 2016, p. 1)*

However vital we feel this work to be, and however urgent it will feel in the live situation of the field, we advocate working at the speed of care in trauma-informed placemaking. Political theorist Joan Tronto states that care includes 'everything that we do to maintain, continue, and repair our 'world' so that we can live in it as well as possible' (1998, p. 17). Care is composed of attentiveness, relationality, responsiveness, and an interdependence with those we are working with. These, in accordance with Tronto (ibid.) are the qualities needed to form and act from a sense of collective, place-based, site-contextual and planetary care. As relational and interdependent, we also have to recognise that care for ourselves as placemakers, as biopsychosocial individuals as well as professional practitioners, works in tandem with care for others and care for places. We take our cue here (fully cognisant of the shoulders we are standing on to say this at all) from the notion of *radical self-care*, the assertion that you have the responsibility to take care of yourself first before attempting to take care of others, from Donna Nicol and Jennifer Yee. Working at the speed of care and with radical self-care, we recognise that as placemakers, our personal and professional obligations are connected to our societal and ecological and environmental ones.

To complement this pathway journey to trauma-informed placemaking we now offer a pathway that extrapolates from the *Six Principles of a Trauma-Informed*

Approach (Trauma Informed Institute, 2023) across to aspects of the placemaking practice journey. This is nascent, and to be extended through the critique of the trauma-informed community of praxis.

A Pathway Through the Six Principles of a Trauma-informed Approach and placemaking

The *Six Guiding Principles of a Trauma-Informed Approach* (Trauma Informed Institute, 2023) have been informative to thinking through the notion of a trauma-informed approach to placemaking. While a direct extrapolation from one across to the other could be reductive and flattening, in the formative endeavour of bringing a trauma awareness into placemaking, we propose a generative reading of the *Principles* into placemaking practice-process.

The *Six Trauma-Informed Principles* were developed by SAMHSA's National Center for Trauma-Informed Care (NCTIC) and CDC's Center for Preparedness and Response (CPR) with the intent to help working professionals better understand how trauma can impact the communities they serve. These *Principles* form a framework for professionals to adopt and adapt to create a trauma-informed workplace culture. As above, no doubt this would be a worthwhile exercise for place-based and place-concerned organisations to turn the lens on their own people culture; we propose though that the *Principles* can inform their externally facing activity and are useful for anyone working with people-in-place. We were first given the opportunity to explore this with Irish Architecture Foundation and its *Reimagine Pocket Guides* (IAF, n.d.) of thought leadership. What is presented below is a further development of what is presented in its *Pocket Guides* series.

Step 1: Develop an Awareness of Trauma, and Trauma-in-place

- in tandem with the Six Principle: 'Safety'

Physical and psychological safety is the first of the *Six Principles*. With safety, when trauma is experienced by the individual, it can be responded to sooner. It involves an environmental risk assessment and the intentional creation of an environment of mutual respect and trust. In the context of people-in-place, trauma is situated and embodied. Traumatic stress-related disorders comprise a specific constellation of symptoms and criteria, including emotional, physical, cognitive, behavioural, social, and developmental reactions. Trauma may also lead people to find they are unable to stop thinking about what happened, a high level of arousal – or feeling alert or 'on guard' – which causes strong sensory reactions in the individual. We have a responsibility as placemakers to inform and educate ourselves, without placing the burden of that exercise onto the traumatised, on the context of the places we work in and the impacts of trauma on the communities of place we work with.

Step 2: Understanding Trauma-in-place as Placemakers

- in tandem with the Six Principles: 'Trustworthiness & transparency'

In the *Six Principles*, trustworthiness and transparency are essential in the (workplace) environment to assure that people are ready for changes that will impact them. The basis of this is intentional communication. To understand trauma-in-place as placemakers we need to be cognisant of the historical precedent – keenly though colonialism, war and natural disaster, for example – and the impact this has on the future precedent through continuing climate and ecological emergency and further pandemics, for example. Think too of the traumatic experience of gentrification and social cleansing; of feelings of menace and threat experienced by many in the public realm; and of the rupture in the people–place relation during the COVID-19 pandemic and the sites of Black Lives Matter protests and police brutality. Engaging with any community we may have the privilege to work with requires meaningfully inclusive practices and an awareness of the basic principles of a trauma-informed approach and to act from this with careful intention and clear communication and mutual expectations.

Step 3: Ask Ourselves What We Can do as Placemakers

- in tandem with the Six Principles: 'Peer support'

The *Six Principles* advocate for 'employee resource groups', firstly, for those with shared identities to connect, and secondly, so that team members feel support, not isolation. As placemakers, we need to be asking ourselves – as a community of praxis – big questions of our practice. Amongst others, how do we conduct ourselves with those we work with? How do we challenge trauma-causing processes? We also need to turn to and learn from each other as a community of praxis – learning through doing, and we need to understand the political implications of our work and strive to keep informed of new conversations and developments in the field. We also need to engage in critical and constructive conversation about what we should not do as placemakers. We need to think of the trauma we have and may cause as placemakers though our practices and processes.

Step 4: Consider a Community of Praxis

- in tandem with the Six Principles: 'Collaboration and mutuality'

In the *Six Principles*, working collaboratively is about creating and working with mutual respect. This requires people to be committed to collaboration as more than an outcome in itself; accepting and working with differences of experience and opinion; and embracing that effective collaboration will offset the standard hierarchy of leadership. In the process of gathering as a group of people with a shared intention and practice to exchange knowledge and problem solve, we are creating action learning sets that signal a step change in the practice of placemaking through the practical application of theory and learning from experience.

Step 5: Become Part of the Community

- in tandem with the Six Principles: 'Empowerment, voice and choice

The objective of this *Principle* is to provide individuals with the skills and resources necessary to speak their truth and make decisions about their experience at work. This impacts the dynamics of leadership and power, no longer using a top-down governance structure by including more stakeholders – specifically, those who will be impacted by the decisions that need to be made. Leaning into collaboration to approach the endeavour of a trauma-informed placemaking community of praxis engenders the forming of a knowledge exchange community.

There is a second dimension to 'community' here – that of the community of place that we work with as placemakers. Firstly, if we are not of the place, we need to be in and with the communities of place that we work with. We need to get to know them from first-hand experience. Secondly, the devolved governance of this *Principle* will accord with many readers as something they recognise with what has become an idiom in placemaking – *the community is the expert in their own lives and place*, and *ipso facto*, should have an equal seat at the placemaking table.

Step 6: Our Call to Action

- in tandem with the Six Principles: 'Cultural, historical, gender issues'

The final *Principle* states that cultural, historical and gender issues should be considered when interacting, responding, and engaging within the workplace. It is important to understand the history and experiences of groups of people. This provides insight into how people of different cultures, identities, and genders may experience events, projects, overall work culture, etc. With this awareness, companies can make better-informed decisions and prevent trauma in the workplace. To parse Emma Geohegan's *Accessibility and Inclusion in Placemaking* (IAF, n.d.) successful placemaking, by holding the community as expert in their own lives and places, should, inherently, be inclusive. Going one step further, as a practice of multivariant partners, successful placemaking should have an inherent disposition to working intersectionally.

Our call to action also asks you to consider the National Institute of Corrections (n.d.) *Four R's of a trauma-informed approach* in your practice:

1. *Realisation*: you and your colleagues have a basic realisation about trauma and understand how trauma can affect people in place at an individual, group and community level.
2. *Recognise*: you and your colleagues are able to recognise trauma signs across identity intersections of age, race, gender, dis/ability etc., and as place/context specific.
3. *Respond*: you and your colleagues are able to respond through your trauma-informed approach awareness.
4. *Resist*: you and your colleagues resist re-traumatisation of people in place through intentional practice.

To close this introduction, we turn once more to the rallying call of Haraway's *staying with the trouble*:

> '*We – all of us on Terra – live in disturbing times, mixed-up times, troubling and turbid times. The task is to become capable, with each other in all of our bumptious kinds, of response… The task is to make kin in lines of inventive connection as a practice of learning to love and die well with each other in a thick present. Our task is to make trouble, to stir up potent response to devastating events as well as to settle troubled waters and rebuild quiet spaces.*'
>
> *(Haraway, 2016. p. 1)*

Action and agency have been shown to increase hope and mitigate increasing eco-/climate anxiety (Haddaway & Duggan, forthcoming; Favretti, 2023; McKeown et al., 2022). If we are to develop solutions through action and agency for climate and ecological crisis, and knowing how trauma can impede both, a trauma-informed lens becomes a critical praxis for placemaking, underpinned by a radical culture of care and compassion. Moving beyond these chapters and the pages of the book, we are entrusted with the mandate – to carry forward the wisdom, insights, and challenges that these chapters present. Our call to action is clear: to be accountable as placemakers, to imbue our actions with sensitivity, empathy, and purpose, and to honour the traumas and healing that are intertwined within our environments. By weaving these principles into the heart and core of our practice, we begin to forge a path forward. A path collectively as practitioners, where each action, each choice, becomes an intention to contribute to the healing of the spaces and places we co-create and to a planetary healing – our lives may depend on it.

References

Ágoston, C., Csaba, B., Nagy, B., Kőváry, Z., Dúll, A., Rácz, J., & Demetrovics, Z. (2022). 'Identifying Types of Eco-Anxiety, Eco-Guilt, Eco-Grief, and Eco-Coping in a Climate-Sensitive Population: A Qualitative Study', *International Journal of Environmental Research and Public Health*, 19 (4), 2461.

American Psychological Association (n.d.). *Trauma* [online]. Available: www.apa.org/top ics/trauma#:~:text=Trauma%20is%20an%20emotional%20response,symptoms%20l ike%20headaches%20or%20nausea. [Accessed: 25 August 2023].

Aravot, I. (2002). 'Back to Phenomenological Placemaking', *Journal of Urban Design*, 7 (2).

Bedoya, R. (2013). 'Placemaking and the Politics of Belonging and Dis-belonging', *GIA Reader*, 24 (1).

Brookins, S. M. (2021). 'Fear of the future', *Futurephobia*. Available: https://fearof.org/futurephobia/ Accessed: 28 August 2023.

Buttimer, A. (1980). 'Home, Reach, And The Sense Of Place' in A. Buttimer & D. Seamon (eds), *The Human Experience of Space and Place*. London: Croom Helm.

Casey, E. (1997). *The Fate of Place: A Philosophical History*. California: University of California Press.

Coburn, A., Vartanian, O., Kenett, Y. N., Nadal, M., Hartung, F., Hayn-Leichsenring, G., et al. (2020). 'Psychological and Neural Responses to Architectural Interiors', *Cortex*, 126, 217–241.

Courage, C. (2022). 'The Relational Museum', *MUSEUM*, Mach-April 2022.

Favretti, M. (2023). *Learning in the Age of Climate Disasters: Teacher and Student Empowerment Beyond Futurephobia*. New York: Routledge.

Geohegan, E. (n.d.). 'Accessibility and Inclusion in Placemaking' in Reimagine Pocket Guide 3, *Irish Architecture Foundation* [online]. Available: https://reimagineplace.ie/pocketguides/placemaking-in-the-irish-town/. [Accessed: 25 August 2023].

Haddaway, N. R., & Duggan, J. (forthcoming). '"Safe Spaces" and Community Building for Climate Scientists, Exploring Emotions through a Case Study' [Accepted manuscript], *Global Environmental Psychology*.

Haiven, M. & Khasnabish, A. (2014). *The Radical Imagination*. London: Zed Books.

Haraway, D. J. (2016). *Staying with the Trouble: Making Kin in the Chthulucene*. Durham: Duke University Press.

Hubbard P., Kitchin R., & Valentine G. (eds.) (2004). *Key Thinkers on Space and Place*. Thousand Oaks, Sage Publications.

Irish Architecture Foundation (n.d.). *Reimagine Pocket Guides*. Available: https://reimagi neplace.ie/resources/. [Accessed: 25 August 2023].

Low, S. (2023). *Why Public Space Matters*. Oxford: Oxford University Press.

Malpas, J. (2006). *Heidegger's Topology: Being, Place, World*. Cambridge MA: MIT Press.

Maté, G. (2022). *The Myth of Normal: Trauma, Illness and Healing in A Toxic Culture*. London: Penguin.

McKeown, A., Hunt, L, Murphy, J., Turner, E., White, R. (2022). *Co-designing for Resilience in Rural Development through Peer-to-peer Learning Networks and STEAM Place-based Learning Interventions*. Available: www.epa.ie/publications/research/epa-research-2030-reports/Research_Report_409.pdf [Accessed 25 August 2023].

National Institute of Corrections (n.d.). *The Four R's of a Trauma-Informed Approach*. Available: https://info.nicic.gov/sites/default/files/Trauma-Informed%20Approach.pdf. [Accessed: 25 August 2023].

Nicol, D. J. & Yee, J. A. (2017). ' "Reclaiming Our Time": Women of Color Faculty and Radical Self-Care in the Academy', *Feminist Teacher*, 27 (2–3).

Relph, E. C. (1976). *Place and Placelessness.* London: Pion Ltd.

Samuel, F. (2023). *Housing for Hope and Wellbeing.* Abingdon: Routledge.

Sexual Trauma & Abuse Care Center (2016). *Neurobiology of Trauma* [online]. Available: http://stacarecenter.org/wp-content/uploads/2015/09/The-Care-Center-Neuro biology-of-Trauma-Nov-2016.pdf. [Accessed: 30 August 2023].

Substance Abuse and Mental Health Services Association (n.d.). *Trauma and Violence.* Available: www.samhsa.gov/trauma-violence. [Accessed: 25 August 2023].

Trauma Informed Institute (2023). *Understanding the 6 Trauma-Informed Principles* [online], 27 February 2023. Available: https://traumainformedinstitute.com/blog/unders tanding-the-6-trauma-informed-principles. [Accessed: 25 August 2023].

Trauma-Informed Care Implementation Resource Center (n.d.). *What is Trauma?* Available: www.traumainformedcare.chcs.org/what-is-trauma/. [Accessed: 25 August 2023].

Tronto, J. C. (1998). 'An Ethic of Care', *Generations: Journal of the American Society on Aging*, 22 (3).

Tuan, Y-F. (1974). *Topophilia: A Study of Environmental Perception, Attitudes, and Values.* Englewood Cliffs: Prentice-Hall.

Tuan, Y-F. (1976). 'Humanistic geography', *Annals of the Association of American Geographers*, 66 (2).

University of Northern Colorado (n.d.) *Neurobiology of Trauma* [online]. Available: www. unco.edu/assault-survivors-advocacy-program/learn_more/neurobiology_of_trauma. aspx. [Accessed: 30 August 2023].

Van Der Kolk, B. (2014). *The Body Keeps the Score: Brain, Mind and Body in the Healing of Trauma.* New York: Viking Press.

Vartanian, O., Navarrete, G., Palumbo, L., & Chatterjee, A. (2021). 'Individual Differences in Preference for Architectural Interiors', *Journal of Environmental Psychology*, 77, Article 101668.

Weinberger, A. B., Christensen, A. P., Coburn, A., & Chatterjee, A. (2021). 'Psychological Responses to Buildings and Natural Landscapes', *Journal of Environmental Psychology*, 77, Article 101676.

Zinner, E. S. & Williams, M. B. (1999). *When a Community Weeps: Case Studies in Group Supervision.* Philadelphia: Brunner/Mazel.

SECTION 1

Understanding and Developing Our Trauma-in-Place Sensitivity

1

TOWARDS TRAUMA-INFORMED PLACEMAKING

Mutual Aid, Collective Resistance and an Ethics of Care

Lynne McCabe

Preface

I am my Irish grandmothers' wildest dreams come true. I grew up in the '80s in an architecturally brutalized and economically beleaguered housing scheme on the outskirts of Glasgow, Scotland, during Thatcher's decimation of the working class. The first person in my father's family to finish high school, I hold a Bachelor of Fine Arts (BFA) Glasgow School of Art and have two master's degrees, one a Master of Fine Arts in socially engaged art and one a Master of Social Work with a concentration in clinical trauma-informed care. I am Irish and Catholic by blood, Scottish by birth, and only fully understood myself as 'white' when I migrated from Glasgow to Houston, Texas, 23 years ago. I am queer, disabled and working class. My body sits at the intersection of multiple privileges and historical oppressions.

As a socially engaged artist my practice focuses on making and holding space for marginalized voices and bodies. As placemaking practice, this has ranged from installing myself as institution-come-artist in museums, teaching classes, hosting panels, and opening my own queer multi-use art space, to the most intimate of spaces, the dinner table. Whether these tables were built from driftwood and perched on the beach of an international border (McCabe & Robbins, 2018), the menu a consumable poem on colonialism and cannibal resistance, or two dismembered dining tables reconstituted to create an imperfect larger table that took over a row house in Houston's Third Ward, the table becoming a physical metaphor, if you will, for the complex nature of socially engaged art making (McCabe & DeBrock, 2007). These intimate dinners were at once mundane and magical sites for sharing culture, history, and love in union with those around me. I have returned to the form of the dinner again and again because the meal provides the most sensual,

DOI: 10.4324/9781003371533-3

equitable way I know for my story, for all our stories, to be transmuted and literally embodied.

And then, COVID happened. Overnight, gathering for a meal became unsafe. I could not ask my community to place their already trauma-soaked bodies in anymore danger, nor could I risk my own immunocompromised disabled body. I realized I had to pivot from host to witness.

And oh, what I have seen. As a disabled immune-compromised individual my trauma response to COVID was to immediately enroll in the relative safety of a virtual master's program at the University of Houston's Graduate College of Social Work. In the pages that follow I share observations from my work as both a socially engaged artist and as a social worker specializing in trauma-informed care. This chapter is an exploration of several trauma responses that I have encountered – Gay, Lesbian, Bisexual, Transgender, Queer, Intersex, and Asexual (LGBTQAI+) Trauma-informed Care, Mutual Aid and Collective Resistance. I will share them the only way I can, in the embodied way that I experienced them, through two vignettes drawn from my first-hand experience. The chapter examines how creative and trauma-informed placemaking practices can support impacted communities by creating secure places in which Mutual Aid and Collective Resistance can happen and generate healing.

From Host to Witness

I have lived on the Gulf Coast of America in Houston, Texas, for over 20 years and in that time it has become impossible not to see how each storm arrives in the region as an unfortunate weather event but also impacts the communities within the Gulf Coast as a set of compounding systems of oppression. The storm of the pandemic hit us just like all the preceding disasters, with resources and health outcomes determined by the ZIP Code people lived in, the historical and well-worn racial inequities of the city made concrete; and the Gulf Coast is made of concrete, to cover up the mud (Bogost, 2017). The converging political, health and weather disasters the residents of the Gulf Coast have endured in only the last three years have led to many, including myself, questioning whether the 'trauma tax' Gulf Coast residents pay to live here, which often shows up in bodies as multiple chronic illness, is worth it (Mankad, 2021).

As I write this in the summer of 2023, members of the LGBTQAI+ community, specifically transgender adults and parents of transgender children, have decided that it is not. We are currently witnessing a migration from the south which echoes 'the great migration' of the '40s (Gavins, 2016) as a direct result of recent legislation that actively criminalises the personhood of transgender individuals (ACLU, 2023).

Living through multiple storms I am keenly aware of how my socially constructed whiteness, (Guess, 2006) and the privileges it has endowed me, has to a great extent protected me from the direst traumatic consequences of the systemic racist and environmental injustices found throughout the Gulf Coast region. Even still I am noticing in myself and in members of my community behaviors that can be categorized into the familiar choices of flight, fight, fawn, or freeze. You see, for me and mine, we have seen this movie before when we grew up queer during the HIV/AIDS crisis, when we had to leave our homes during our *An Gorta Mor* (Gaelic for The Great Famine of 1845–1849). We know how fast plagues strike. We've lived through the blight once and we won't watch our children starve again. We know what we know.

What Our Bodies Know

The trauma response is characterized by a lightning-fast over-correction driven by a body wired for survival. In the trauma state the prefrontal cortex, where all our executive thinking and decision making takes place, literally goes offline and our emotional brain, the amygdala, where our fight, flight, freeze and fawn (Evans et al., 2023) response resides, takes over.

Our brain makes sense of the world from the bottom up. Our 'lizard brain' (O'Sullivan, 2022) at the base of our skull registers input from our sensory neuroprocessors, sight, sound, touch, smell and taste, these signals are then passed along our synaptic highways until they reach the prefrontal cortex where they can be deciphered and organized. A brain in trauma skips this processing stage and the amygdala is in charge. There is a saying in neurobiology, 'what fires together wires together' (Perry & Winfrey, 2021). Essentially this means that with each new traumatic experience the brain develops stronger and stronger neural pathways to the amygdala making it increasingly more difficult to engage our higher ordered thinking (O'Sullivan, 2022).

Time and the differentiation between short- and long-term memory exists in the prefrontal cortex (Van der Kolk, 2014); subsequently, from a neurobiological standpoint, when someone suffering from post-traumatic stress disorder (PTSD) is triggered, they experience their historical trauma in the present. An important neurobiological study conducted during the pandemic has shown that whether you were infected with the COVID virus or not, your neural pathways have literally been 'rewired' by the collective trauma we have all lived through (Brusaferri et al., 2022).

Mutual Aid / Collective Resistance in Placemaking

Taking these professional and methodological contradictions and shortcomings into account how does one then embark on a trauma-informed approach to practice? I propose that to do so ethically we must acknowledge and embrace the tensions inherent to this work. For this potentially transformative practice to have efficacy it must define the structure of the intervention, not be applied to existing structures as an additive flavor. I would argue that Mutual Aid efforts offer us an array of examples of possible trauma-informed collective care. In his 2020 book, '*Mutual Aid: Building Solidarity During This Crisis (and the Next)*', Dean Spade describes Mutual Aid as 'a term used to describe collective coordination to meet each other's needs, usually stemming from an awareness that the systems we have in place are not going to meet them' (Spade, 2020). These efforts, long practiced by marginalized communities across the globe, stand in contrast to social justice actions designed to pressure powerholders to do the right thing.

Mutual Aid must be the site from which the placemaker, the artist, the social worker, the community organizer or the architect, whomever, is invested in a practice that intentionally engages in the ethics of care. The lesson a disaster teaches us, be it natural or political, is that when the collective is harmed, the response and repair must also be collective.

As previously discussed, the historically oppressed communities of black, brown and indigenous peoples of the Gulf Coast have been one of the hardest hit by accelerating climate, health and socio-political disasters in America (Noffsinger, et al., 2012). These communities sit at a challenging intersection of history, geography, industry, and political power. The history of slavery, Jim Crow laws, immigration and systemic racism and voter suppression within the state governments of Louisiana, Texas, Mississippi, and Alabama have left the inhabitants' fate in the hands of Republicans friendly to the petrochemical industry (Jensen, 2016). The resultant lax regulations and negligent enforcement has combined to heighten the risk of disasters for the poor communities living and working in proximity to industry sites (Bullard, 1999). Adding insult to injury federal evacuation plans do not cover these poor communities, despite their increased risk due to proximity to petrochemical sites and floodplains, as the federal government prioritizes securing property over people (Ait Belkhir & Charlemaine, 2007). Instead, people look to their state governments, swollen with callous Republican politicians beholden to the petrochemical industry to save them.

Vignette One: LGBTQAI+ Trauma-informed Care

In the second year of my master's program, I was awarded a fellowship in the emergent liberal discourse of trauma-informed care and received a grant to conduct research on how trauma showed up within the LGBTQAI+ community, concentrating on the health disparities particular to transgender and non-binary members of the community. While I was learning to examine the multiple and intersecting systems of oppression in order to craft therapeutic interventions, in my social work program, the trauma-informed fellowship offered little to no research or analysis on community-wide trauma responses. In fact, I made it through one of the nation's preeminent trauma-informed psychotherapy fellowships (Amtsberg, 2023) without encountering a single example in the pedagogical materials of queer trauma. This is shocking considering the extensive research demonstrating that LGBTQAI+ youth are disproportionately affected by multiple forms of childhood trauma, including rejection by their families, which leads to increased homelessness and higher risks of practicing survival sex and being trafficked (Roberts et al., 2012). Due to a lack of trauma-informed gender-affirming transitional housing, members of this community experience homelessness for longer periods of time; consequently, presenting to services with greater mental and physical needs (Vandenburg et al., 2021).

This disconnect between the words and actions of those at the forefront of professionalizing trauma-informed care became emblematic of tensions I began to identify within this burgeoning field. I came into the trauma fellowship program understanding the neurobiology of trauma through the lens of my own chronic illness. Trauma-informed care originated in the United States in the 1970s as behavioral health professionals struggling to adapt to the very specific emotional and physical 'injuries' soldiers were presenting with as they returned from the Vietnam War (Hyde, 2015). After observing this population, doctors with the department of Veterans Affairs coined the term 'post-traumatic stress disorder' replacing the First World War term 'shellshocked'. In 2001 the US government created the National Center on Trauma-Informed Care, spearheaded by Susan Salasin, a project officer at Substance Abuse and Mental Health Services Administration (SAMHSA) (Hyde, 2015).

When confronted with these open wounds, behavioral health professionals such as phycologists, psychiatrists, and psychotherapists, could adopt an ethics of care that centers social justice and systems theory (Walker, 2012) in their patients healing journey. Instead, I witnessed my professors and the scholars I studied cling to 19th-century individualistic ideologies that located the success and failure of healing from trauma within the individual, not the traumatizing systems within which we all exist (Dalal, 2018). It seemed to me that the primary goal of trauma-informed care was to address it head-on and take care of it, and in doing so ensure the comfort of neo-liberal society.

Trauma in the Temporal

Given the way multiple injustices overlap and compound the effects of a disaster, the only just response to the traumas disasters leave in their wake must be grounded in a strength-based community-led (West-Olatunji & Goodman, 2011) approach. A just trauma response must center the survivor as they lead their own recovery process (Pyles, 2016). Some disaster responses in Europe are already allowing grassroots organizations to leverage governmental funds and infrastructure to get resources to communities in need (Stokols et al., 2009).

One important element that must be considered in this work is time. Trauma-informed care practices are crafted to remind an over-reactive nervous system that it is safe, and that can only happen effectively if the body *is safe,* and community is established. Right now, queer families, especially those with transgender and gender queer children are fleeing the South. Those with resources to relocate themselves are doing so individually, others are building coalitions and Mutual Aid societies inspired by the 'Rainbow Railroad' (Powell, 2023). TikTok has been a vital organizing platform for fundraising and training during this exodus. This digital placemaking is led by Mutual Aid funds such as A Place for Marsha (an organization working towards helping transgender refugees in the US find safe homes) (Emanuel, 2023). I am tempted by the urgency of the situation and, the immediacy of the issue as I write this, to look for trauma-informed solutions and responses to include in this text. However, now is not the time for that. Families are still gathering resources to move or establishing homes and communities in a new state. We can't convince a traumatized nervous system that it is safe before the boxes are unpacked. It is easy to confuse urgency to help with the need to produce something for those outside the trauma, who are ready to see solutions.

I am heartened by Alice Elliot's compelling argument for the equivalence of *hope* as an engine for forced migration (Fiddian-Qasmiyeh et al., 2020) and how this might hone and expand the scope of trauma-informed practices. As a queer cis-gender (denoting or relating to a person whose gender identity corresponds with the sex registered for them at birth (Human Rights Campaign, 2023)) woman living in Texas I know how centering hope and joy in the face of oppression can have the effect of a soothing balm on my understandably traumatized and hyper vigilant sympathetic nervous system. I want to make an argument for trauma-informed practices to center beauty, hope, and joy as a foundational element of its ethics of care. A practice that deploys and shores up the community organizations and Mutual Aid, funds drag shows, that harness the radical power of queer beauty (Vaid-Menon, 2022) as a tool of resistance to the ugly, hateful and terrifying anti-LGBTQAI+ laws that southern Republican state legislators have passed during the 2023 legislative session (ACLU, 2023). I'm interested in how this shift changes the internal narrative and neurobiological experience of the self-proclaimed 'transfugee' (Tranniemom, 2023) community that makes up the majority of the people in the forced migration we are currently witnessing.

Vignette Two: A Place of Belonging

Writing about trauma is often a re-traumatizing experience. As a person with multiple learning disabilities, I have always approached writing as a trauma-informed practice. My dyslexia, attention deficit hyperactivity disorder (ADHD) and dysgraphia demand that my writing is made in concert with my community. This is comprised of trusted friends in academia, most commonly my wife, the artist and educator, Ruth Robbins, who not only edits my texts to remediate the spelling errors that are a defining expression of my dyslexia, but also on occasion allows me to utilize her working memory by typing as I dictate.

However, during the writing of this essay this writing partnership with my wife fell apart. As I talked to her about my ideas and research, she became frustrated and irritated. I had triggered her amygdala and our harmonious working partnership, where she in effect allowed me to timeshare her prefrontal cortex, was off the table. Her 'lizard brain' was now in charge. You see, Ruth had been living and working in New Orleans for almost seven years when Katrina hit, and was a part of the forced migration west to Houston that filled the astrodome and our Houston schools.

Ruth was invited to be a teaching artist at what became known as New Orleans West Elementary (Dart, 2015) by the charter school organization Knowledge Is Power Program (KIPP). KIPP is a network of charter schools that is based in Texas, operating in cities around the country. The charter organization approached the Houston School Board and proposed to run a short-term school serving evacuated children. They already had contact with teachers and administrators who had been working for KIPP in New Orleans and were able to reach out to these evacuated employees and invite them to come to Houston, to teach. When New Orleans West opened every person in the building was a Katrina evacuee.

Leveraging the resources of the Houston School District and funneling them into this facility is a compelling example of what an exterior organization can do to provide structure and support in time of crises. As the school was beginning to take form the administrators began to explore how they would reach the displaced students of New Orleans who had been evacuated to Houston and were still living in a makeshift refugee camp in the Astrodome (Harris, 2021). They utilized a colloquial culturally specific vehicle to recruit students and families, they made *Mardi Gras* beads with a medallion with all the enrollment information attached. As Ruth tells it the staff simply 'walked through the crowds in the Astrodome and started handing out beads at which point the actual customer you need when you are trying to start a school – the children – started running towards them jumping and waving their hands and saying throw it to me.' This simple, but deeply felt and recognized, cultural gesture signaled that to some degree this school could be a place where the children were understood. This is exemplar of a grassroots culturally competent approach to trauma-informed practice.

The shift I made from socially engaged artist to social worker was in hindsight a trauma-driven response to COVID-19 laying bare the precariousness of my financial and, by default in America, healthcare situation as a professional artist. My art practice is rooted in a collaborative intersectional social justice-focused approach, the switch to social worker seemed like a logical extension of this work. I was aware that after years of making work with diverse communities, ultimately the cultural capital the works accrued within the art world had not transferred to the kind of political capital that could be leveraged for real substantive investment in these communities. I naively thought that in the role of social worker I might be able to change that.

Ethics of Care

One thing I have learned is that every trauma response is specific and embodied, and each body responds from its precise location and with its own nature. Factors as small as bio-chemical reactions and as large as geo-political forces intersect in each of us to create a unique landscape on which large-scale traumas (hurricanes, forced migration, pandemics) unfold.

The intersecting roles of trauma-certified licensed Social Worker/artist is what frames my interrogation of the tension between the idea of trauma-driven and trauma-informed placemaking, i.e., the reactionary and the considered. I would even go so far as to state that the essential goal of adopting an ethics of care for we who engage in what has been termed 'trauma-informed work' is primarily harm reduction, figuring out how to move away from the unthinking, unconscious response and bring it to consciousness. This is a slow, iterative process that must be 'led from below' (Klein, 2015). That's the only way it works. Although trauma theorists invoke systems theory (Walker, 2012) to position the subject (Gulf Coast resident/Irish/queer) at the nexus of intersecting oppressive systems, a disheartening amount of the discourse and pedagogical materials instructive in the development of trauma-informed practices continue to center individualistic neo-liberalism when, as I have stated; *if the collective is harmed, the response and repair must also be collective.*

It is my assertion that anyone wanting to lay claim to the title of trauma-informed practice must employ cultural humility and a strength-led approach to the community they are working with (Jacobs, 2018). Open to appreciating how community grassroots organizations conceptually and structurally refute the neoliberal individualism that many in the behavioral health professions, at the forefront of this emergent field of trauma-informed practice, continue to cling to.

The concept of Mutual Aid as a voluntary, reciprocal, sharing of resources, skills, and ideas, is not only well understood in communities we as practitioners may find ourselves invited to work with, but, I want to stress that as these organizations become more established, they take on rich cultural traditions as well as offer community financial support and can become hubs for education, preparedness and peer-to-peer support (Mutual Aid Houston, 2022). Large-scale traumas are unique in the way they inspire action, moving us to participate in collective action to assist the impacted communities. This impulse to respond to the vulnerable and engage in an ethics of care manifests in a diverse range of actions, attending a vigil, donating online, volunteering, offering space in one's home. Often, the urgency of the situation discourages the mindfulness and reflection that are necessary to safeguard the interests of those involved. As we witness the wave of escalating geo-political climate disasters that has barely begun to crest, I am convinced that the work of understanding how trauma works in our bodies and in our communities is urgent and necessary for those who are called to act collectively.

References

Ait Belkhir, J. & Charlemaine, C. (2007). 'Race, gender and class lessons from Hurricane Katrina', *Race, Gender & Class*, 14.

American Civil Liberties Union (2023). *'Mapping attacks on LGBTQ rights in U.S. state legislatures'* [online]. Available: www.aclu.org/legislative-attacks-on-lgbtq-rights. [Accessed: 12 August 2023].

Amtsberg, D. (2023). *Trauma education program*. University of Houston. Available: https://uh.edu/socialwork/academics/msw/unique-opportunities/trauma-education-program-tep/#:~:text=The%20GCSW%20educates%20Trauma%20Fellows,the%20needs%20of%20adult%20clients. [Accessed: 11 August 2023].

Bogost, I. (2017). 'Houston's Flood is a design problem', *The Atlantic* [online]. Available: www.theatlantic.com/technology/archive/2017/08/why-cities-flood/538251/. [Accessed: 9 July 2023].

Brusaferri, L. et al. (2022). 'The pandemic brain: Neuroinflammation in non-infected individuals during the COVID-19 pandemic', *Brain, Behavior, and Immunity*, 102.

Bullard, R. D. (1999). 'Dismantling environmental racism in the USA', *Local Environment*, 4(1).

Dalal, F. (2018). 'Introduction Hyper-rationality', in *CBT: The cognitive behavioural tsunami: Managerialism, politics and the corruptions of science*. Boca Raton: Routledge.

Dart, T. (2015). 'New Orleans West': Houston is home for many evacuees 10 years after Katrina', *The Guardian* [online]. Available: www.theguardian.com/us-news/2015/aug/25/new-orleans-west-houston-hurricane-katrina. [Accessed: 2 April 2021].

Emanuel, E. (2023). *A Place For Marsha*. Available: www.aplaceformarshaofficial.org/. [Accessed: 10 July 2023].

Evans, O. G. et al. (2023). 'Fight, flight, freeze, or fawn: How we respond to threats', *Simply Psychology*. Available: www.simplypsychology.org/fight-flight-freeze-fawn.html. [Accessed: 4 August 2023].

Fiddian-Qasmiyeh, E. & Elliot, A. (2020). '7. Mediterranean distinctions: Forced migration, forceful hope and the analytics of desperation', *Refuge in a moving world tracing refugee and migrant journeys across disciplines*. London: UCL Press.

Gavins, R. (2016). *The Cambridge Guide to African American history*. New York: Cambridge University Press.

Guess, T. J. (2006). 'The social construction of whiteness: Racism by intent, racism by consequence', *Critical Sociology*, 32(4).

Harris, C. (2021). 'Educating the uprooted', *GovTech*. Available: www.govtech.com/em/disaster/educating-the-uprooted.html. [Accessed: 14 August 2023].

Human Rights Campaign (2023). 'Cis-gender' in *Glossary of terms*. Available: www.hrc.org/resources/glossary-of-terms. [Accessed: 5 June 2023].

Hyde, P. S. (2015). 'Foreword', *Quick guide for clinicians based on tip 57: Trauma-informed care in Behavioral Health Services*. Rockville, MD: U.S. Department of Health and Human Services, Substance Abuse and Mental Health Services Administration, Center for Substance Abuse Treatment.

Jacobs, F. (2018). 'Black feminism and radical planning: New directions for disaster planning research', *Planning Theory*, 18(1).

Jensen, M. (2016). 'Oil industry pollutes Texas politics, too', *The Daily Texan*. Available: https://thedailytexan.com/2016/07/22/oil-industry-pollutes-texas-politics-too/. [Accessed: 23 April 2021].

Klein, H. (2015) 'Beyond Chiapas', *Compañeras: Zapatista Women's Stories*. New York: Seven Stories Press.

Mankad, R. (2021). 'Living in Houston comes with 'the hidden tax' of trauma from disasters', *The Houston Chronicle*, 21 February 2021.

McCabe, L. & DeBrock, T. (2007). *Pot luck*. Houston, TX.

McCabe, L. & Robbins, R. (2018). *On Cannibalism: a poem on migration and colonialism in six courses*. American Camp National Park, San Juan Island, WA.

Mutual Aid Houston (2022). Available: https://mutualaidhou.com/. [Accessed: 15 April 2023].

Noffsinger, M. A. et al. (2012). 'The burden of disaster: Part I. Challenges and opportunities within a child's social ecology', *International Journal of Emergency Mental Health*, 14(1): 3–13.

O'Sullivan, K. (2022). *The Neurobiology of Trauma, The Neurobiology of Trauma | Danielle Rousseau*. Available: https://sites.bu.edu/daniellerousseau/2022/04/25/the-neurobiol ogy-of-trauma/. [Accessed: 25 May 2023].

Perry, B. D. & Winfrey, O. (2021). *What happened to you?: Conversations on trauma, resilience, and healing*. New York: Flatiron.

Powell, K. (2023). *Rainbow railroad, Rainbow Railroad*. Available: www.rainbowrailroad. org/. [Accessed: 16 May 2023].

Pyles, L. (2016). 'Decolonising disaster social work: Environmental justice and community participation', *British Journal of Social Work* [Preprint].

Roberts, A.L. et al. (2012). 'Elevated risk of posttraumatic stress in sexual minority youths: Mediation by childhood abuse and gender nonconformity', *American Journal of Public Health*, 102(8).

Spade, D. (2020). *Mutual aid: Building Solidarity during this crisis (and the next)*. London: Verso.

Stokols, D. et al. (2009). 'Psychology in an age of ecological crisis: From personal angst to collective action', *American Psychologist*, 64(3).

Tranniemom (2023). 'Transfugee', *ticktok.com*. Available: www.tiktok.com/t/ZT8NyeDsY/. [Accessed: 28 April 2023].

Vaid-Menon, A. (2022). *I'm fighting for beauty, ALOK*. Available: www.alokvmenon.com/ blog/2020/12/4/im-fighting-for-beauty. [Accessed: 2 November 2021].

Van der Kolk, B. (2014). *The body keeps the score: Brain, mind, and body in the transformation of trauma*. New York: Viking.

Vandenburg, T., Groot, S. & Nikora, L. W. (2021) '"This isn't a fairy tale we're talking about; this is our real lives: Community-orientated responses to address trans and gender diverse homelessness', *Journal of Community Psychology*, 50(4).

Walker, S. (2012). *Effective social work with children, young people and families: Putting systems theory into practice*. London: SAGE.

West-Olatunji, C. & Goodman, R. D. (2011). 'Entering communities: Social justice oriented disaster response counseling', *Journal of Humanistic Counseling*, 50(2).

2

(EM)PLACING TRAUMA

The Wounds Among Us

Jacque Micieli-Voutsinas

The Wounds Among Us

The discourse around trauma remains heartbreakingly relevant. In this COVID era, we are experiencing compounding crises as ecological vulnerability, economic precarity, and structural violence rippled across the globe, leaving our communities, our families, even our own selves, deeply wounded, psychologically, emotionally, and physically. As we learn to live with, and alongside, the virus and its consequences, we are seeing an explosion in trauma-informed care practices as well as renewed interest in the management of chronic post-traumatic stress throughout much of the world. Recent years have taught us that a crisis can radically interrupt our lives at any moment. Together, this forms a context in which trauma happens more often than not, and the risk of being traumatized remains high, an omnipresent reality always looming above us, or waiting just around the corner. It's no wonder that rates of anxiety leapt up during the pandemic (World Health Organization, n.d., 2022) – exacerbating a long-term trend, with rates doubling from 2008 to 2018 (Goodwin et al., 2020).

Perhaps we are right to feel anxious. As Lauren Berlant highlights in her 2011 book, *Cruel Optimism*, trauma isn't rare. It is an everyday part of our contemporary world. Therefore, when discussing trauma, we shouldn't frame it as something extraordinary, but something, well, ordinary. Twelve years later and Berlant's thesis rings as clear as ever.

Berlant's thesis sought to describe the social and anthropogenic structures that undergird our society and how they are prone to producing traumas, both big and small. While that certainly rings true in the context of neoliberal global capitalism, the novel corona virus illuminated the economic and political fallout of decades of deferred economic, racial, and climate justice. As such, discussions of trauma

DOI: 10.4324/9781003371533-4

have shifted over the past few years to include more historic and systemic forms of trauma – intergenerational land dispossession and race-based traumatic stress – to the insidiousness of gendered trauma and chronic resource scarcity. Discussions of trauma have therefore never been more prevalent, and perhaps, most importantly, pertinent to our lives.

Though we find trauma everywhere around us, we rarely think of it in terms of place, or the places where trauma happens. This growing spatial approach is understood, in some circles, as the 'geographies of trauma', and began as a series of conversations at the 2013 annual meeting of the American Association of Geographers, in Los Angeles, California. There, researchers gathered to discuss spatial approaches to trauma and the ways it interacts with – and exists within – places and spaces (Micieli-Voutsinas & Coddington, 2017). Such approaches are imperative to understanding our current historical moment and are further outlined in the sections below.

(em)Placing Trauma

Trauma takes place in places. It exists somewhere, geographically speaking. Sometimes it even permeates the very grounds upon which we walk. In the footprint of New York City's World Trade Center, for example, one can feel the immense trauma that happened there, in that very spot, in 2001. Standing in the site's memorial space can awaken something profoundly visceral in visitors, an acknowledgment that a wound persists there. Acknowledging trauma's existence in wounded spaces allows for a profound collective reckoning and working through of trauma's unresolved pain (Micieli-Voutsinas, 2021).

As Till explains (2012, p. 7), 'wounded cities' are engaged in an insidious process of healing that is neither confined to catastrophic events or those communities most intimately impacted. Rather, the 'mo[ve]ment of the wound' is non-linear as an intergenerational public reckons with the historical experience of trauma as it continues to reverberate beyond the present, dictating what and how an appropriate, collective response should be. Unacknowledged trauma – trauma that is intentionally expunged from the landscape or denied the right to exist – however, persistently begs for our engagement by haunting places and their descendants.

Recent memorial debates across the globe demonstrate how the unresolved trauma of settler colonialism, slavery, and racial violence, continue to haunt places and their communities. The residue of chattel slavery drapes itself over the US South, but its remnants are unavoidable in the Central Florida landscape where I write this paper from. When I moved to the US South during the pandemic, I desperately tried to avoid living in places named after plantations, real or imagined. I would soon come to learn, however, that such romanticized placenames functioned primarily as aesthetic distinctions in a landscape intimately constructed around and impacted by slavery. In an area that once consisted of an antebellum economy of nearly 50 plantation sites, it is impossible to avoid living on land that isn't tainted by the legacy of slavery or haunted by the enslaved and their memory. In fact, I found myself 'stepping in trauma' most places I went – at home, at work, at my children's school, and so on. Despite my concerted efforts to avoid it, I ended up with one of the largest area plantations as my next-door neighbor.

Places of difficult heritage like the 9/11 Memorial & Museum and the Whitney Plantation teach us that although trauma can never be fully understood, it can be felt. Trauma is, first and foremost, something that is experienced on a cellular level; it is deeply felt by the peoples and places impacted, and those feelings can be emplaced within the ruins of witnessing landscapes. Even without a memorial, sometimes trauma can ensnare a place; it can permeate it, seeping into the soil and into the very lives of inhabitants spanning generations.

That's a reality Gail Adams-Hutcheson points to in her 2017 study of Christchurch, New Zealand – the site of a 7.1 magnitude earthquake in 2011. Here, the natural disaster produced both physical scars and psychic shockwaves throughout the community. It upended many people's fundamental sense of safety and understanding of home – including their sense of place in the world – so much so that the aftershocks of the trauma seemed to become part of the landscape, as much as any mountain or tree. As the community continued to suffer seismic aftershocks and chronic post-traumatic stress disorder (C-PTSD) in the months following the initial quake, the cumulative effect of the event exceeded the resiliency threshold of some residents and cemented their relocation from the area. But as the residents relocated, so too did their memories of trauma; a traveling, living memory of the event etched inside of the very bodies of the people who lived through it.

Research into epigenetics, for example, demonstrates how traumatic memory is stored in the body as a function of its own innate resilience. The lessons of survival associated with traumatic events are literally encoded within our DNA and genetically passed down to subsequent generations in utero (Rosner, 2017). Such memories are inherited in order to prepare the next generation of offspring to biologically adapt to dangers in the environment as a means of intergenerational survival. Akin to the generational transmission of trauma described by children of Holocaust survivors, trauma, in other words, moves within and with us, as the next section considers.

Trauma's Mobility

'Trauma piles on trauma!' writes Art Spiegelman in his graphic novel, *In the Shadow of No Towers* (2004, p. 5). The child of Auschwitz survivors, Spiegelman recalls his father's inability to describe the smell of the Nazi death camps as he himself struggles to define – and wishes to forget – the smell engulfing lower Manhattan in the aftermath of the 11 September 2001 terrorist attacks. Highlighting the ability of trauma to disrupt linear notions of time and connect disparate spaces, the cartoonist recalls the cultural event known as '9/11' in relation to familial histories and memories of the Jewish Shoah. The temporal distance separating past trauma from present is shattered in Spiegelman's scent memory of Lower Manhattan as trauma compounds the spaces of here and there, then and now, self and other, entangling both traumatic events – the Holocaust and 9/11 – and rendering the historic past indistinguishable from the unfolding historical present.

Although trauma is experienced within specific geographical places – and at specific moments in time – trauma is also on the move. And, as it moves, it reverberates outwardly, generating radically new constellations of interconnectivity between peoples, places, and histories, forcing us to change how we both experience and remember trauma's affective afterlife.

This is illustrated in vicarious trauma – a well-known phenomenon among therapists, researchers, and advocates. Vicarious trauma is a term primarily used to describe the experiences of mental health providers who, in their professional capacity to support survivors of trauma, 'take on' symptoms of post-traumatic stress, including physical, psychological, and emotional distress, as a result of their repeated exposure (Branson, 2019). Social scientists have even argued that conducting research in a traumatic setting can lead the researcher to become traumatized in the field (Drozdzewski & Dominey-Howes, 2015). As such, repeated exposure to someone else's trauma can end up triggering trauma inside of your own body, even if the traumatic event didn't happen to you directly. Today, vicarious, or secondary trauma is a major topic among professional organizations in psychology and family therapy as a result of its commonplace occurrence.

In 2017, Kate Coddington pushed this idea further, suggesting the existence of contagious trauma. Rather than the direct transfer of trauma from patient to therapist, contagious trauma is much more pernicious and adaptable. In this form, trauma seeps into most social relations, casually spreading from person to person. Not as an exact replica of someone else's experience of trauma, but as a newly formed experience, triggered by past events, memories, and experiences within the secondary person. And, as it takes root in new individuals, it meets a whole other host of traumatic memories to create new webs of social relations – and reactions – to the external world.

Trauma has a way of moving through space independent of the people who have suffered the original event, but not unrelated to their cultural memories, as Spiegelman's memories at the onset of the section demonstrate. This is where trauma really starts to reveal its power as something that readily moves across space and time. This is a central theme in a 2017 study on migrant detention centers conducted by Alison Mountz. In her interviews with detained persons, harrowing descriptions of war-related trauma, political violence, and forced migration, were rooted throughout the conversations she recorded within the carceral setting of the camps. Although the trauma of confinement and detention created new bouts of trauma for those detained, it did not prevent their personal stories and experiences of trauma from escaping confinement and spreading to others. As stories of trauma traveled across the camp, they morphed as they came into contact with other displaced persons, humanitarian aid workers, and detention workers themselves.

Such complex transmission means that the trauma that is spread is also much more complex.

It's a terrifying thought, but it also makes sense when we think about the ways mass traumas continue to echo throughout entire societies well-after the original survivors are no longer with us. From the Holocaust to the Vietnam War, events of mass trauma reverberate through entire societies and ethnic groups, and will continue to do so, if not meaningfully confronted.

The removal of Confederate monuments by anti-racist activist and their supporters, for example, signifies the onset of a broader campaign of truth-telling and reconciliation in the United States. As the country contends with histories of racial terrorism, the removal of Confederate monuments marks the beginning, not the ending point, of anti-racist efforts to tell more truthful histories about the past and how the trauma of racial terrorism is both embedded and erased from the American landscape.

Pamela Moss and Michael Prince address similar issues of shame and erasure in their 2017 study of traumatized war veterans in Canada. Navigating both the political and cultural context in which traumatic memories are mediated, their study reveals where and how the trauma of military service is actively hidden from view or addressed publicly. Here, military service is actively remembered by Canadian officials as valorous and heroic at the expense of acknowledging the ongoing suffering and vulnerability of returning service members experiencing PTSD. This outright denial of emotional distress enables governmental agencies to dismiss widespread reports of disproportionately high suicide rates amongst returning servicemembers and delay the systematic policy changes desperately needed in veterans' healthcare. Shame is an extremely important motivator here as various stakeholders engage in a series of political debates around the cultural meaning of national identity, military service, gender, and healthcare. As Erica Doss importantly reminds us in her work on American Memory, every society has its own truths, its own certain way to think about right and wrong, and its own narratives about traumatic events, their meanings, and their uses. Societal factors can, as a result, compound certain traumas, just as they could potentially mitigate and heal them.

In 2017, for example, researchers Geraldine Pratt, Caleb Johnston, and Vanessa Banta, working in the same cultural context, looked at how migrant caregivers from the Philippines handled their trauma as international laborers residing in Canada. The community in question suffered at the hands of international trade agreements and global economic exploitation, leading them to take up employment opportunities like the ones that brought them to Canada to work as nannies for Canadian families, while living thousands of miles away from their own families and children.

To express these experiences, the women engaged in community theater. It was a bold move, one that highlighted the truly transnational nature of their physical journeys and emotional wounds – wounds caused by far reaching multinational corporations and sweeping trade agreements, yet still suffered on the most personal and interpersonal level: their role are caregivers within their own families. The women's stories so aptly revealed the gendered exploitation underpinning the global caregiver economy, but they also revealed deeply personal narratives of loss and shame. Here, the theatre provided the women with a space to process the unresolved trauma of being forced to leave their own family and children behind in order to care for them financially from afar. They found that publicly communicating their stories in this way opened up moments of profound witnessing by their mostly Canadian audience.

The shared sense of humanity seemed to bridge the divide of cultural, geographic, and experiential distance, while on the stage. But there were risks, too. Much of the trauma on display was related to the structures of power and trade that left Canadians audiences much wealthier on the whole than the Philippine performers coming to serve in caregiving roles. Therefore, if common humanity could not be palpably felt, there was always the threat of deepening the trauma. This risk was even more sticky when the play was performed back in the Philippines: how would their community, their family, feel about the play and its contents; would it deepen interpersonal connections and foster understanding and healing, or would it deepen the fractures of already distant social relationships?

The approaches above show a variety of ways that we can think about trauma spatially and geographically. But in the end, how does that really help us? There is something hopeful in understanding trauma and the way it moves through time and space. By recognizing it, we gain a richer appreciation for how it bends our actions and understanding of places and peoples. We can see that so much of how we relate to the world is a direct response to trauma – even, or maybe especially, when we don't even realize it. To see trauma as it moves across the room, across the continent, across the ocean, means we can engage it where it is. And we shouldn't underestimate the power of this. Trauma is a kind of knowledge (something we learn from experience) that is embodied in and recalled through our emotions. That makes it very hard to deal with in dialogue, but it can be done, if you know where to look. Healing will require us to deepen our emotional intelligence and spatial fluency as we navigate the inherently complex terrains of our individual and collective wounds.

References

Adams-Hutcheson, G. (2017). 'Spatialising skin: Pushing the boundaries of trauma geographies' in Micieli-Voutsinas, J. & Coddington, K. (eds.) (2017). Special Issue: 'Geographies of trauma'. *Emotion, Space and Society* 24(1).

Berlant, L. (2011). *Cruel Optimism*. Durham: Duke University Press.

Branson, D. C. (2019). 'Vicarious trauma, themes in research, and terminology: A review of literature', *Traumatology*, 25(1).

Coddington, K. (2017). 'Contagious trauma: Reframing the spatial mobility of trauma within advocacy work' in Micieli-Voutsinas, J. & Coddington, K. (eds.) (2017). Special Issue: 'Geographies of trauma' *Emotion, Space and Society* 24(1).

Drozdzewski, D., Dominey-Howes, D., (2015). 'Research and trauma: understanding the impact of traumatic content and places on the researcher', *Emotion, Space and Society*, 17(1).

Goodwin, R., Weinberger, A., Kim, J., Wu, M., & Galea, S. (2020). 'Trends in anxiety among adults in the United States, 2008–2018: Rapid increases among young adults', *Journal of Psychiatric Research*, 130.

Micieli-Voutsinas, J. & Coddington, K. (eds.) (2017). Special Issue: 'Geographies of trauma', *Emotion, Space and Society*, 24(1).

Micieli-Voutsinas, J. (2021). *Affective Heritage and the Politics of Memory after 9/11: Curating Trauma at the Memorial Museum. Interventions: Cultural Approaches to International Relations Series*. New York: Routledge.

Moss, P. & Prince, M. (2017). 'Helping traumatized warriors: Mobilizing emotions, unsettling orders' in Micieli-Voutsinas, J. & Coddington, K. (eds.). Special Issue: 'Geographies of trauma' *Emotion, Space and Society*, 24(1).

Mountz, A. (2017). 'Island detention: Affective eruption as trauma's disruption' in Micieli-Voutsinas, J. & Coddington, K. (eds.). Special Issue: 'Geographies of trauma' *Emotion, Space and Society*, 24(1).

Pratt, G., Johnston, C., & Banta, V. (2017). 'Filipino migrant stories and trauma in the transnational field' in Micieli-Voutsinas, J. & Coddington, K. (eds.) (2017). Special Issue: 'Geographies of trauma' *Emotion, Space and Society*, 24(1).

Rosner, E. (2017). *The Survivor Café: The Legacy of Trauma and the Labyrinth of Memory*. Los Angeles: Counterpoint Press.

Spiegelman, A. (2004). *In the Shadow of No Towers*. New York: Pantheon Books.

Till, K. E. (2012). *Wounded cities: Memory-work and a place-based ethics of care*. Political Geography, 31(1), p. 3–14.

World Health Organization, n.d. (2022). '*COVID-19 pandemic triggers 25% increase in prevalence of anxiety and depression worldwide*' [online]. Available: www.who.int/news/item/02-03-2022-covid-19-pandemic-triggers-25-increase-in-prevalence-of-anxiety-and-depression-worldwide. [Accessed: 13 August 2023].

3

THE DMZ AND THE LAUNDRY

Lessons From K-drama for Trauma-Informed Placemaking

Ayako Maruyama, Laura Van Vleet,
Molly Rose Kaufman, Liam Van Vleet and
Mindy Thompson Fullilove

Introduction

This chapter investigates the role of the Korean Demilitarized Zone (DMZ) and a public laundry in a South Korean television show, *Crash Landing on You*, created by Studio Dragon, originally broadcast in South Korea on TvN between December 14, 2019–February 16, 2020 and then streamed on Netflix. This internationally popular Korean drama – or *K-drama* as such shows are called – tells the story of a South Korean businesswoman who accidentally paraglides into the DMZ and lands in the arms of a North Korean army officer. He and his men commit to getting her home safely. While they are working on that, she spends a month in a very poor North Korean village. The life of the village women centers around a public laundry, an institution which has disappeared in the developed world, and which represents both the distinction in development between North and South and the communitarian lifestyle that the heroine has never had.

Korea was one country at the start of the 20th century, ruled by the Joseon dynasty which had been in power since 1392. The 20th century was marked by vast changes, including: the imposition of Japanese colonialism, 1910–1945, which attempted both to 'modernize' Korea and to wipe out its language and culture; the imposition of partition at the end of World War II; and the brutal Korean war, 1950–1953 (Cumins, 2005; Hong, 2014; Tudor, 2018). The war, which ended in an armistice rather than a peace treaty, continues to haunt both countries: they are locked away from one another, connecting only inside the DMZ at the Joint Security Area, commonly known as Panmunjon.

Since 1953, the Republic of Korea, popularly known as South Korea, has pursued a path that began with capitalist dictatorship, rapid industrialization and a rise in its standard of living, and transitioned to democracy and an ongoing effort to

DOI: 10.4324/9781003371533-5

undo the patriarchal systems that impose terrible penalties on women's lives. North Korea followed a different path, with a communist dictatorship, the establishment of a ruling family, militarization, and limited economic development.

South Korean dramas like *Crash Landing on You*, reflect the traumas that have resulted from this long series of dislocations. For example, the Korean War left 3,000,000 dead and 100,000 orphans, and this trope is woven into many K-dramas. It is the alleviation of some of the fallout from this history of trauma that forms the heart of the shows. In this paper, we investigate the ways in which the contrasting, yet interdependent, spaces of the DMZ and the public laundry offer lessons for trauma-informed placemaking.

Method

In 2021 the Cities Research Group of the University of Orange (Fullilove, 2020) launched a *Popular Media Project*, which involved watching an array of television shows and discussing their import for the COVID pandemic, collective recovery and other social processes then underway. We were all 'sheltering in place' because of the COVID-19 pandemic. We agreed to watch the television show *Crash Landing on You* on our own and met weekly to discuss what we had seen.

Crash Landing on You is an internationally successful example of the popular Korean drama form we call 'situation story'. These dramas center on a love story, using that as a base from which to explore issues in contemporary Korean culture, such as the traps of social hierarchy and arranged marriage, as well as the path to enlightenment. The shows draw on deep archetypes, like Rapunzel in the tower (*Rookie Historian*), the rude person who insults the powers-that-be and gets sent on a journey (*Live Up to Your Name, Dr. Heo*) and many versions of the Brave Maiden (*100 Days My Prince* and *Cinderella and the Four Knights*). In *Crash Landing on You*, the archetypal situation is that of the 'poor little rich girl', the young woman with wealth but no love who falls into a situation short on wealth but long on human kindness.

The Story

Crash Landing on You is a story about the relationship and adventures of Yoon Se-ri, a successful businesswoman from a prominent South Korean *chaebol* family, and Captain Ri Jeong-hyeok who serves in the Korean People's Army in North Korea. Because of the division between North and South Korea, they met against impossible odds. Yoon Se-ri, CEO of a beauty company, *Seri's Choice*, is testing some new leisurewear by wearing it to paraglide. A tornado carries her some distance and she lands in a tree. She calls for help and Captain Ri arrives. He orders her to come down and she falls into his arms. Stunned to learn she is in North Korea, she runs away. He calls after her that she should take the right path at the fork, but she doesn't trust him. Yoon Se-ri's choice is to take the left fork which leads her to a North Korean village just outside the DMZ. She watches the people exercise together in the public square and call their children in for the night. To Yoon Se-ri's surprise, then the lights go out. It happens that the village is where Captain Ri lives when not patrolling the DMZ. He finds her and whisks her to safety at his house.

Having established this premise, the show devotes the next nine episodes to following several threads: Yoon Se-ri's adjustment to life in North Korea; the efforts of Captain Ri and his men to get her back to South Korea; Yoon Se-ri and Captain Ri falling in love; the recovery from earlier trauma of both Captain Ri and Yoon Se-ri; and the efforts of the villain Cho Cheol-gang, a North Korean State Security Department officer, who had killed Captain Ri's brother several years before the events depicted in the show, to destroy Captain Ri and his family. The second part of the show, episodes 11–16, takes us to South Korea where the confrontation between our heroes and the villainous Cho Cheol-Gang plays out.

Yoon Se-ri's adjustment is characterized by both her shock at the state of affairs in North Korea and her ability to adapt. In Episode 2, Captain Ri is lighting a charcoal stove to the admiration of a young soldier from a poorer village further north. Captain Ri says kindly to the young man, 'Modernization will come to your village'. Yoon Se-ri mutters under her breath, 'If that is modernization, what would they say to a gas stove?' Captain Ri calls her to task for her scorn. What she is to learn in that encounter and many that follow is to appreciate what her hosts have and are sharing with her, as opposed to despising them for what they do not have. Yoon Se-ri, for the first time in her life, lives in community, tasting simple pleasures, and laughing at ordinary games. Her ability to play drinking games (Episode 4) and her ability to make a plain dress fashionable (Episode 4) are talents that support her in adapting to this different kind of life.

The team – Captain Ri and his men – cannot easily return Yoon Se-ri to South Korea. They come up with a series of plans. Escape by sea, escape by paragliding, and escape by joining an athletic team under a false passport all fail. It is Captain Ri's understanding of the dynamics of North Korean life that enables them to survive the failures and eventually get her home again.

The adventures surrounding her return involve Captain Ri and Yoon Se-ro in a difficult and time-consuming project through which they come to appreciate each other and fall in love. They also slowly uncover that they had actually encountered each other some years earlier in Switzerland. Captain Ri, then a music student, saved Yoon Se-ri from suicide and she planted doubt in his mind that his arranged marriage was right for him.

Yoon Se-ri was in Switzerland because that country has euthanasia and she wanted to die. She was her father's illegitimate daughter and was hated by her stepmother, whom he forced to raise her. The stepmother tried to kill her by abandoning her on a beach when she was a young girl. Yoon Se-ri is haunted by the memory of this, just as Captain Ri is haunted by grief for his brother. He suspects foul play, but has no proof, so he is in a constant state of alert. In the process of their growing relationship, both are able to re-negotiate those earlier traumas.

However, as viewers of K-drama know, there's always a 'snake in the garden'. Cho Cheol-gang is a man who survived a difficult childhood and longed for power and prestige. He found that he was able to achieve these through the underground system of graft and corruption. The show suggests that many people are engaged in the system. One of Captain Ri's men, for example, is a devoted fan of K-drama, which is highly illegal. Captain Ri and all the villagers visit the local market which has products brought in from South Korea, including some of the cosmetics made by Yoon Se-ri's company. But Cho Cheol-gang is working at a different point in the system, getting power over high-ranking men because he can get them things they want, like apartments in Pyongyang. We get the impression that killing is acceptable to him, and perhaps pleasurable.

Captain Ri's father is a very high-ranking officer, whom Cho Cheol-gang would like to depose so that people he controls might accede to power. Captain Ri's brother had evidence of his perfidy, which, we learn, is why he was killed (Episode 10). Cho Cheol-gang wants to destroy the whole family and he is in avid pursuit of this goal from Episode 1 to the end of the show.

The DMZ and the Laundry: The Spaces of Life

Of course, in addition to this tangle of stories, the show is always examining the relationship between North and South Korea. The current partition, with the rise of dictatorship in the North and robber baron capitalism in the South, is the latest in this long series of threats to the integrity of the Korean peninsula and its life, culture and dignity (Hong, 2014). Yoon Se-ri's landing in the DMZ – as off-limits a space as exists in the world today – symbolizes the impermanence of partition. We are shown the North Korean soldiers going about their routines of patrolling the area, hearing the South Korean anthem, encountering grave robbers, and visiting a deserted home. These moments in the drama open the space of the DMZ to our imagination and hope.

At a more intimate level of scale, in the scenes in the public spaces of the village, the women exercise together in the village square, make kimchi on the banks of a nearby waterway, and do laundry at the communal laundry in the center of town. The village laundry exemplifies life built around such a public facility, designed to provide scrubbing surfaces and water at which a number of people can wash clothes (Roddier, 2003). As these facilities have disappeared in the developed world, its appearance as part of a North Korean village is an important signal of the state of the nation. On the one hand, North Korea does not provide washing machines to each household or even a laundromat for each village. On the other hand, the laundry is not on the riverbank: it is made of concrete and is adroitly positioned in the middle of the village, making it possible for women to attend to washing clothes and other social functions, like welcoming children on their way home from school.

Communal laundries can be found in many parts of the world. In France in the 1850s government subsidies helped municipalities build public laundries away from the riverbanks. These proliferated rapidly, making full use of the flexibility of concrete and newly installed municipal water systems. There was a rapid evolution of their designs, making the washing process easier and easier. For example, outdoor washing on riverbanks required a great deal of stooping, while the new *lavoirs* – the public laundries – provided waist-level sinks. The long sinks allowed many women to work at the same time. The *lavoir* was a place of gathering and part of the social information system. Blood – or lack of it – on a newlywed's sheets, for example, was a matter of great collective interest and speculation. The outdoor *lavoirs* quickly fell into disuse in the 20th century and have been abandoned or demolished.

In Captain Ri's village, the laundry is the heart of the rhythm of life: we see women together beating the clothes, wringing them out, hanging them to dry. The village laundry exemplifies the saying, 'pain shared is pain divided'. Washing clothes at such a laundry is harsh labor. While the work is going on, however, we see the women chatting and sharing news and ideas. Gossip is exceptionally important in K-dramas. In one dramatic scene in Episode 7, a nurse's gossip reveals to Captain Ri the depth of Yoon Se-ri's feelings for him. In the space of the laundry, the combination of two kinds of vital work eases the burden either alone might impose (Figure 3.1).

The introduction of Yoon Se-ri into these spaces acts like a stone thrown in a pond: the ripples are far-reaching, as are the echoes. One of the ripples is that Yoon Se-ri lives in the village for a month. The village women are a constant presence as

FIGURE 3.1   Women's work at the public laundry. Drawing by Ayako Maruyama.

the trials of Yoon Se-ri's stay unfold, and she is slowly drawn into their community. In Episode 4, she goes to a party and proves herself by taking a simple dress and making it state-of-the-art – after all, the viewers are reminded, she is the tycoon of a fashion and lifestyle business in South Korea. In Episode 5, the village women take her side when it seems that Captain Ri has betrayed her for another woman: they help her update her wardrobe to fit in in Pyongyang and win Ri back. When she wants to buy gifts for them prior to her return to South Korea, in Episode 8, they help her pawn a ring. The village women's 'Magnificent Five' strut back from the pawn shop is the pinnacle of their embrace of Yoon Se-ri.

Another of the ripples is that the central bad guy, Cho Cheol-gang, finds Yoon Se-Ri at Captain Ri's house (Episode 2) and believes that he can use her to destroy Ri and his family. This puts him in direct conflict with Captain Ri, in the course of which Cho Cheol-gang's evil network is exposed, disturbing the village. Two episodes in, that process reveals the underlying structure of women's solidarity in the face of threat. The senior woman in the village is Ma Young-ae, who is married to a colonel in the army. She has great power over him, and therefore over the lives of everyone in the village. She is fawned over by all the women. Her husband, however, had accepted graft from Cho Cheol-gang and was arrested. Afterwards she is alone in her house. In Episode 12, the other women discuss the situation at the laundry, but the stern advice is to stay away from Ma Young-ae so as not to fall under suspicion. As it turns out, the women go secretly one-by-one to help Ma Young-ae and run into each other there. Ma Young-ae is, at first, scolding them to go away so as to avoid trouble, but then cries with relief at their help.

The second incident related to Cho Cheol-gang is an attempted kidnapping (Episode 15). The women are called by the children as two men, posing as government officials, attempt to capture the wife and child of a key player on Ri Jeong-hyeok's team. The women all arrive on the scene with whatever tool was at hand: Ma Young-ae arrives with one of the wooden sticks used to beat the laundry. She cleverly penetrates the disguise of the kidnappers and saves Go Myeong-eun and her son, U Pil.

These ripples trigger further echoes. Captain Ri's falling in love with Yoon Se-ri triggers the end of his arranged marriage to Seo Dan. But before the engagement ended, Seo Dan's mother had secured her daughter a wedding home in the village and got to know the village women. Later it is Seo Dan's mother who is able to find and share Yoon Se-ri's tribute to the village women, a special line of products with drawings of their faces. The product line is called *Saudade*, from the Portuguese, meaning 'melancholic longing'.

The final echo is in Yoon Se-ri's life in South Korea. She had lived in profound isolation from other people, including her own family, before she crash landed in North Korea. On her return, she misses being 'in community'. With help from Captain Ri, she begins to create a community in the high-rise building in which her company is located. In the final scene there, we see her seated with her team working on ideas together, 'beating' the ideas into shape, 'washing off' the waste, and 'cleaning up' their materials and messages. Yoon Se-ri has made a laundry of her own.

Lessons for Trauma-Informed Placemaking

While the women in *Crash Landing on You* were washing their clothes and fighting Cho Cheol-gang, we who were watching at home had our own troubles: a badly managed pandemic that was costing too many lives, a fraught election followed by an invasion of the US Capitol, and a rapid acceleration in the harms and processes of global warming. In that context of isolation coupled with a series of fearsome events, we took several lessons from the show and our discussions.

First, the show highlighted for us the importance of community. Community is an abstract noun, used in many ways. We are using it here in the sense of a place-based, self-conscious group that offers care and concern. Our group, based in the 'place' of Zoom, offered us time and space to articulate our joys and concerns, breaking our isolation and protecting us from being prey to the dark thoughts that arise in isolation. Community, as we've defined it here, can exist in many places – real and virtual – and is a fundamental asset to human wellbeing. For Yoon Se-ri, the community she finds in North Korea not only saves her life by getting her safely back to South Korea and protecting her from Cho Cheol-gang, but also eases the deep wounds left by her early abandonment and later suffering from the jealousy of her stepbrothers. Similarly, Captain Ri, in protecting Yoon Se-ri and learning to love, lets go of his prolonged grief and comes to life again. Joining in the work and life of community offers all of us those possibilities of protection, support and love.

Second, the show taught us about public space. The laundry in the public space of the village was in stark contrast to our own meetings on Zoom as well as to Yoon Se-ri's life in South Korea, where the privatization of services and experiences gave a sense of control at each step of the way and filtered out the adventurous experiences of public space. The public is a place to encounter the unexpected: the homeless, the young and old, the infirm, the gorgeous, the startled pigeon. Creating a public space that is welcoming and safe, that offers potential for passing through or lingering, and that is inclusive is a task for all societies. Such public spaces contribute to the building of community that extends far beyond 'like-minded people' to include everyone.

Third, the show's emphasis on the DMZ reminded us that world peace conditions all of our lives. Established in 1953 with the signing of the armistice that ended the Korean War, the DMZ is a space so impenetrable that it has 'rewilded' to the astonishment of ecologists (Brady, 2020). The contrast between the intimacy of the laundry and the antagonism of the DMZ is stern. The threat of missiles ranged along the two sides of the DMZ looms over peaceful places in both the North and South. The public laundry is a site for convening and, sometimes, a site for calibrated spreading of news or information. But what it means to have a calculated threat and frequent shows of power is that no serene, attractive and functional public space is

FIGURE 3.2   Reunification statue in the DMZ. Photo by Mindy Fullilove.

truly safe. And because of the domino effects of international aggression, an attack in one place can precipitate a worldwide disaster.

In the DMZ, at a site that tourists can visit, there is a statue of a sphere broken in half, with people trying to push the two halves – one representing the North and one representing the South – back together (Figure 3.2). That is the work that lies ahead if we are to be safe. The ancient Roman philosopher Publius Flavius Vegetius Renatus's (1996) idea that preparing for war is the path to peace drives much public policy, but we would argue from this analysis of *Crash Landing on You* that peace is 'made' by making peace. At the local level peace is 'made' by shared activities in the public space, such as the women's work that took place around the laundry. While the tasks that 'make' peace at the international level may not be as clear to us, they exist. Our placemaking at the international level of scale must uncover these tasks and set us to accomplish them. The most powerful lesson for trauma-informed placemaking is that (Hanh, 2021).

References

Brady, L. M. (2020). 'From war zone to biosphere reserve: the Korean DMZ as a scientific landscape', *Notes and Records*. 75: 189–205.

Cumings, B. (2005). *Korea's place in the sun: A modern history* (updated edition). New York: WW Norton & Company.

Fullilove, M. T. (2020). *Main Street: How a City's Heart Connects Us All.* New York: New Village Press.

Hanh, T. N. (2021). *Zen and the Art of Saving the Planet.* New York: Random House.

Hong, E. (2014). *The Birth of Korean Cool: How One Nation Is Conquering the World Through Pop Culture.* London: Picador Publishing.

Renatus, F. V. (1996). *Military Art and Science.* Liverpool: Liverpool University Press.

Roddier, M. (2003). *Lavoirs: Washhouses of Rural France.* New York: Princeton Architectural Press.

Tudor, D. (2018). *Korea: The impossible country: South Korea's amazing rise from the ashes: The inside story of an economic, political and cultural phenomenon.* Clarendon, Vermont: Tuttle Publishing.

4

FLYING, FLEEING, AND HANGING ON

Trauma Baggage at Airports

Rezvaneh Erfani, Zohreh BayatRizi and Samira Torabi

'A diaspora is a network of people, scattered in a process of nonvoluntary
displacement, usually created by violence or under threat of violence or death.
Diaspora consciousness highlights the tensions between common bonds created
by shared origins and other ties arising from the process of dispersal and the
obligation to remember a life prior to flight.'

(Gilory, 1997, p. 328)

Introduction

Airports are known as 'emotion-laden environments' (Malvini Redden, 2013,
p. 122). They are places of joy and pain, reunion and separation, gateways to dream
vacations and departing points toward disaster and doom. For immigrants and
refugees, the story of airports is slightly different from others. This chapter opens
a conversation about how diasporic populations (see Gilroy, above) experience
airports as traumatic places both in normal times – as a place of separation and
uprooting – and in extraordinary times – in the aftermath of air disasters and during
war and conflicts. We specifically focus on Tehran Imam Khomeini International
Airport (IKA) and the Ukrainian Flight 752 tragedy in 2020, while also making
brief references to the Kabul's Hamed Karzai International Airport (KBL) in the
aftermath of the American military withdrawal in 2021, to discuss airports as
disaster areas where collective traumatic stories and experiences are repeatedly
(re)shaped. In our interviews with people impacted by the Ukrainian Flight 752
tragedy, some of our participants shard their fears of the mental pain and emotional
heaviness they would experience when they arrive at IKA for the first time after
the incident: it was after all, where the families of victims hugged their loved ones
for the last time and the place where they continue to gather to commemorate the

DOI: 10.4324/9781003371533-6

victims. Further, even in ordinary times, many diasporic populations, including from Iran and Afghanistan, experience airports as places of involuntary separation from loved ones. These and other incidents point to the fact that airports are places of major trauma, especially in destabilized regions in the world.

This chapter begins by outlining a theoretical approach to studying airports at the intersection of emotions, space, and politics, intersections often of that diaspora that Gilroy talks of above. We will then discuss IKA in the context of Iranian modern political history and move on to present findings from our research on the Flight PS752 tragedy to contextualize the theoretical framework introduced here. We will also describe images of people hanging from American planes at Kabul's airport, to discuss their resonance for many refugees and immigrants especially those from conflict-zone countries. We will conclude by bringing these threads on airports, trauma, and diasporic experiences together and consider how traumatic experiences might influence people's decisions to fly or not, or go back 'home', or not.

Airports at the Intersection of Emotions, Space, and Politics

Traveling is an emotionally diverse experience depending on the purpose of leaving and the process of arriving at a destination (Wang, Hou & Chen, 2021, p. 1085). At the most basic level, some of us experience fear of flying (aerophobia or aviophobia), which is one of the most common phobias and is experienced more by those with anxiety disorders (Flasbeck et al., 2023). This fact alone colours our emotional experience of going to or being at an airport. Airports are themselves places of layered and complex emotions, originating in the personal, familial, historical, cultural, and political terrains. If you live far from friends and family, the airport is the last and first place you see your loved ones. Upon arrival, much that was hidden behind cameras during fun family video calls becomes visible and tangible in the airport when you see your family and friends in person: their increased frailty, the first glimpse of the cane helping your father walk, the sorrow that they can no longer hold back over the death of loved ones that they had been hiding from you. In the departure gate at the airport, you wonder if this is the last hug, the last goodbye, or the last picture. Despite the multitude of emotions, you may find yourself trying to appear put-together, as it might be that last picture.

Emotions move throughout the airport with passengers, making airports excellent candidates for studying 'emotion management', especially in regard to security and sense of safety (Malvini Redden, 2013). Airports are highly secured places with multiple security measures and performances requiring numerous compulsory interactions between passengers and employees (ibid.). In addition, political factors and power relations add to the complex emotional experience of airports. In *Discipline and Punish* (1977), Michel Foucault, discusses two theoretically distinct models of power: 'sovereignty', which relies on the power to kill, torture, take away (life, taxes, property), and exclude (from land, from certain spaces, or from rights) versus 'discipline', which relies on supposedly gentler methods such as surveillance, training the bodies, channeling people and their life forces into certain directions, and distributing bodies in certain formation in specific spaces. Airports are places where both of these forms of power are simultaneously exercised (Salter, 2007), impacting our bodies and emotions. On the one hand, sovereignty is exercised at airports through the 'power to ban or exclude' (p. 51). Immigration officers, as representatives of the government, are empowered to question people, revoke their visas, detain them, and to use biometric scanners to help determine who belongs and who does not, making our bodies a direct site of power relations. Customs officers are similarly empowered to question travelers and search their belongings. Airports might be described as a space of flows (Castells, 1989; Wilhoit, 2018), but the flow is subject to a combination of privilege, autonomy, and politics: the economic privilege to buy tickets and fly; privileges of holding citizenship and travel documents; and the social and cultural capital of visiting another country. This encounter with sovereign power, exercised through immigration officers, is often stressful for people who are routinely targeted for extra scrutiny due to the colour of their skin, nationality, place of birth, religion, etc.

'Disciplinary power' is also on display at airports. Ubiquitous cameras, security checkpoints, body scanners, and manual searches of our bodies and luggage in airports are material reminders of what Foucault (1977) has described as 'disciplinary' powers that govern our bodies and our movement through space. While Foucault's argument was specifically developed with reference to prisons, workhouses, and factories, we can paraphrase him to argue that airports too are spaces of 'articulated and detailed control to render visible those who are inside it' (p. 172). The architecture of airports helps 'to transform individuals: to act on those it shelters, to provide a hold on their conduct, to carry the effects of power right to them, to make it possible to know them, to alter them' (ibid.). The spatial arrangements of airports can be analyzed through a Foucauldian perspective as demonstration of political and power relationships (Salter, 2007). Multiple lines and security queues work as spatial ordering devices to manage body movements and population mobilities. All the while, we are 'tied within a web of individuation and deindividuation marked by perpetual surveillance' (Wood, 2003) and even become temporarily disconnected from our bodies as 'the traveler.'

By playing our part in the security theater, we engage actively in constructing airports (Putz, 2012; Wilhoit, 2018). And yet the controllability of our bodies and body movements gives us the sense that we lack agency over the process and over what will happen next, especially during body scanning, passport checks, and crossing the border. By 'being scanned, having our identification checked, or submitting our luggage for X-rays, we become accustomed to security—and (dis) comforted by its presence' (McHendry, 2016, p. 551). Later in this chapter, we will also discuss how problematic and traumatic such an experience can be for those of us with political baggage when entering and exiting home and how the airport can become a 'locus of anxiety and interrogation' (Salter, 2007, p. 49). Yet, in general, a lack of agency and autonomy over the process of security checks and how our bodies will be treated is experienced at least to some extent by all of us. Anxiety, fear, and stress are therefore experiences that many people report in airports (Malvini Redden, 2013).

The above discussion helps create an integrative framework that captures the interplay of emotions, space, and politics in the context of airports. Such a framework allows us to understand the experience of fleeing, flying, and hanging on to the trauma as an immigrant. In the following, we examine this framework in the context of IKA after providing a short description of this airport and its history, which is meant to explicate the overlapping political, emotional, and spatial forces that shape people's experience of it.

IKA's Emotional Geography

As a major city of 12 million inhabitants and the capital city of Iran, Tehran is served by two main civilian airports. Mehrabad Airport (THR), nestled in the western neighborhoods of the city, opened in 1938 and served as the main international hub of the country until 2004, when the Imam Khomeini Airport (IKA) began operations and took over as the main hub of international flights. Unlike Mehrabad's intimate location within the city with its views of the Alborz mountains and its proximity to major landmarks and the hustle and bustle of Tehran, IKA is located in a desolate desert setting 60 kilometers from the city. On the way to the airport, one drives on a crowded and fast-moving highway that snakes through low-income and poverty-stricken neighborhoods in the southern outskirts of the city before it reaches Tehran's famous Behesht-e Zahra cemetery, home to over one and a half million graves and the mausoleum of Ayatollah Khomeini, the founder of the Islamic Republic, as well as the tombs of many victims of war and political violence (BayatRizi & Ghorbani, 2019). The highway then stretches further into the desert before it reaches IKA. The location of IKA in the thinly populated southern outskirts of Tehran puts it in proximity to the metropole's underbelly: industrial sites and large-scale cattle farms, the foul smell of which assaults the nose for kilometers on end. Thus, one's first embodied impressions of IKA can include distance (traveled on highways), noise (of cars and airplanes), smell (of air pollutants), death-anxiety (provoked by the cemetery, the desolate desert, and highway travel in a country known for its high rate of road accidents), and darkness (as major international flights are scheduled for late night and early hours of the morning).

As a major transportation hub, IKA is highly politicized. Its opening during the administration of reformist President Mohammad Khatami in 2004 was contentious and controversial. The Islamic Revolutionary Guard Corps (IRGC) of Iran laid siege to the airport by land and air in an attempt to wrest control of airport services from the duly elected government, on the pretence of national security. The brazen showdown shocked the nation and delayed the airport's opening, but the IRGC finally got what they wanted and continues to have a visible presence at the airport, controlling security screening and occasionally interrogating or even arresting dual citizens or academics on trumped up charges of 'espionage', 'collaboration with enemies', and so on.

As the country's main international hub, IKA is a place of separation and reunion. In the last few years, with a strong pick up in the country's emigration statistics, IKA has served as a gateway for students and immigrants bound for Western, and more recently Eastern European and Asian, destinations. According to Iran Migration Observatory (2022) it is estimated that 65,000 Iranians are migrating to foreign countries every year. On the other hand, for those who leave the country to study abroad, only 10% are willing to return as opposed to 90% prior to the 1979 revolution (Azadi, Mirramezani and Mesgaran, 2020). For these departing emigrants IKA is a gate to their dreams. But as we will see below, for some, it has become a place of nightmares.

Flying in Dangerous Skies: The PS752 Tragedy

On 8 January 2020, Ukrainian Flight PS752 left IKA at 06:19 local time. Only three minutes later, when the plane was still close to the airport, it was shot down by two missiles of the Islamic Revolutionary Guard Corps, killing all 176 passengers and crew. Iranian officials later claimed that an air defence operator had mistaken the plane for a foreign missile (Jones & Burke, 2021) while many among the families of victims and the general Iranian population maintain that the shooting was targeted at the airplane. Most of the passengers were Iranian immigrants or Iranian international students going back to Canada after visiting their families and friends over the winter break (Government of Canada, 2023). The rest were bound for Western European countries.

The downing of Flight PS752 is arguably the most painful tragedy the Iranian-Canadian community has experienced in the past decades and has affected the Iranian diaspora in many corners of the world. Soon after this tragedy, we designed a sociological research project to understand and document how the Iranian community was affected by this tragedy in Edmonton, Canada. We focused on the community's experience of grief as well as their reflections on their immigrant identity and sense of belonging to Iran and Canada in the aftermath of the tragedy (BayatRizi, Erfani & Torabi, 2022).

Some of our participants shared their feelings toward traveling back to Iran, especially through IKA, which is forever connected to this tragedy. IKA is not only the airport where the PS752 passengers last said their goodbyes and boarded the plane, it is also the site of an IRGC security operation in the form of a second security checkpoint that everyone has to go through, on top of the one operated by IKA management. Study participants used specific wording to describe or mention IKA, such as 'the airport of miseries', used by the 31-year-old Parsa, and expressed their hatred towards the place and 'its walls [that] smell like blood' (Maryam, 42).

Sima, 33, wished she could be transported directly to home as soon as she stepped out of the plane, so she would not have to go through IKA. Hamed, 42, regretted that when at the airport one has to pray for not getting hit by missiles, rather than think about loved ones, review the pleasant memories, or take time to feel how one might miss 'home.' Many participants expressed anxiety and fear about getting on a plane, any plane, after the downing of Flight PS752. What made this tragedy even harder on many study participants was the fact that their life trajectories as immigrants mirrored those of the victims, which made our participants imagine themselves or their loved ones on that plane. We described this experience as 'vicarious death' (BayatRizi, Erfani & Torabi, 2022).

Fleeing with an Emotional Baggage

In addition to the study discussed above, we also conducted auto-ethnographies in our visits to Iran, via IKA, after the tragedy. Reflexively, as passengers, as immigrants, as dual citizens of Iran and Canada, and as academics, we have several layers of emotions at this airport. As passengers, we worry about missing our flights, flight delays and cancellations, luggage weight and dimensions, etc. As immigrants, we experience IKA as a place of reunion and a threshold of separation and self-imposed exile. As dual citizens of Iran and Canada and as academics, we fear being interrogated, detained, and even arrested upon arrival or departure by the IRGC which has for political purposes arrested and imprisoned several high-profile academics and journalists, especially at or on their way to IKA. Dual citizenship also puts scholars even at more risk as they can easily be accused of espionage, 'endangering national security', 'associating with foreign diplomats', and 'collaborating with a hostile government' (Wilson Center, 2007; Human Rights Watch, 2018; BBC News, 2022). Among dual nationals arrested at IKA are several Iranian-Canadians, including anthropologist Homa Houdfar and philosopher Ramin Jahanbegloo. Journalists Parnaz Azima, Mehrnoush Solouki, and Nazanin Zaghari-Ratcliffe who also had dual citizenships (with the USA, France, and Britain, respectively) and were arrested at IKA (Human Rights Watch, 2006; 2018; BBC News, 2022). Researchers residing in Iran may also face an exit ban, a tool widely used by the Iranian judiciary and the IRGC to punish and control civil society activists and journalists. For example, Saeed Madani, criminologist and sociologist, was interrogated at IKA and detained by the IRGC's intelligence branch when flying for a sabbatical at Yale University in January 2022 and sentenced to nine years in prison. Madani, who is accused of suspicious foreign relationships, has published extensively on prostitution, addiction, civil society and social movements in Iran.

As Iranian-Canadian academics, we face multifaceted negative emotions when traveling through this politically charged and emotionally fueled space: fear of getting detained and held hostage, sadness over leaving family and friends behind, worry that we may never see them alive again, anxiety over their wellbeing, and guilt over saving ourselves and leaving our families behind to deal with the current economic and political circumstances of Iran.

Hanging On

Sadly, IKA is not unique in its trauma-filled baggage. Many other airports across the world have become scenes of invasion, coups, disasters, and exodus. Take Kabul's KBL, which became the focal point of the disastrous American military withdrawal from Afghanistan in 2021. Twenty years after its invasion of Afghanistan, the United States withdrew its last forces from Afghanistan in August 2021. The withdrawal was accompanied by chaos in Afghanistan as Taliban forces marched on the capital, the government fell, and many people flocked to the airport seeking to escape Taliban rule. Video footage of people clinging on to an American Air Force plane to flee the country and falling off the plane was shared widely on social media and was described as the 'defining images' of the 20 years of Western military intervention and invasion (Aljazeera, 2021). Among several people who died trying to cling to the American plane as it took off from Kabul airport was Fada Mohammad, a 24-year-old dentist who had seen on Facebook that 'Canada and the United States were airlifting anyone who wanted to leave out of the Kabul airport' (Shih, Masih & Lamothe, 2021). The image of people hanging from the plane resonated with many immigrants and asylum-seekers trying to flee a conflict zone. As diaspora we hang on to our home even when it is full of traumas. Hanging onto hope while carrying our trauma baggage everywhere we go becomes part of our identity. This is where the alliance between place and identity intertwines with trauma for refugees and immigrants (Cox & Connell, 2003).

Leaving the country for the first time in pursuit of a better life is a heavily traumatic experience for many immigrants and refugees, even those who immigrate with their whole family unit. Re-entering the country after a traumatic national event (for example, the PS752 tragedy) also creates a unique experience for those living in exile as it delays the tangible encounter with loss and grief. Upon return, the airport becomes the space where the loss materializes for the traveler. In this chapter, we drew on our interviews and our auto-ethnographies in order to open a conversation about how diasporic populations might experience airports as traumatic places both in normal times (as a place of separation and uprooting) and in extraordinary times (in the aftermath of air disasters and amidst global conflicts).

References

Aljazeera (2021). 'Afghans cling to moving US Air Force jet in desperate bid to flee', Aljazeera [online], 16 August 2021. Available: www.aljazeera.com/news/2021/8/16/afgh ans-cling-to-plane-defining-image. [Accessed: 15 May 2023].

Azadi, P., Mirramezani, M. & Mesgaran, M. B. (2020). 'The struggle for development in Iran: The evolution of governance, economy, and society'. California: Stanford University Press. Available: www.sup.org/books/title/?id=33646

BayatRizi, Z., Erfani, R. & Torabi, S. (2022). 'Grieving immigrants: Emotions at the intersections of politics and race' in *Reading Sociology: Decolonizing Canada* (4th edition), in J. JeanPierre, V. Watts, C. E. James, P. Albanese, X. Chen, M. Graydon (eds.) Oxford: Oxford University Press.

BayatRizi, Z., & Ghorbani, H. (2019). 'Risk, mourning, politics: Toward a transnational critical conception of grief for COVID-19 deaths in Iran', *Cuurent Sociology*, 69(4). Available: https://journals.sagepub.com/doi/full/10.1177/00113921211007153

BBC News (2022). Who are the dual nationals jailed in Iran? [online]. Available: www.bbc. com/news/uk-41974185. [Accessed: 15 May 2023].

Castells, M. (1989). *The informational city: Information technology, economic restructuring, and the urban-regional process*. Oxford, England: Blackwell Publishing.

Cox, J., & Connell, J. (2003). 'Place, exile and identity: the contemporary experience of Palestinians in Sydney', *Australian Geographer*, 34(3).

Flasbeck, V., Engelmann, J., Klostermann, B., Juckel, G., & Mavrogiorgou, P. (2023). 'Relationships between fear of flying, loudness dependence of auditory evoked potentials and frontal alpha asymmetry', *Journal of Psychiatric Research*, 159.

Foucault, M. (1977.) *Discipline and Punish: The Birth of the Prison*. (trans. Alan Sheridan). New York: Vintage.

Gilory, P. (1997). 'Diaspora and the detours of identity', Woodward, K. (ed.) *Identity and difference*. London: Sage.

Government of Canada (2023). *Canada's response to Ukraine International Airlines Flight 752 tragedy* [online]. Available: www.international.gc.ca/world-monde/issues_deve lopment-enjeux_developpement/response_conflict-reponse_conflits/crisis-crises/flight-vol-ps752.aspx?lang=eng. [Accessed: 15 May 2023].

Human Rights Watch (2006). Iran: Top *scholar detained without charge*. Available: www. hrw.org/news/2006/05/04/iran-top-scholar-detained-without-charge. [Accessed: 15 May 2023].

Human Rights Watch (2018). *Iran: Targeting of dual citizens, foreigners prolonged detention, absence of due process* [online]. Available: www.hrw.org/news/2018/09/26/ iran-targeting-dual-citizens-foreigners. [Accessed: 15 May 2023].

Jones, R. P., & Burke, A. (2021). Flight PS752 shot down after being 'misidentified' as 'hostile target', Iran's final report says [online]. *CBC News*, 17 March 2021. Available: www.cbc. ca/news/politics/final-report-flight-ps752-1.5953340. [Accessed: 15 May 2023].

Malvini Redden, S. (2013). 'How lines organize compulsory interaction, emotion management, and 'Emotional Taxes': The implications of passenger emotion and expression in Airport Security Lines', *Management Communication Quarterly*, 27(1).

McHendry, G. F. (2016). 'Thank you for participating in security: Engaging airport security checkpoints via participatory critical rhetoric', *Cultural Studies-Critical Methodologies*, 16(6).

Putz, O. (2012). 'From non-places to non-events: The airport security checkpoint', *Journal of Contemporary Ethnography*, 41(2).

Salter, M. B. (2007): 'Governmentalities of an Airport: Heterotopia and Confession', *International Political Sociology*, 1(1).

Shih, G., Masih, N., & Lamothe, D. (2021). 'The story of an Afghan man who fell from the sky', *The Washington Post* [online], 27 August 2021. Available: www.washingtonpost. com/world/2021/08/26/story-an-afghan-man-who-fell-sky/. [Accessed: 15 May 2023].

Wang, L., Hou, Y., & Chen, Z. (2021). 'Are rich and diverse emotions beneficial? The impact of emodiversity on tourists' experiences', *Journal of Travel Research*, 60(5), 1085–1103.

Wilhoit, E. D. (2018). 'Space, place, and the communicative constitution of organizations: A constitutive model of organizational space', *Communication Theory*, 28.

Wilson Center (2007). *Reports from Tehran Indicate that Haleh Esfandiari Has Been Formally Charged with Espionage and Endangering Iranian Security* [online], 11 June 2017. Available: www.wilsoncenter.org/article/reports-tehran-indicate-haleh-esfandiari-has-been-formally-charged-espionage-and-endangering. [Accessed: 15 May 2023].

Wood, A. (2003). 'A rhetoric of ubiquity: Terminal space as omnitopia', *Communication Theory*, 13(3).

5

TRAUMA

The Counterproductive Outcome of the Land Restitution Program

Juan David Guevara-Salamanca and Gina Jimenez

Introduction

One of the current tendencies relating to healing trauma is the repetition of narrating the traumatic events in question to expose violence and seek justice. However, remembering and retelling can be a revictimization process that contravenes the healing process (Das, 2007, p. 103). In this chapter we firstly consider the Land Restitution program, a Colombian State initiative that seeks to return to the victims of the armed conflict the land they lost, as a narrative/discourse that depicts the tension created by the Colombian State in the constitution of the category of victims of the armed conflict. Secondly, we consider narratives as vehicles used to make sense of the world and tell stories about human experiences. We consider narratives to be excellent research tools that facilitate the understanding of the reality depicted by them (De Fina & Johnstone, 2015).

Our chapter works on the dyad narrative/discourse inspired on Livholts and Tamboukou's 'narrative as/in discourse' (2015, p. 40). *Narrative as/in discourse* has three main characteristics: first, the need for tracing their power/knowledge effects; second, their surrounding processes of power and desire; and third, the productive property of narrative as/in discourse for shaping human subjectivity (Livholts & Tamboukou, 2015).

Narrative as/in discourse illuminates the linguistic and authoritarian processes and forces of naming and authoring. These processes are relevant in this chapter since naming the other not only produces and incorporates the existence of other voices, rules, regulations and a different social landscape, but a subjectivity that striates and excludes relations that are real but usually not actual, i.e., virtual (Livholts & Tamboukou, 2015; Shields, 2003).

DOI: 10.4324/9781003371533-7

In our case study, *narrative as/in discourse* provides the personal experience, life stories and the description of the events of suffering and trauma, while the Land Restitution program captured and reshaped them through its institutional disposition and linguistic enclosing of its discursive practices, personified by the agents involved throughout the process. Such a purview permits us to identify and highlight the productive interconnections of power, desire and knowledge (Foucault, 2003), while opening up the possibilities of language and its performance intra-activity (Barad, 2008). Our understanding of the *narrative as/in discourse* of the Land Restitution program does not exhaust with the denouncing of its possible revictimization and power force over the victims of the armed conflict but seeks to provide a diffractive and ontoepistemological reading to uncover the potentialities and possibilities for healing that are hidden but contained in the program.

This chapter presents our insights and critiques of our experience working for the Land Restitution program for over six years. We strongly believe the program's relevance for Colombian society, but with some rethinking and improvement to consolidate its main goals and objectives. The chapter is organized by: 1) a brief presentation of the Land Restitution program; 2) a short problematization of institutional and collective trauma; 3) the narrative/discourse of the Land Restitution program; 4) the Land Restitution program discursive materiality as placemaking that is counterproductive to its aims by contributing to trauma. Finally, we conclude with a list of possible elements to consider a diffractive and ontoepistemological approach to the Land Restitution program and its potentialities for healing and placemaking.

1.   Land Restitution Unit and Its Institutional Mandate

The Land Restitution Unit (LRU) is the Colombian State actor that develops the Land Restitution program through administrative and judicial tools to formalize land ownership and make reparation to the rights of the armed conflict victims. Its main objective is to support and advocate for the recognition of property rights on abandoned or dispossessed land and facilitate the safe return of displaced farmers to their lands (Congreso de la República de Colombia, 2011). The program comprises three stages: the administrative, the judicial and the post-court ruling.

Administrative Stage

In this stage, the LRU is responsible for recognizing the category of victim of the claimant and the legitimacy of the property rights claimed (Congreso de la República de Colombia, 2011). The social, cadastral and legal teams collect the evidence required to prove that the claimant has the category of victim, and their property relationship with the land. While the social and legal teams collect primary and secondary information, the cadastral team is responsible for identifying, measuring and georeferencing the land claimed. Legally, this process has four months to systematize, analyze and include the data collected in the applicant/victim case report that will determine if the land claimed can be registered in the Land Restitution program database. Once the acceptance or rejection is informed to the applicant/victim, the first stage finishes, starting the second stage with the presentation of the case report to a Land Restitution Judge.

Judicial Stage

The judicial stage seeks to resolve the claim with a judicial decision recognizing the property right of the applicant/victim over the land claimed. This stage allows the LRU to represent the victims before a Land Restitution Judge. This judge must decide the property rights and further benefits the applicant/victim will receive as part of the reparation process led by the State. These benefits aim to generate productive processes on the land claimed that strengthen the private property on the land (Ministerio del Interior de la República de Colombia, 2012). This stage should take four months to be completed. If the land recognized as owned by the applicant/victim is not occupied by a third person, the applicant/victim can return to it and start a 'productive project.' However, if a third person occupies the land, the judge must decide regarding this occupation and how to make effective the rights of the applicant/victim.

Post-court Ruling Stage

The final stage aims to accompany the process of the new proprietor/victim to recover its land, facilitate their safe return and compliance with the sentence of the Land Restitution Judge. In this stage, the intervention of various State programs is relevant, and their coordinated action must accomplish the restitution of the land and the reparation for the damage caused by the armed conflict (Ministerio del Interior de la República de Colombia, 2012). The post-court stage lasts another four months. For the applicant/proprietor/victim, this is the most relevant stage of the process due to their possibility to return to the land. The sentence can mandate the construction or renovation of the house, the cancellation of debts, and the development of a 'productive project' supervised by the LRU. At this time, the applicant/victim should consolidate the healing process by returning to the land with a new life project.

The structure of the law, the concept of land, the individual reparation, alongside the administrative and judicial procedures, have resulted in a program that generates trauma instead of healing. Our approach disentangles the discourse that forms the restitution process in Colombia, making evident how an institutional and governmental program can contravene its own goals.

2. Collective and Institutional Trauma

Collective and institutional trauma is a transformative process with historical and sociocultural connotations. It results from a series of actions committed against individuals' wellbeing and will. Trauma can be deployed after certain events unexpectedly influence the wellbeing of individuals and collectives and requires mediation of attributions and discourses that create a master narrative of social suffering in which recollection of events and pain are socially constructed (Alexander, 2012). The acknowledgment of aspects of trauma and possible responsibility are explicit and socially accepted, and a social connection and recognition of the trauma socially widespread. Trauma can also be a social mobilizer and generator of collectivity. Overcoming societal trauma can empower a socio-political movement that provides a voice to individuals and groups to admit and confront their victimization and to provide alternative solutions (Ball, 2000).

However, the paradoxical nature of trauma and its mediation is not transparent nor neutral but crossed by institutions and hierarchies that control the said and the unsaid (Foucault, 2003; Alexander, 2012). Thus, institutional trauma has two features. First, legal institutions shape discourses on trauma through legally bound responsibilities, material reparations and punishments (Alexander, 2012). Second, the operation of State bureaucracy delimits and controls processes, contents, meanings and subjects regarding reconciliation, truth and reparations, 'channelling the spiral of signification that marks the trauma process' (Smelser, 1962, in Alexander, 2012, p. 23).

In this sense, the representation of trauma through narratives/discourses opposes healing and the consolidation of a social collective via different mechanisms. In various cases, the social circulation of the traumatic narrative generates a saturation directed against the victims due to its capacity to empty trauma's 'psychological and rhetorical force' (Ball, 2000, p. 16). This saturation nullifies the traumatic narrative through a fascination with the grotesque (Bakhtin, 1984) and the individuals desire to both produce and consume spectacle (Debord, 1967). As a result, there is a gap between private and public registers of experience (Ball, 2000, p. 17), alongside the creation of an anonymous and observing public audience (ibid.).

Additionally, traumatic narratives/discourses controlled and commodified by political and social actors can domesticate and exploit the traumatic impact 'of widespread suffering upon a complicit, complacent, and privileged class of spectators' (Ball, 2000, p. 16). Due to the difficulties in connecting, reconstructing and representing the traumatic experience, the weight of the recollection lies upon the memories that reconstitute the traumatic situation. This memorial shift allows a disentangling of the events that produce the trauma while at the same time extends the possibility of interpretation of memories and collective recollection (Ball, 2000, p. 8).

Das (2007) suggests that trauma cannot simply heal through time. A traumatic event only sits in the consciousness as an experience through nightmares and repetitions that the survivor takes from the experience (ibid.). In this sense, trauma behaves as a memorial device with affective responses, creating possibilities, restrictions and a field of dispute to narrate the traumatic memory (Guevara Salamanca & Jiménez Espinosa, 2022), a narration that constantly actualizes a past that is not fully accessible (Shields, 2003; Das, 2007). This problematization of trauma highlights how collective trauma and memory can be composed and shaped by institutions, generating a field of dispute with sociopolitical components as a narrative as/in discourse, and causing further harms and declining their potential power.

3.   *The Narrative/Discourse of the Land Restitution Program*

Collective and institutional trauma contain challenges and potentialities of narratives/discourses for healing. The Land Restitution program, through the LRU, develops a discursive logic that constrains individuals, and the materiality that co-constitutes them in the manner of which we argue, produces an overlapping and saturation of trauma in its conception and materialization of land. As a process of placemaking by the State, there can be a revictimization process that contravenes the healing process with various operational aspects.

i.

Abuse of power, as a modus operandi of action, imposes the one over the other. It is exercised indirectly and in action. Power operates in the realm of the possible, as actions organize to be possible upon other actions. Power is a development of a managerial technique that constrains and makes possible actions. A governing activity that manages the conduct of individuals and groups legitimately subjected to the power regulates modes of action in a calculated manner. This manner of governing generates a structure to forecast and enclose others possible actions to regulate them (Foucault, 2003, p. 138).

ii.

Discourses seek to generate specific practices that shape individuals. Discourses act as self-disciplinary technological *dispositifs* (Foucault, 2003) that structure the agencies of individuals. Judicial systems amalgamate relations of power, space and body and create the subjects they represent (Foucault, 2003).

The Land Restitution program produces a particular individual actor with specific subjective characteristics: the applicant/victim. This specific configuration of subject is a product of a 'complex web of relations and significations of power' (Foucault, 2003, p. 127) and shows how the rationalization of the State acts upon individuals. For instance, the category of applicant/victim develops next to a bureaucratic process that is usually an excessive demonstration of political power.

The applicant/victim subject overlays other subjectivities that are purposely hidden in the application of the law and the work of the Unit due to the need to homogenize and control the restitution process.

This strategic relationality makes invisible historical subjectivities that suffered violence and attempts their annihilation. The categories of claimant, applicant and victim reposition social categories such as farmers or communal leaders, for example. The applicant/victim dyad paradoxically reinforces and diminishes the categorical narratives of victims due to their effective vanishing in the Land Restitution program. As victims, they are relevant as testimonial actors in the conflict and perform according to the legal category of applicant/victim to access the program; however, they must also wait for this recognition by the same institutions that seek to represent and advocate for them. Thus, the category of the victim is being devoured by the category of applicant. This is a concomitant process to the bureaucratic rationalization that permeates the relationship of the LRU with applicants/victims (Bauman, 2007).

The technique and the struggle that emerges from this subjectification correspond to the consolidation of the State, which is happening due to the culmination of the armed conflict, the occupation of national territories previously ruled by guerrillas and the consolidation of the monopoly of violence (Hobbes, 2008). Socio-historico-political discursive practices have positioned certain interests aligned with the circulation of capital and its significance to the land. We suggest that this consolidation process is reinforced by the 'modern matrix of individualization' as a referent of the discourses about peace, reparation and land restitution (Foucault, 2003, pp. 131–33).

The 'modern matrix of individualization' describes how individuals seek redemption and salvation in the present, a particular production and expenditure of their happiness (Foucault, 2003, 2010). The logic of calculation is a relevant element of this matrix in that individuals should look to diminish the risks of living and securing survival (health, property, security). This form of novel 'pastoral power' is executed by different agents of socialization such as public servants, the family, the education system, among others. The creation of scientific knowledge is also relevant to comprehend and inform the agencies of these actors and to reinforce the tools and processes of individuation under this matrix (Foucault, 2003).

Annually, the managerial actors of the LRU define goals. The need to reduce the risks of living moves the Restitution teams to focus on the products that correlate with the applicant's category, and not on the healing process for the victim. Any calculation (as an element of rationalization) that seeks to reduce their risk of living installs a productive logic that contravenes the relationality of the Land Restitution program (reparation to victims). On the one hand, the neoliberal logic of effectiveness rules the work of the Restitution teams, measured by products to define their continuity. This effectiveness creates high levels of pressure, accompanied by the difficulties of working in specific territories where the State is in the process of consolidation and other internal issues that diminish the relevance of the victims during the process. It is common to observe how certain cases are considered problematic due to the multiple issues involved being left behind to resolve later while 'easier' ones have priority since any case counts as a product. Also, a dynamic of withdrawals and rejections of applications serve to bump up the products, increasing the likelihood of achieving the numerical goals requested (Comisión Colombiana de Juristas, 2018).

A paradoxical situation arises detached from the effectiveness logic, by leaving behind the complicated cases. This paradox of effectiveness against waiting times has a detrimental effect on victims who have long processes that exceed the law and are still waiting for the resolution of their cases. The paradox of the narrative as/in discourse of the Land Restitution program also relates to the constant turnover of personnel and the implications of achieving goals that increase the likelihood of making mistakes while collecting and analyzing the information, which reinforces a revictimization and voiding narrative due to the constant repetition of the traumatic events by the applicants/victims, creating action with damage (Comisión de la Verdad Colombia, 2022). The exceeding of times relates to the depiction of relations between citizens and the State in Auyero's (2011) concept of 'patients of the State.' The waiting suffered by welfare recipients is full of uncertainty, confusion, and arbitrariness, despite the significant socialization encountered during the waiting. This waiting can shape a political strategy that compels applicants/victims to prolong this category and their trauma (De Certeau, 1996).

The bureaucratization, rational-calculation and the paradoxes of the effectiveness of the Land Restitution program generate a saturation and overlapping of trauma that confronts the expectations of the process and the perpetuation of the category of applicant/victim that labels citizens in their specific relations with the State.

4. The Land Restitution Discursive Materiality and Its Placemaking

Our final remark focuses on the co-constitutive character and materiality of narrative as/in discourse on the land restituted, making visible the material-discursive practices as place makers in the Land Restitution program. Our approach aims to produce a diffractive reading of the program using the post-humanist approach of Karen Barad. Barad (2003, 2008) proposes an onto-epistemology that considers the co-constitutional intra-activity of Humans and Non-Humans. This diffracting reading illuminates the (re)configurings of the applicants/victims in their intra-activities to land (ibid.).

Barad's onto-epistemology argues that the relationship of Humans and Non-Humans enacts a co-constitutive process in which no predetermined or aprioristic entities exist. Material discursive practices are specific materializations of Human and Non-Human bodies; they compel us to consider their productive aspects and ongoing becomings (Barad, 2003). Discursive Materiality considers that meaning is a 'material (re)configuring of the world, a semantic indeterminacy that is locally resolvable in specific intra-actions' (Barad, 2003, p. 819). The main difference between the conceptualization of discourse and Discursive Materiality is that matter matters in the apparatus of enactment and meaning in Discursive Materiality. Barad (2003, 2008) argues that Discursive Materiality in the (re)configuring of the world enacts boundaries, properties and meanings in its ongoing agential intra-actions. In this sense:

> meaning is not a property of individual words or groups of words but an ongoing performance of the world in its differential intelligibility. In its causal intra-activity, a 'part' of the world becomes determinately bounded and propertied in its emergent intelligibility to another "part." Discursive practices are boundary-making practices with no finality in the ongoing dynamics of agential intra-activity
>
> *(Barad, 2003, p. 821)*

Neoliberal Discursive Materiality disregards that territory entails a complex web of forces, entities and agencies, leaving them behind and obscured. This complex web of intra-actions and becomings has also suffered the violence exerted by the armed actors affecting other co-constitutive Non-Humans. Also, Human and Non-Human intra-activities such as farms, roads, and crops are put aside by an imposing Discursive Materiality that encloses their performativity in a homogenizing pattern of placemaking. The 'matrix of individualization' which influences the Neoliberal Discursive Materiality, blurred communitarian processes of intra-activity and their possibilities of emergence due to the misrecognition of the multiple factors that intervene in their constitution and their multiplicity and deterritorializations (Deleuze & Guattari, 2004). Communities were multiplicities through which individuals related to other communities and entities, there were different intra-actions of farmers, entities and forces with socio-historical places that shape forms of property and relationality between one and the other.

The Discursive Materiality of the Land Restitution program relies on an intra-activity to the land that engages with some properties while hiding others. For the program, the land is its ultimate goal and a passive entity in placemaking; its passivity is performed as another commodity in the neoliberal puzzle in which it constrainedly acts. Thus, the land is bounded, limited and appropriated by the intra-action of the Land Restitution program as part of a trade in which the applicant/victim is also subject to a (re)configuring that requires them to perform as private proprietor of the land, enacting as individual clients available to produce and consume its now owned land.

The Neoliberal Discursive Materiality of the Land Restitution program energized the boundaries created by their intra-actions leading to the reinforcement of saturating and overlapping traumas. For instance, prior to being applicants/victims, farmers used to connect and intra-act through an interwoven web of knowledge/wisdom, traditions, customs and spaces of representations (Lefebvre, 1991; Shields, 1991). The program's Discursive Materiality places these elements under the rug and prioritizes individuals in singular to be repaired. Farmers and their families may return to the land as if the traumatic events were evaporated and impossible to actualize (Shields, 2003), and as if their affective relationship of entities and forces will never release its effects (Deleuze & Guattari, 2004). This avoidance of comprehending the particular intra-actions and connections of trauma may respond to the State's codified imposition of the Land Restitution program.

Conclusion – Transforming the Future Land Restitution Program

A possible diffractive and ontoepistemological performance and enactment of the *narrative as/in discourse* of the Land Restitution program will have to consider the incorporation of Non-Humans as equal partners/agents of Humans. The Colombian jurisdiction recently recognized the territory as victim of the armed conflict, which implies the acceptance that Non-Humans have rights. For us, this is a great advance but still limited. We propose the following elements to overcome a restrictive *narrative as/in discourse* in our diffractive reading of the Land Restitution program: 1) actors, agents and entities (Humans and Non-Humans) are defined and perform in their co-constitutive relationship; 2) local enactments are necessary to allow the expression of boundaries, multiplicity and diversity of territories and their relationships; 3) state is another actor in intra-activity and cannot impose hierarchical significances of reparation, restitution and healing; 4) state requires a local engagement with the actors, agents and entities in intra-activity; 5) Non-Human agents, entities and forces are co-constitutive elements of healing and placemaking; and 6) restitution, healing and placemaking are ongoing and intra-active processes that require responsibility to one another.

References

Alexander, J.C. (2012). *Trauma: a social theory.* Malden: Polity.

Auyero, J. (2011). 'Patients of the State: An Ethnographic Account of Poor People's Waiting', *Latin American Research Review*, 46(1), pp. 5–29.

Bakhtin, M. M. (1984). *Rabelais and his world* (1st Midland book ed.). Bloomington: Indiana University Press.

Ball, K. (2000). 'Introduction: Trauma and Its Institutional Destinies', *Cultural Critique*, 46.

Barad, K. (2003). 'Posthumanist Performativity: Toward an Understanding of How Matter Comes to Matter', *Signs: Journal of Women in Culture and Society*, 28(3).

Barad, K. (2008). *Meeting the Universe Halfway: Quantum Physics and the Entanglement of Matter and Meaning.* Durham: Duke University Press.

Bauman, Z. (2007). *Modernity and the Holocaust* (Reprint). Cambridge: Polity Press.

Comisión Colombiana de Juristas (2018). 'Cumplir metas, negar derechos: balance de la implementación del proceso de restitución de tierras en su fase administrative, 2012–2017'.

Comisión de la Verdad Colombia (2022). Enfoque de acción sin daño – Glosario. Available: https://web.comisiondelaverdad.co/transparencia/informacion-de-interes/glosario/enfo que-de-accion-sin-dano. [Accessed: 14 April 2023].

Congreso de la República de Colombia (2011). *Ley No. 1448*. Available: www.centrodem emoriahistorica.gov.co/micrositios/caminosParaLaMemoria/descargables/ley1448.pdf. [Accessed: 12 August 2023].

Das, V. (2007). *Life and words: violence and the descent into the ordinary*. Berkeley: University of California Press.

De Certeau, M. (1996). *La invención de lo cotidiano 1: Artes de hacer*. México: Universidad Iberoamericana.

De Fina, A. & Johnstone, B. (2015). 'Discourse Analysis and Narrative', in D. Tannen, H. E. Hamilton, and D. Schiffrin (eds.) *The Handbook of Discourse Analysis* (2nd edition). Chichester: Wiley Blackwell, pp. 152–67.

Debord, G. (1977) [1967]. *The Society of the Spectacle*, translation by Fredy Perlman and Jon Supak (Black & Red, 1970; rev. ed. 1977).

Deleuze, G. & Guattari, F. (2004). *Mil Mesetas. Capitalismo y esquizofrenia*. 6a edn. Valencia, España: Pre- textos.

Foucault, M. (2010). *The birth of biopolitics: lectures at the Collège de France, 1978–79*. M. Senellart (ed.) Translated by G. Burchell. New York: Palgrave Macmillan.

Foucault, M., Rabinow, P. & Rose, N. S. (2003). *The essential Foucault: selections from essential works of Foucault, 1954–1984*. New York: New Press.

Guevara Salamanca, J. D. & Jiménez Espinosa, G. P. (2022). 'Configuraciones y encuentros espaciales: convivencialidad y memoria', *Documentos de Trabajo* (Unicevantes), 4(4).

Hobbes, T. & Gaskin, J. C. A. (2008). *Leviathan*. Oxford: Oxford University Press.

Livholts, M. & Tamboukou, M. (2015). *Discourse and narrative methods*. London: Sage publication.

Ministerio del Interior de la República de Colombia (2012). '*Decreto No. 4829*'. Available: www.centrodememoriahistorica.gov.co/micrositios/caminosParaLaMemoria/descargables/ley1448.pdf. [Accessed: 14 April 2023].

Shields, R. (1991). *Places on the Margin: Alternative geographies of modernity*. Abingdon: Routledge.

Shields, R. (2003). *The Virtual*. Abingdon: Routledge.

Smelser, N. (1962). *Theory of Collective Behaviour*. London: Routledge.

6

LANDSCAPES OF REPAIR

Creating a Transnational Community of Practice with Sheffield- and Kosovo-based Researchers, Artists and Civil Society on Post-traumatic Landscapes

Amanda Crawley Jackson, Korab Krasniqi and Alexander Vojvoda

Introduction

In this chapter, we will discuss a 2021 knowledge exchange project between researchers from the University of Sheffield and forumZFD Kosovo, in cooperation with Kosovo artists, researchers, cultural workers and civil society activists. The collaboration – *Landscapes of Repair* – occurred through an online exhibition featuring the work of six Kosovo-based artists and architects and a programme of international cross-sector talks and masterclasses. This collaboration gave space to questions on the role of informal networks of communities, cultural workers, artists and civil society activists in tangible, transformative interventions in dealing with traumatic pasts in public space. Questions such as what does repair mean in the context of a wounded built environment? How might repair be performed at a physical and a symbolic level? What forms might it take? Working proactively through a self-reflexive and action-research lens, we now share insights from *Landscapes of Repair.* Our insights include exploring – the reclamation, reoccupation and re-imagining of sites shaped by difficult, violent or traumatic histories; reflections on how processes of memorialisation and creative artistic and architectural practice might contribute to repairing landscapes scarred by trauma; critiques of the tools of knowledge exchange as a means of building a transnational forum for trauma-informed reflection and praxis with regard to the repair and re-appropriation of public space in post-conflict Kosovo; and we present our mutual learning and critical self-reflection on the community of practice that has emerged through *Landscapes of Repair*. We conclude with a reflection on how projects of this kind are *de facto* embedded in power imbalances and inequalities. From this we show how we can continue to work towards building trust and respect, as well

DOI: 10.4324/9781003371533-8

as sharing common experiences, practices, lessons learned and strategies in the context of post-conflict and post-traumatic placemaking.

This chapter was written collaboratively by three members of the *Landscapes of Repair* community of practice. Writing together, in a dialogical and self-reflexive mode, has been a valuable way for us to hold space for each other and proactively assess and reflect on our continued efforts to embed equity in our partnership working. It builds on and further contributes to the meta-cognitive and ethical work that we believe must underpin any sustainable, ethical collaboration around trauma-informed placemaking.

From Post-Traumatic Landscapes to *Landscapes of Repair*

In March 2020, the Graves Gallery (Sheffield, UK) opened an exhibition co-curated by Amanda Crawley Jackson with Museums Sheffield, *Invisible Wounds: Landscape and Memory in Photography*. It included works by seven internationally acclaimed artists whose works variously encapsulated the visible and invisible scars inflicted by trauma on the landscape and explored how we remember and deal with the past in places we occupy in the present.

The landscapes depicted by the artists addressed the temporal complexity of post-traumatic landscapes. While the prefix 'post-' suggests that the trauma referred to is located in the past, the term 'post-traumatic' is typically used in clinical and everyday discourse to describe the lingering, disruptive resonance of the past in the here and now. To speak of the 'post-traumatic' is, therefore, to describe a trauma which is ongoing, still live and is not yet addressed or accounted for in the present. It describes also an expanded 'colonial present' (Gregory, 2004), in which intersectionally marginalised communities continue to be harmed in the *longue durée* of systemic injustice and disproportionately exposed to cycles of trauma and re-traumatisation (Deming & Savoy, 2011, p. 3).

Although *Invisible Wounds* was forced to close prematurely due to COVID-19, several significant themes emerged and were amplified in the context of the pandemic and the 2020 Black Lives Matter protests. These were further explored in the edited book, *Invisible Wounds: Negotiating Post-Traumatic Landscapes* (2020) and in a two-day online symposium, which dealt with issues such as how we bear witness to often unmarked and contested spaces of trauma; the various ways in which memory is materially and symbolically inscribed – or deliberately erased – in landscapes and communities; how and why post-traumatic landscapes might be neglected, avoided or suppressed; how systemic injustice manifests and is experienced in place; and how artistic interventions might begin to acknowledge, remember, reclaim and repair post-traumatic landscapes in creative and socially just ways.

The reparative and memorial functions of art and civil society interventions were at the heart of the first iteration of *Landscapes of Repair* (2021), a collaborative transnational project on post-traumatic landscapes in Kosovo. The partnership emerged directly from forumZFD Kosovo's encounter with the *Invisible Wounds* book, building on its emergent reflections on trauma-informed placemaking and the role played by creative practice and the arts in a more localised geographic context. *Landscapes of Repair* brings together Crawley Jackson's research on post-traumatic landscapes with forumZFD's expertise in dealing with the past in the Western Balkans, foregrounding the role of Kosovan artists, organisations and civil actors in thinking about, as well as enacting, community- and practice-led repair.

The project began in 2021, with key outputs including an online exhibition co-curated by Amanda Crawley Jackson and then University of Sheffield student Faye Larkins, and doctoral researcher Emily-Rose Baker, in discussion and collaboration with the participating artists; an interactive website (www.landscapesofrepair.org) and a two-day international, cross-sector symposium hosted and organised by forumZFD Kosovo staff Alexander Vojvoda and Korab Krasniqi in cooperation with the University of Sheffield and various Kosovo-based researchers, academics, artists and civil society activists in the field of memorialisation practices, transitional justice and dealing with the past. The symposium was delivered online in English, Serbian and Albanian and created a transnational platform for sharing experience, insight and learning, bringing together a range of expertise in post-traumatic and post-conflict placemaking, with speakers including academic researchers, activists, architects, creative practitioners, artists, representatives from Kosovo-based NGOs and civil society. The online exhibition, featuring works by Brixhita Deda, Lulzim Hoti, Argjirë Krasniqi, Korab Krasniqi, Vildane Maliqi and Dardan Zhegrova, explored questions around the practices of memorialisation, reoccupation and repair in post-conflict landscapes; what repair might look like and what forms it might take in Kosovo and the Western Balkans; the role of creative practice in repairing, repurposing and reimagining damaged topographies; and the possibility of co-creating landscapes of care and architectures of peace. Many of the artworks displayed showed buildings scarred by the visible and invisible traces of multiple past conflicts, bringing us 'closer to an understanding of how art is implicated in repairing localised legacies of the past inscribed within the present' (Landscapes of Repair exhibition, 2021).

forumZFD Kosovo has continued to grow the *Landscapes of Repair* website, with recorded talks by researchers and activists in Kosovo, commissioned papers, special issues of journals, a multi-media mapping tool and an archive featuring site-specific interventions into post-traumatic landscapes via the collection of soundscapes, images, texts and videos. In 2022 and 2023, the online exhibition was co-curated *in situ* by Alexander Vojvoda in cooperation with local civil society, cultural actors including Anibar and Kinema Jusuf Gërvalla in Peja, the University of Prishtina, the Centre for Narrative Practice in Prishtina and the artists involved in the exhibition. An exhibition catalogue, featuring texts by the exhibited artists, was published in Kosovo in 2023. The collaboration, conversations and activities have thus continued and expanded since the first edition of the project came to an end in summer 2021. The present chapter marks an important stage of reflection in the collaboration as forumZFD Kosovo, in collaboration with the *Landscapes of Repair* collective and Kosovan and international partners, continue to plan future activities.

Landscapes of Repair is part of forumZFD Kosovo's long-term approach to reflect on public spaces and space-creation in post-conflict settings as discursive arenas for competing narratives on the violent and traumatic past in the Western Balkans region. forumZFD Kosovo has assessed space-making, public spaces, post-traumatic landscapes and the documentation of memorial sites as a relevant context with high mnemonic potential and an important field for discussions about dealing with the past. Exemplary of forumZFD's work in this are projects like *MOnuMENTI* (2014), *Kosovo Memory Heritage* (2017-), *Memory Mapping Kosovo* (2014–16), *Memory Lab* (2014–15), *Inside-out, Outside in* (2017–18) and *Landscapes of Repair* (2021-). These activities have served as a forum for critical reflection on the (violent) past in Kosovo, the region and beyond. Ultimately, the way in which Kosovo deals with its past in public spaces has a profound impact on the country's culture of remembrance. It seeks to challenge one-sided narratives and supports fact-based, inclusive practices of memorialisation.

Kosovo: The Politics of Violence

With the constitutional amendments of 1974, which effectively provided *de facto* self-government, Kosovo, although not entirely equal to other republics, became an autonomous region within the Federal Republic of Yugoslavia. The population of Kosovo consisted of 82.2% ethnic Albanians, approximately 10% ethnic Serbs and Montenegrins, and the remaining 7.8% made up of other ethnic minorities (Clark, 2000, p. 215). With growing nationalism and the rise of Serbian nationalist leader Slobodan Milošević in the federal government, Kosovo's autonomy was revoked in 1989, sparking a series of violent protests. At the beginning of the 1990s, Albanians were totally segregated from educational, social, political and cultural life. The Kosovo local government was dissolved, ethnic Albanian workers were fired, students were segregated from the educational system, thus forcing ethnic Albanians to unilaterally declare independence from Serbia and to establish parallel state structures, with the government in exile financially supported by the Albanian diaspora (Clark, 2000, p. 103).

In the early 1990s, as the Federal Republic of Yugoslavia began to dissolve, Albanian resistance remained largely peaceful and non-violent. The Albanian parallel government organised democratic elections, with Ibrahim Rugova, a pacifist, writer, academic and founder of the political party Kosova Democratic League, elected as president. Rugova's strategy of peaceful resistance attracted widespread support from the Kosovan ethnic Albanians. As the work of parallel state structures proved not entirely effective, Kosovan ethnic Albanians began engaging in peaceful, non-violent protests in public and urban areas, though they often faced excessive use of force by the police and severe political repression and persecution. Non-violent, peaceful resistance became a political mantra of Ibrahim Rugova and his ethnic Albanian followers. As Clark notes, the turn to non-violence and the establishment of parallel state structures (Clark, 2000, p. 95) also marked a general shift, as 'old traditions' were replaced and 'the aspiration to be 'modern' and 'European' took hold of a younger generation who hoped they could emulate people power movements elsewhere in Eastern Europe' (ibid., p. 46).

Meanwhile, as Slobodan Milošević consolidated repressive measures in Kosovo and Rugova attempted to improve social cohesion and the political position of Kosovo, an armed group calling themselves the Kosovo Liberation Army emerged, launching armed and violent resistance that resulted in countermeasures from the federal army and Serbian police. By the end of 1998, NATO demanded that the oppression and violence inflicted on civilians end, and that Kosovan autonomy should be restored, with the withdrawal of the Yugoslav Army and Serbian forces from Kosovo. In March 1999, following Russia and Serbia's refusal to sign the Interim Agreement for Peace and Self-Governance in Kosovo, NATO launched a bombing campaign that would last almost three months.

War and Trauma

Between 1 January 1998 and 31 December 2000, a total of 13,518 people were killed in Kosovo, including 11,661 men and 1,857 women. Among them were 10,794 Albanians, 2,197 Serbs, and 527 from other ethnicities, with. 8,662 were civilian Albanians and 2,132 guerrilla soldiers; 1,197 were civilian Serbs and 1,000 were soldiers of the Yugoslav army, Serbian police officers, and/or paramilitaries (Kosovo Memory Book). Around 20,000 women and men were raped (Karic et al., 2022). Hostage taking, imprisonment, forced disappearances and mass expulsions were part of the strategy and apparatus of war (ibid.), as was 'urbicidal violence' (Graham, 2011, p. 378), with buildings looted and demolished, land registry records stolen, and public infrastructure, sites of cultural heritage and monuments destroyed (UNEP & UNHCR, 1999, p. 68; Vojvoda, 2021; Krasniqi, 2022, p. 13). Around 120,000 houses were damaged, of which approximately one third were beyond repair (UNEP & UNHCR, 1999, p. 70). The estimated cost of the material damage amounted to approximately 22 billion euros (Ahmeti, 2018; Krasniqi, 2022, p. 13).

The use of violence in the context of war aims to inflict harm on bodies, landscapes and architecture as an act of political intimidation and cultural delegitimisation. The deliberate destruction of religious, historical and cultural architecture has been well documented and discussed, mainly through the prism and discourse of violence and materiality. The destruction, deliberate damaging and/or erasure of landscapes, architecture, monuments and archaeological objects as cultural and mnemonic repositories, qualify as war crimes according to international human rights laws and as such are punishable (see Protocol Additional to the Geneva Conventions of 12 August 1949, and relating to the Protection of Victims of International Armed Conflicts [Protocol I], 8 June 1977 to the Geneva Convention, Article 53).

Between Memory and Oblivion

When the war ended, fewer than a million people who had fled and sought refuge in Albania, North Macedonia, Montenegro, Serbia and elsewhere, returned to their homes. Upon their return, they were confronted with grave architectural, infrastructural and natural resource destruction and the post-conflict political context nurtured and maintained the culture of violence against buildings, sites and architecture. On 17 and 18 March 2004, violent rioting took place throughout Kosovo, incited by sensationalist and inaccurate media reports that ethnic Serbs living in North Mitrovica had been responsible for the drowning of three Albanian children in the Ibar River. The violent riots posed a serious threat to security in the place, targeting ethnic minorities, their (550) homes and properties, as well as infliction of harm and destruction on built cultural heritage, (27) Orthodox churches, and monasteries (Human Rights Watch, 2004). Deliberate attempts to destroy socialist memorials were a common practice as well. The vandalism we witnessed in Prishtina reflects an attempt to erase the memory of co-existence and anti-fascist struggle, a common ground that united people in Yugoslavia. Examples include the degradation and vandalisation of the Partisan Martyrs' Cemetery, which commemorates those who fought in the Second World War (see Caka, 2021); at the memorial complex in Landovica, the removal of the bust of Boro Vukmirović (an ethnic Montenegrin) from its position next to that of Ramiz Sadiku (an ethnic Albanian), friends who fought against the Italian occupation of Kosovo and Albania in 1941 (see Krasniqi, in Vojvoda et al, 2022, pp. 48–52); the planting of dynamite at the base of the Monument to Heroes of the National Liberation Movement and the painting of partisan fighters carrying flags of the allies who supported the liberation of Kosovo.

Towards Repair

As James Joyce noted in the margins of the manuscript of *Ulysses* (1922), 'Places remember events.' The trauma of conflict scored and scarred the natural and built environment in Kosovo, shaping modes of sociality and co-existence, spatial policies and public memorialisation in the present (Vojvoda, 2021). Over the last two decades in Kosovo and the region, the question of dealing with the past has preoccupied researchers, civil society and the state. Martti Ahtisaari's *Comprehensive Proposal for the Kosovo Status Settlement* (UN Security Council, 2007) set out the obligation for Kosovo to 'promote and fully respect a process of reconciliation among all its Communities and their members' and to 'establish a comprehensive and gender sensitive approach for dealing with its past, which shall include a broad range of transitional justice initiatives.' European Union (EU) institutions and enlargement policies invite and suggest that Kosovo 'should develop an overarching strategy for transitional justice, including a comprehensive approach to addressing its past' (EU Commission, Kosovo Report 2022).

The prism through which civil society, state institutions and international mechanisms in Kosovo commonly address past human rights violations is the *Conceptual Framework for Dealing with the Past* (2012), drafted by Swisspeace, an action-oriented peace research institute with headquarters in Bern, Switzerland. The framework promotes a holistic approach to dealing with the past that aims at developing a culture of accountability, strengthening the rule of law and fostering reconciliation through four key principles approved by the UN Human Rights Commission 1997: the right to know, the right to justice, the right to reparations, and the guarantee of non-recurrence.

In other terms, post-conflict societies engage in transitional justice processes in ways that encompass independent legal or normative frameworks as well as grassroots reconciliation initiatives, known also as retributive or restorative justice mechanisms. The first seeks to prosecute and hold accountable those who committed gross human rights violations in the context of war; and the former seeks to restore and/or repair relationships, trust and cohesion in society. Restoring and repairing are not only about fixing what is broken or restoring what is lost. They are also about acknowledging what is hurt or missing and finding ways to cope and heal. This approach also reaches towards the creation of new meanings and possibilities from the fragments and traces of memory, through means of symbolic activities, the re/creation of public spaces and spacemaking.

In the context of restorative justice and repair, memory constitutes a key societal dynamic that enables the past to be remembered and inscribed in the present. It is a means of negotiating the future through the meanings of the past and present. Restorative justice positions memories as dynamic and multivarious – co-existing rather than competing, and thereby producing a form of collective memory that can be understood as a moral framework for building social consensus among affected people following a period of brutality (Karamanić, 2012).

In contrast to traditional understandings of dealing with the past and transitional justice practice, which deal with the legacy of conflict and trauma through a limited temporal and spatial perspective and are strongly related to normative practice, the works exhibited in the *Landscapes of Repair* exhibition sought to accommodate the multitude of memories, promoting a horizontal approach to memorialization and repair. They made space for multiple and sometimes conflicting narratives, representing and remembering the pain and trauma of different social groups through creative praxis in public space and engaged mnemonic practice. The exhibition remains online, encouraging sustained discussion and continued reflection on the complexity of post-conflict placemaking.

Korab Krasniqi

Since 2017, Korab Krasniqi has been working at the intersection of research and photo-documentary to create *Kosovo Memory Heritage*, a project which documents a multitude of architectural features and functions, sites and memory landscapes, which, as explained earlier, were damaged or destroyed during the war. The seven photographs shown in the *Landscapes of Repair* exhibition call for reflection and constructive social dialogue about the entangled layers of historical, cultural, and political pasts. Interrogating and countering the homogeneity of narratives, they enable interpretive reconfigurations of realities and fictions, problematizing memory as fluid and transformative, and collapsing normative meanings of time and space. The photographs position memory as participation and anticipation, rather than recollection, updating the past in the present to respond to a vision for what lies in the future.

Lulzim Hoti

The closed Ibar bridge in the ethnically divided city of Mitrovica has become the focus and symbol of the physical divide between the city's Serb-dominated north and the Albanian-dominated south. Born and raised in Mitrovica, artist Lulzim Hoti challenges perceptions of the Ibar bridge and the use of public spaces in Mitrovica and Kosovo, raising questions about the (re-)appropriation of spaces 'that re-awaken memories, emotions and insecurities' (Hoti, cited in Vojvoda 2022, p. 109). With the video installation 'How you throw, I will throw it back', he lays open in a striking way the lines of fragmentation symbolised by the bridge and finds ways to build grounds for common perspectives in a divided city. The work shows two people using the bridge as their playground, throwing a frisbee back and forth. This playful, almost childish intervention in public space, which draws the attention of the police, shines light on the mnemonic potential of the Ibar bridge in different contexts; and calls also for new ways of thinking about public spaces as places for repair, reconciliation and radical openness.

Argjirë Krasniqi & Brixhita Deda

In 2020, architect and artist Argjirë Krasniqi led a project in the depopulated town of Janjeva, working with local residents to map and co-design new futures for approximately 600 empty buildings. What, she asked, could they be used for? How might their creative repurposing bring life back to Janjeva and benefit its residents? In another project, Krasniqi and architect Brixhita Deda explored how the difficult pasts of buildings associated with conflict trauma have been dealt with in the present. In one of their case studies, a drawing depicts atrocities committed outside the former House of the Yugoslav People's Army (Vojvoda et al., 2022, p. 121), which is now the Kino Armata, a vibrant hub of cinematographic culture. Exhibited next to the drawing is a colour photograph of the building's contemporary interior. Everywhere, there are signs of cultural activity, purposeful organisation and life. This building is presented as an example of healing transformation. However, other buildings associated with violence and trauma have been abandoned, falling into a state of such dereliction that demolition is now the only solution.

Slow Collaboration as a Sustainable Framework for Practices of Repair

The rationale for our opting for a 'slow' collaboration was to address preliminary questions regarding the safety of participants and contributors, fairness, referencing and the (re-)appropriation of results, transparency and openness, and cross-cultural/ social/economic commitment to empowerment; and to lay the foundations for a longer-term cooperation before launching projects about collaboratively repairing places. Without laying these groundworks, a classic project-focused cycle would not lead towards a community of practice with emphasis on action-based and actors-focused processes. These preliminary, resource-intensive processes and discussions resulted in an interdisciplinary, cross-sectoral cooperation between academia, civil society and artists to develop a community of practice which platforms its work, activities, exhibitions and conversations about placemaking in post-conflict contexts. Our approach to the agreed outputs of the first edition were conceived less as endpoints than as generative way markers, learning from which continues to inform our understanding of both the values and challenges of sharing knowledge across international borders.

Key insights from this deliberately slow approach to working together for better outcomes include recognising the value and importance of:

1. Collaborative and participatory development processes. The funding application for the first edition of the project was co-authored by forumZFD Kosovo and researchers from the University of Sheffield, in consultation with Kosovo-based researchers, civil society organisations and cultural workers. The costings and implementation process involved financial remuneration for all Kosovo-based participants and UK students working on the project. Institutional finance systems can, however, in their complexity and opacity, inhibit equitable collaborations. It was crucial for the funding application to be co-written by the Sheffield academic team and partners in Kosovo, in order to ensure that a) all parties were clear about each other's respective engagement and commitments, and b) no labour was invisible, and that funding was equitably allocated.
2. Co-creating an online platform for civil society actors, artists and researchers from the UK, Kosovo and the Western Balkans. This platform is intended to facilitate transparency, access and the sharing of new approaches and innovative methods on post-traumatic landscapes, memorialisation, capacity building, networking and trauma-informed practices of repair.
3. Sharing unique knowledge and perspectives on local and regional socio-political dynamics, especially in a post-conflict environment. The consideration of local and diverse perspectives is key, as is the co-design of activities. Various academics and researchers, civil society activists and artists were invited to co-curate panels, presentations and other inputs during the two-day symposium in June 2021. In the case of the symposium, each partner organisation partook

in the overarching programme design and was invited to curate a panel, with funding made available to support the engagement of non-academic speakers, including artists and filmmakers.

4. Establishing transnational learning processes, including critical self-reflection and collaborative writing, which inform further activities and potential projects.

5. Creating potential trajectories for common project development and fundraising for projects about dealing with the past.

6. Language. With regard to cultural pre-conditions, English as the main project language was questioned as it does not reflect the realities of the participants and the socio-political environment in which *Landscapes of Repair* is working. Language barriers were discussed as, while English may be the first language for many of the UK-based researchers, this is not the case for the participants, researchers and artists working in the Western Balkans. This was addressed by dedicating funds and resources to the translation of project documents and the website, as well as simultaneous interpreting across three languages (Albanian, English, and Serbian) during the symposium.

Conclusion: 'Slow' Cooperation, Lessons Learned and Trauma-informed Communities of Practice

The *Landscapes of Repair* project team sought not only to co-create a platform for exchange but was also aware of potential imbalances between various actors in such exchanges. In this context, the University of Sheffield team and forumZFD Kosovo surfaced certain risks and sought to create a space of transparency and fairness in which mitigation strategies could be developed; therefore, the paced approach was adopted in developing a common framework.

During the *Landscapes of Repair* symposium on 13–14 July 2021, Nehari Sharri (Country Director, forumZFD Kosovo) stressed the importance of implementing non-extractive policies and principles in collaborative endeavours, 'because we all have or [have] had experiences when your knowledge or your experiences were exploited – even hijacked in some cases. So, I think it is very important to emphasise that in whatever way we [the *Landscapes of Repair* project team] continue and with whomever we continue, in whatever setting, the process should be or will be fair and transparent and everybody's involvement, experience, and work will be acknowledged.' Sharri thus addresses questions of power imbalances and equity in cooperations between different actors from different fields, equipped with different resources. As the Arts and Humanities Research Council (AHRC) *Common Cause Report* of 2018 observed, UK universities' ambitions around knowledge exchange, impact and public engagement have often, and in many ways, reinforced, rather than dismantled, barriers to social justice. A key obstacle is the difficulty experienced by partners and stakeholders from underrepresented groups and communities as they attempt to navigate institutional finance systems. The council also emphasises that knowledge exchange projects and collaborations can reveal power imbalances, biased and one-sided structures that benefit UK academia, to the disadvantage of intersectionally marginalised communities – in our case, Kosovo civil society activists, artists and cultural workers. As the report states:

> *There is important and urgent action that needs to be taken in particular by funders and by university leaders, to address the current situation in which the knowledge, expertise, interests and needs of diverse communities are not being reflected in the research landscape. [...] Alongside the necessary work to address structural inequalities in other areas—such as widening participation and decolonising the curriculum—these research collaborations will begin to transform universities into powerful spaces for mutual learning, dialogue and the enrichment of our collective knowledge base.*
>
> *(p. 98)*

Furthermore, the exchanges and inputs during the symposium and in panel discussions raised questions regarding ethics and socio-political dynamics when dealing with the past, especially when it comes to sensitive topics in the context of victims of violence and trauma. These discussions and iterative meta-learning around the co-creation of more equitable transnational partnerships continue to inform and structure the workings of the collaboration. This strategy is in line with Donjeta Murati's observation during the symposium on the context of transnational exchanges:

> *A precondition for [transnational exchanges] would be a kind of agreement or a common project in order to serve the commons and the common needs beyond stratification which might be based on economic conditions or any other condition. That's how I see it, coming from here [Kosovo] for the importance of transnational exchange.*

The common agreement that emerged in conversations between the *Landscapes of Repair* project team, as well as during the discussions, exchanges and symposium, was that transnational knowledge exchange needs to formulate clear political aspirations and demands that are embedded in and reflect the local and regional socio-political environment and dynamics, especially when discussing public spaces as reflections of trauma, war and past violence.

References

Ahmeti, N. (2018). 'Over 22 billion Euros of war damages' in Radio Free Europe [online], 23 September 2018. Available: www.evropaelire.org/a/mbi-22-miliarde-euro-deme-te-luftes-/29505183.html. [Accessed: 29 May 2023].

Brumund, D. & Pfeifer, C. (eds.) (2014). *MOnuMENTI: the changing face of remembrance*. Belgrade: forumZFD.

Caka, F. (2021). '"Martyrs" Cemetery in Velania: A memorial park in despair or a potential landscape for repair?' in *Dealing With the Past* [online], 2 November 2021. Available: https://dwp-balkan.org/martyrs-cemetery-in-velania-a-memorial-park-in-despair-or-a-potential-landscape-for-repair/. [Accessed: 7 May 2023].

Clark, H. (2000). *Civil Resistance in Kosovo*. London: Pluto Press.

Deming, Alison Hawthorne and Savoy, Lauret E. Savoy. (2011). *The colors of nature: Culture, identity, and the natural world*. Milkweed Editions.

EU Commission, Kosovo Report (2022). Chapter 23 - Judiciary and fundamental rights, Domestic handling of war crime cases, 24.

Graham, S. (2011). *Cities Under Siege: The New Military Urbanism*. London: Verso.

Gregory, D. (2004). *The Colonial Present: Afghanistan, Palestine, Iraq*. Oxford: Blackwell.

Human Rights Watch (2004). Failure to Protect: Anti Minority Violence in Kosovo. Available: www.hrw.org/report/2004/07/25/failure-protect/anti-minority-violence-kosovo-march-2004. [Accessed: 25 August 2023].

Karamanić, S. (2012). 'Truth and Reconciliation: a new political subjectivity for post-Yugoslavs?', Open Democracy [online], 6 November 2012. Available: www.opendemocracy.net/en/truth-and-reconciliation-new-political-subjectivity-for-post-yugoslavs/. [Accessed: 20 May 2023].

Karic, H. & Domi, T. L. (2022). 'We need a better way to prosecute sexual assault in conflict', *Foreign Policy* [online] 9 March 2022. Available: https://foreignpolicy.com/2022/03/09/rape-sexual-assault-war-crime-justice-kosovo/. [Accessed: 29 May 2023].

Kosovo Memory Book (1998–2000). Available: www.kosovomemorybook.org. [Accessed: 7 May 2023].

Krasniqi, K. (2022). *Hijacked Childhoods: Accounts of Children's Wartime Experiences*. Prishtina: forumZFD.

Krasniqi, K. et al. (eds.) (2017). *Kosovo Memory Heritage*. Prishtina: forumZFD.

Landscapes of Repair exhibition (2021). www.forumzfd.de/en/landscapes-repair-exhibition-and-symposium#:~:text=The%20Landscapes%20of%20Repair%20symposium,conflict%20place%2Dmaking%20in%20Kosovo.

Swisspeace (2012). *A Conceptual Framework for Dealing with the Past: Holism in Principle and Practice*. Bern: Swisspeace.

UNEP & UNHCR (1999). *The Kosovo Conflict: Consequences for the Environment and Human Settlements*. Geneva: UNEP & UNHCR.

UN Security Council (2007). Letter dated 26 March 2007 from the Secretary-General addressed to the President of the Security Council, Addendum, Comprehensive Proposal for the Kosovo Status Settlement, Article 2, Human Rights and Fundamental Freedoms, 2.5, 3.

Vojvoda, A. (ed.). (2021). 'Violence taking place – Trauma and war in architecture and public spaces' in *Balkan Perspectives Magazine* #17, forumZFD. Available: https://dwp-balkan.org/balkan-perspectives-2/. [Accessed: 25 August 2023].

Vojvoda, A. et al. (eds.) (2022). *Exhibition Catalogue: Landscapes of Repair*. Prishtina: forumZFD.

7

THE FILIPINO SPIRIT IS [NOT] WATERPROOF

Creative Placeproofing in Post-Disaster Philippines

Brian Jay de Lima Ambulo

As someone born and raised in the Philippines, the frequency of experiencing natural hazards is nothing new to me. My earliest, faintest memory of this 'interaction' was in 1995 with Typhoon Rosing, one of the earliest categorised as a *super typhoon*. I was four years old at that time and seeing some parts of our roof and ceiling effortlessly carried by a 290 km/h howling, sustained wind was something to be remembered. We were fortunate to be situated in a high-elevated mountain area, far from flood-prone locations. We neither evacuated nor got displaced, but it was one of the longest eight hours of my life.

State in Calamity

The Philippines' unique geographic positioning renders the archipelago highly susceptible to the adverse effects of natural hazards, which significantly shape the collective experiences of its inhabitants. It is also placed along the typhoon belt, with an average of 17 to 20 tropical cyclones entering the Philippine Area of Responsibility (PAR) annually, according to the Philippine Atmospheric, Geophysical and Astronomical Services Administration (PAGASA) (Tropical Cyclone Information, 2020). Five of these cyclones are now usually classified as super typhoons (Barber, 2013, p. 5). The recurring onslaught of typhoons and other catastrophic events has resulted in a distinctively Filipino resilience that permeates deep into its cultures. Whilst these calamitous events are undeniably detrimental, it is also essential to consider their potentialities to serve as spaces for community healing, resilience, creativity, and transformation. Framing within its spatial, temporal, and relational dimensions, this analysis investigates disasters as spaces of suffering and trauma, inculpability, and co-optation.

DOI: 10.4324/9781003371533-9

According to McFarlane and Norris (2006, p. 4), a disaster is 'a potentially traumatic event that is collectively experienced, has an acute onset, and is time delimited; disasters may be attributed to natural, technological, or human causes.' Experiencing disasters, agency and scale notwithstanding, is always a phenomenon of all fronts, challenging social, environmental, economic, cultural, political, and psychological dimensions of an individual and a community – 'all-encompassing occurrences' with 'multiplicity of interwoven, often conflicting, social constructions' (Oliver-Smith, 2002, pp. 23–4). It disrupts ways of living but simultaneously creates a culture of meaningful interactions, relationships, creativity, and innovation. It blurs the lines between being an *event* and the *processes* that go through it. It offers a unique context to the field and its seemingly 'unique' characteristic to 'occur at the intersection of nature and culture' (ibid., p. 24).

As the 21st century is confronted with an escalating frequency and severity of disasters and crises, such as climate change and environmental degradation, global pandemics, economic and financial risks, geopolitical tensions, technological disruptions, historical decline, social inequalities, and political violence (Klein, 2007; Meiner & Veel, 2012; Sassen, 2014; United Nations Intergovernmental Panel on Climate Change, 2019), the interrelation of the natural and cultural realms demands further academic exploration. Comprehending this intricate relationship is vital, as it can reveal the intricate ways in which human endeavours, both communal and individual, are moulding the global ecosystem and perpetuating disparities (Haraway, 2016). Moreover, Latour (2018) contends that the complex interweaving of nature and culture within the Anthropocene has led to a dissolution of the demarcation between human and non-human entities, thereby necessitating a reassessment of how we perceive and tackle environmental challenges.

For the Philippines, the constant exposure to both recorded and unrecorded disasters and hazardous events, big or small, ceaselessly influence the daily lives of Filipinos. From 2000 to 2020, 'natural' disasters in the Philippines affected almost 149 million people, with 25,374 deaths and 385,611 injured (Guha-Sapir, 2020). Socioeconomic damages accounted for $21.3 billion USD, averaging $1.2 billion USD annually (Guha-Sapir, 2020). These data were only the recorded ones. Incontestably, these scenarios affect its long-term growth and developmental effects. According to Bankoff (2003, p. 178), what makes the Philippines different is that hazards are 'a frequent life experience, and consequently one where the chronic threat of such events has been normalised as an integral part of culture.'

Filipinos are exposed, laid bare, and susceptible to natural hazards in whichever shape and form, encountering the little shocks and troubles that were uncovered but unreported, unconsciously entering the banality of everyday life. On the other hand, the country 'has devoted significant resources to build disaster capacity and reduce population exposure and vulnerability' (Alcayna et al., 2016, p. 2). Ironically, due to the commonness of disasters in the Philippines, the resilience that made the country more informed in drafting 'a seemingly well-crafted disaster management plan' (Santiago et al., 2016, p. 644) is the same resilience it has accumulated from experiencing these hazards. Nonetheless, such events should not be utilised to rationalise a country's disaster response and management methods. Despite this understanding, studies show that a typhoon of similar intensity could lead to a death toll in the Philippines that is 17 times higher than in Japan (Ginnetti et al., 2013, p. 3).

Embracing a context-specific and comprehensive approach that considers the various social, economic, environmental, and political elements contributing to disaster risk and vulnerability is pivotal to effectively confronting these challenges and nurturing resilience. This involves fostering an environment of readiness and adaptation, boosting cooperation among different stakeholders, and employing novel strategies and technologies that enable communities to handle the intricacies of disaster risk reduction and management.

Recognising the interconnected nature of these challenges and adopting a comprehensive viewpoint allows us to build more resilient, adaptable, and inclusive communities. These communities will be better armed to endure and rebound from the harmful impacts of disasters and other crises. Through such efforts, we can transform the way communities and individuals face these challenges, ultimately striving towards a more sustainable, equitable, and robust future for all.

A pressing need exists to reimagine the bond between nature, culture, and creativity, fostering an environment that encourages healing, regeneration, nurturing, preservation, and overall resilience. Guided by this frame of reference, I propose *creative placeproofing* as one of the alternatives.

Culture of Resilience

'*The Filipino spirit is waterproof*' – but it is not. For many years, this unofficial national battle cry has been employed and empowered by the government and the media as a living testament to the survivalist attitude of the Filipino people – a seeming unbreakable resilience in the face of disasters and other adversities (Fig. 7.1). This resilience narrative, whilst seemingly acknowledging the strength and adaptability of the population, has been criticised for obscuring the government's responsibility to provide effective disaster management and risk reduction measures. Simply put, it overlooks the structural inequalities and inadequate policies that exacerbate the impact of disasters on vulnerable communities.

The concept of 'proof' is wide-ranging, from validating propositions in philosophy to demonstrating theorems in mathematics, adapting to change in business, and designing resilient infrastructures in urban planning. 'Proof' also extends to policymaking, with evidence-based policymaking integrating scientific evidence with practical experience. In the context of the Philippines, the narrative of the 'waterproof' Filipino spirit, symbolising resilience against natural disasters, requires critical scrutiny. It is essential to demand more accountability from the government in addressing these challenges rather than romanticising resilience and overlooking systemic issues.

FIGURE 7.1   Amidst the azure allure of Siargao Island, a family seeks solace while the scars of super typhoon Odette's wrath – uprooted coconut trees – serve as a sombre backdrop. Image credit: the author.

As disaster risk reduction, response, and preparedness are shaped by cultural values and how we perceive resiliency and vulnerability, it is crucial to have a closer look into how disaster informs resilience, as it sheds light on the 'public and its problems' (Chandler, 2014, p. 147). It is necessary. Now more than ever, dissecting resilience in a disaster setting unveils and raises a manifold of themes and contemporary issues that shall give more body to socio-environmental knowledge. Institutions must be aware that the granularities of the culture of resilience make way for well-informed crafting of disaster risk reduction and other disaster mitigation policies, plans, and programmes.

International organisations and humanitarian agencies define resilience as 'the ability of a system, community or society exposed to hazards to resist, absorb, accommodate to and recover from the effects of the hazard in a timely and efficient manner, including through the preservation and restoration of its essential basic structures and functions' (UNDRR, n.d., 85). Resilience is also widely used within policymaking and development work circles vis-à-vis the environment and natural resources field. Acting as a paradoxical boundary concept (Uekusa & Matthewman, 2017, p. 186), or as a metaphor (Norris et al., 2008, 127), resilience as a term has been used as a point of inquiry in relation to varied themes, including culture (Bankoff, 2003); a sense of community/ies (Aldrich & Meyer, 2015; Paton et al., 2001; Quarantelli, 1978); vulnerability (Pelling, 2003; Wisner et al., 2004); and governance (Chandler, 2014; Gaillard, 2015).

Undoubtedly, there is a considerable degree of veracity to the assertion made by experts that, through enduring disasters, individuals and communities develop resilience. Yet, whilst it is widely accepted that resilience accumulates with each successive challenge surmounted, the distinctive spatio-temporal context of communities within the Philippines warrants closer examination and contemplation.

Talking about the same observation, Shakira Sison, an award-winning Filipino essayist, critiques the narrative of Filipino resilience. She notes that while Filipinos are often praised for their cheerfulness and self-sufficiency, even in the face of disasters, this has led to a troubling immunity to such adversities. Sison expresses concern that Filipinos have become accustomed to severe conditions and often resort to self-help, knowing that external assistance may not arrive (Sison, 2014).

Some Filipino experts – and most of the Filipino population – succumb to the undesirable but inevitable claims that 'we are prone to disasters because we are located in the tropics and the so-called 'ring of fire' (Vicente & Villarin, 2005, p. 1). The media and politicians take advantage of 'naturality' as an alibi, 'distracting national governments from the root causes of vulnerability' (Gaillard, 2015, p. 8).

The Philippine government treats natural hazards as *the* 'enemy', and the root cause of disaster that must be stopped and fought against, relying on military chains of command (Alexander, 2002). Such a level of strategy is reflected and somehow explains why the National Disaster Risk Reduction and Management Council (NDRRMC), the lead council 'responsible for ensuring the protection and welfare of the people during disasters or emergencies' (NDRRMC, 2014), functions under the Office of Civil Defense of the Department of National Defense. This rather defensive, combative, and reactive nature must shift to proactive measures, trusting that 'the underlying social, cultural, economic, and political causes of vulnerability are not usually viewed as military or civil-defense matters' (Wisner et al., 2004, p. 209). Building resiliency, from a public service perspective, ought not to rely on the sporadic tendencies of a disaster but rather on providing accessibility to regular social protection and standardised support from the government. The failure to do such regularisation leads survivors to believe that they have become resilient since they have gone through a disaster. And due to its frequency in Philippine territories, Filipinos may conclude that they are even more resilient after surviving all these disasters – a dangerous mode of social conditioning from the national and local governments. However, surviving does not equate to resilience.

Creative Placeproofing

The espousal of the 'waterproof' resilience narrative may ultimately serve to evade accountability for its role in mitigating and addressing the impacts of natural hazards on vulnerable populations. Weaponising the resilience of the Filipino spirit, the power dynamics divert attention from structural inequalities and inadequate policies that exacerbate disaster impacts and discourage critical examination of its actions. Whilst acknowledging the genuine resilience of Filipino communities, I am concerned that the 'waterproof' metaphor may inadvertently contribute to a depoliticisation of disaster risk, concealing the structural vulnerabilities and incompetent public administration that exacerbate the impacts of natural hazards on marginalised populations (Gaillard, 2007; Bankoff, 2019). In such instances, the emphasis on resilience may be superficial, tokenistic, or even harmful, as it can shift the burden of responsibility onto affected communities, obscuring systemic failures or injustices that underlie the challenges they face.

By integrating creative and trauma-informed placemaking, communities can harness the power of arts, culture, and creativity to reimagine their futures, shape their environments, and forge new pathways towards collective wellbeing and sustainability. Genuine resilience is built on social cohesion, collective action, equitable resource allocation, and a deep understanding of local contexts and vulnerabilities. It is about acknowledging these underlying intentions, processes, and outcomes while simultaneously urging a culture of transparency and accountability, advocating for equitable policies and actions, and appealing for inclusive and participatory decision-making.

I propose creative placeproofing as an emergent pivotal approach in fostering community resilience and resistance. I envision it as a union between creative placemaking and trauma-informed placemaking. This amalgamation provides a unique opportunity to cultivate community resilience, healing, and empowerment in response to challenging circumstances. This emphasises the significance of comprehending the repercussions of traumatic experiences on individuals and communities whilst capitalising on the transformative potential of arts, culture, and creativity to encourage recovery and foster a shared sense of identity.

Creative placeproofing is an approach that acknowledges the intricate ways in which trauma influences individuals and communities, moulding their interactions with the built environment and the social dynamics of their localities. This method aspires to establish safe, inclusive, and nurturing spaces that facilitate healing, social cohesion, and wellbeing. Creative placeproofing understands that physical spaces and experiences, along with the processes through which they are conceptualised, can either intensify or alleviate the effects of trauma on individuals and communities. Moreover, creative placeproofing encompasses a profound commitment to healing, regeneration, and cultivating a community of care. This entails not only the physical renewal of urban spaces but also the psychological and emotional wellbeing of inhabitants. Implementing creative strategies, such as participatory design processes and incorporating green spaces, can facilitate restorative environments that promote mental health and overall quality of life. Importantly, no singular approach to creative placeproofing can be universally applied across diverse contexts.

Integrating creativity and trauma-informed through creative placeproofing offers a comprehensive and symbiotic approach to community development and place planning. This collaborative framework recognises the importance of addressing both the psychological and socio-cultural aspects of resilience and healing following traumatic experiences while utilising the power of arts and creativity as catalysts for transformation.

At its core, creative placeproofing prioritises integrating creative solutions, which are deemed 'creative' by the community and stakeholders in the area, to ensure that evidence-based practices support places and spaces. It assumes critical importance in ensuring that places and spaces undergo a creative transformation endorsed and cultivated by local communities and stakeholders. This process ultimately leads to the emergence of resilient, resistant, and evidence-based environments adaptable to change and conducive to collective wellbeing.

A key tenet of creative placeproofing is the transformation of vulnerability into strength. This process involves harnessing the unique features of a community or space and capitalising on them to bolster resilience. For example, a neighbourhood with a high population of artists may utilise their creative skills to develop public art projects that inspire local pride and foster a sense of identity, thereby strengthening community bonds and mitigating the impacts of potential threats. Placeproofing strategies that harness the power of creative approaches can transform spaces prone to socio-economic, environmental, or cultural challenges into thriving, self-sustaining communities.

Community is Necessary

One significant reason attributed to Filipino survival and seeming resilience amidst all the disasters is portrayed through the concept of *bayanihan*. It cannot be directly translated into English, but it can come close to the idea of 'mutual support.' It is a deeply rooted socio-cultural value harnessing the power of collectivity, selflessly offering service to those in need, be it for noncritical or emergency reasons. Furthering *bayanihan* from its supportive and responsive nature, I encourage that this concept may be expanded to an oversight function that leads to increased accountability of the government in narrowing vulnerabilities. Within territorial and relational boundaries are human attachments and interactions that disasters can severely disrupt. How a community responds to and bounces back from a disaster also relies upon these interactions and is contingent on a community's pre-existing vulnerabilities and capacities.

Studies suggest that 85% of post-disaster survivors are responded to and rescued by families, friends, neighbours, and other community members (Quarantelli, 1978). Local communities must be engaged and heard in all disaster mitigation processes – risk reduction, response, relief, recovery, and rehabilitation. Community members must be empowered to recognise and scrutinise their own needs and vulnerabilities to hazards; access ubiquitous but accurate and timely means of information; devote time to familiarising neighbours and networks of assistance, and develop problem-solving capacities (Aldrich & Meyer, 2015, p. 256; Norris et al., 2008, p. 143). Motivating the frontliners and those who suffer from the initial point of contact with hazards to share their observations to inform improved disaster risk reduction shifts the policy and governance model to 'increased people power' (Uekusa, 2018,

p. 191), actor-centred and oriented (Drury et al., 2013, p. 35; Uekusa, 2018, p. 184) approach.

Indigenous knowledge has been transmitted from generation to generation, 'developing mechanisms to survive and to adapt to such hazards for centuries… relating to the environment and how to live in harmony with it' (Shaw et al., 2009, p. 7). Resilience evolves, accumulates, develops, and sharpens with quotidian experiences, skills, resources, networks, and capital.

However, it must be noted that these aforementioned sets are neither stable nor fixed. Oftentimes they are enhanced, reduced, or completely lost. The only constant about disasters is uncertainty, on the grounds that we get to measure the ramifications after the fact. Communities must then acknowledge these uncertainties regarding disaster preparedness – planning for not having a plan (Norris et al., 2008, p. 143). This flexibility allows for innovation and creativity to persist and breaks away from chains of command when not necessary.

After the crucial first few hours post-disaster, and expecting communities mitigated some risks, external assistance must have continued its functionalities. Local risk reduction councils, government counterparts, aid agencies, humanitarian organisations, and other private partners must deploy relief and rescue support and assistance. Communities need a trusted, consistent, and continuous flow of information on the rapidly changing environment and its impact on past disasters, present hazards, and future risks to readjust adaptably. This is important in truly drawing on someone's resilience.

Transformative Recovery and Resilience Building

In a society where resilience narratives directly affect the psyche, discourse, governance and public administration, disaster risk reduction and preparedness have deeply suffered. Hence, there is a need to reposition the concept of resilience, amongst other factors. Disasters are caused by hazards and human failure to enact the root causes of vulnerabilities. Gaillard (2015) elaborates that until the Philippines 'seriously consider that disasters are the consequences of unequal access to resources and poor governance illustrated by the oligarchic nature of the Philippine state, disasters will unfortunately continue to happen on a regular basis' (p. 38).

The configurations surrounding this rhetoric and the power at play must be transformed. Espousing resilience under the guise that Filipinos can cope and survive anyway is not resilience. Romanticising resilience for the lack of accountability must not be normalised. No one becomes resilient by solely surviving a disaster. Nothing improves by enduring the status quo. In terms of governance, there must be an enabling political and socio-cultural environment that does not resist change and is willing to ensure local knowledge is well-blended in its policies and programmes.

Slogans such as *'The Filipino spirit is waterproof'; 'The Filipino spirit is stronger than any typhoon'; 'Where I'm from, everyone's a hero'; 'Bagyo ka lang, Pinoy kami! (You're just a mere storm, we're Filipinos!)* – whilst I believe that these inspirational lines and slogans do mean well and truly play a role in boosting morale, it results in the improper valorisation of local efforts and sanitises accountability of large-scale mitigation and preparedness measures of the state. In addition, it gives both the individual and the community a false sense of safety and security. This illusion of safety (Weick & Sutcliffe, 2007) makes one think one has confidently planned and anticipated a natural hazard. Still, inversely, they become more 'prone to chance' (Dürrenmatt, 1998, in Lorenz, 2010, p. 13).

This same sense of submission and passivity desensitises people to yearn for necessary policy and institutional changes in swiftly executing response, relief, recovery, and rehabilitation efforts during post-disasters. And it is of utmost importance to address this rhetoric as 'how we view adversity and pain have important implications on how society engages with the threats that it faces' (Furedi, 2007, p. 485) and 'how people are likely to cope in an emergency or a disaster are shaped by prior experience but also by a cultural narrative that creates a set of expectations' (ibid., p. 485).

Disasters must be 'considered as the extension of permanent emergency situations and should not be considered as accidents in society or the consequence of the occurrence of rare and extreme natural hazards' (Gaillard, 2015, p. 65). It does not only need the global, national, and institutional prescriptions or a 'top-down imposition seeking to direct, manage or assert control over things' (Chandler, 2014, p. 13). It is imperative to make sense and take meaning from community knowledge – their sets of socio-cultural beliefs, practices, and resources. Communities must be the locus and 'point of engagement, without which there will continue to be gaps in integrating knowledge, with science viewed as far superior due to its more global nature' (Gaillard & Mercer, 2012, p. 100).

The nuances around the concept of resilience must be scrutinised and deconstructed, as it plays a major role in how we look at an all-encompassing phenomenon like disasters and how it influences individuals and communities. It builds a strengthened, self-reliant network for disaster readiness. Communities must be engaged in all disaster mitigation steps; be empowered to recognise their networks, skills, and resources; and readjust to fluctuations and exercise creative and flexible problem-solving approaches.

Resilience needs an ecosystem with a changed mindset on disasters; non-weaponisation measures; an empowered community; and accountable government institutions committed to building back better. However, it really does take more than just proclaiming these recommendations. De-centring resilience is a good start.

Critiques and Final Thoughts

As I position creative placeproofing in its exploratory stage, it is, without a doubt, open to critique and feedback. One of the major insights would be the potentially inherent paradox of the concept itself, as the term 'proofing' implies a finite and stable process, whereas 'place' is dynamic, growing, evolving, and unpredictable.

Furthermore, the increasing reliance on empirical data and rigorous evaluation methodologies in informing decision-making processes is a testament to the recognition of the value of evidence-based practices in enhancing the efficacy of policies and interventions. By grounding decisions in the best available evidence, these entities seek to optimise the use of scarce resources and ensure greater accountability and transparency. Nonetheless, the obstacles and criticisms linked to this approach, including the 'cherry-picking of evidence to promote political interests' (Parkhurst, 2017, p. 7). This likelihood of selective and manipulative use of evidence and the danger of sidelining qualitative and context-sensitive perspectives must be met to guarantee the rigour and significance of evidence-based approaches in influencing policy and implementation.

A central issue in this context is the propensity to concentrate on quantifiable aspects among evidence-based policy advocates, which ought to be regarded as a deliberate or implicit political preference that prioritises specific values over others. This highlights the need for a more nuanced and critical understanding of the implications of such choices, particularly given the complex nature and issues surrounding and embedded within places and their interaction, say, to climate change, political conflict, gentrification, and poverty, amongst others.

The notion of 'proofing', as seen in terms like climate-proofing or planet-proofing, urges us to enhance and broaden our worldview, carrying significant consequences for human life and our existential state. Evidently, there is a communal obligation to create an environment that establishes certainty, safety, reliability, and resilience in the face of challenges and uncertainties – both tangible and intangible. This pursuit of proof, present across many domains, embodies the human endeavour to secure knowledge, ensure safety, and promote justice. As we persist in navigating the intricacies and unknowns characterising our modern world, the notion of 'proof' can be a crucial reference point. It can guide our joint undertakings to enhance comprehension, establish significant relationships, and craft a fairer, more sustainable future.

Regardless of the evident importance and potential consequences of the 'proofing' concept, a void persists in our current comprehension and exploration of this idea. This gap warrants further exploration and analysis by scholars and practitioners in the field and possible reframing to be utilised as a broader model.

References

Alcayna, T., Bollettino, V., Dy, P. & Vinck, P. (2016). 'Resilience and Disaster Trends in the Philippines: Opportunities for National and Local Capacity Building', *PLoS Currents*, 8.

Aldrich, D. P. & Meyer, M. A. (2015). 'Social Capital and Community Resilience', *American Behavioral Scientist*, 59(2).

Alexander, D. (2002). 'From Civil Defence to Civil Protection – and Back Again', *Disaster Prevention and Management: An International Journal*, 11(3).

Bankoff, G. (2003). *Cultures of Disaster: Society and Natural Hazard in the Philippines*. London: Routledge Curzon.

Bankoff, G. (2019). 'Remaking the World in our Own Image: Vulnerability, Resilience and Adaptation as Historical Discourses', *Disasters*, 43(2).

Barber, R. (2013). *Localising the Humanitarian Toolkit: Lessons from Recent Philippines Disasters*. Melbourne: Save the Children Australia.

Chandler, D. (2014). *Resilience: The Governance of Complexity*. London: Routledge.

Drury, J., Novelli, D. & Stott, C. (2013). 'Representing Crowd Behaviour in Emergency Planning Guidance: 'Mass Panic' or Collective Resilience?', *Resilience*, 1(1).

Dürrenmatt, F. (1998). *Die Physiker. Anhang: 21 Punkte zu den Physikern*. Zürich: Diogenes.

Furedi, F. (2007). 'The Changing Meaning of Disaster', *Area*, 39(4).

Gaillard, J. C. (2007). 'Resilience of Traditional Societies in Facing Natural Hazards', *Disaster Prevention and Management: An International Journal*, 16(4).

Gaillard, J. C. (2015). *People's Response to Disasters in the Philippines: Vulnerability, Capacities and Resilience*. New York: Palgrave Macmillan.

Gaillard, J. C. & Mercer, J. (2012). 'From Knowledge to Action', *Progress in Human Geography*, 37, pp. 93–114.

Ginnetti, J., Dagondon, B., Villanueva, C., Enriquez, J., Temprosa, F. T., Bacal, C. & Carcellar, N.L. (2013). 'Disaster-Induced Internal Displacement in the Philippines: The Case of Tropical Storm Washi/Sendong', *Internal Displacement Monitoring Centre Report*. Geneva: Internal Displacement Monitoring Centre. Available: www.internal-displacement.org/sites/default/files/publications/documents/2013-ap-philippines-DRR-country-en.pdf. [Accessed: 2 February 2023].

Guha-Sapir, D. (2020). *The Emergency Events Database of the Center for Research on the Epidemiology of Disasters*. Brussels: Université catholique de Louvain. Available: https://public.emdat.be/. [Accessed: 10 May 2020].

Haraway, D. J. (2016). *Staying with the Trouble: Making Kin in the Chthulucene*. Durham: Duke University Press.

Klein, N. (2007). *The Shock Doctrine: The Rise of Disaster Capitalism*. Toronto: Knopf Canada.

Latour, B. (2018). *Down to Earth: Politics in the New Climatic Regime*. Polity Press.

Lorenz, D. F. (2010). 'The Diversity of Resilience: Contributions from a Social Science Perspective', *Natural Hazards*, 67(1).

McFarlane, A. C. and Norris, F. H. (2006). 'Definitions and Concepts in Disaster Research', *Methods for Disaster Mental Health Research*, pp. 3–19. New York: Guildford Press.

Meiner, C. and Veel, K. (2012). *The Cultural Life of Catastrophes and Crises*. De Gruyter.

National Disaster Risk Reduction Management Council. (2014) *Situational Report Re Effects of Typhoon Yolanda (Haiyan)*. Available: www.ndrrmc.gov.ph/attachments/article/1329/FINAL_REPORT_re_Effects_of_Typhoon_YOLANDA_HAIYAN_06-09NOV2013.pdf. [Accessed: 2 May 2020].

Norris, F. H., Stevens, S. P., Pfefferbaum, B., Wyche, K. F. & Pfefferbaum, R. L. (2008). 'Community Resilience as a Metaphor, Theory, Set of Capacities, and Strategy for Disaster Readiness', *American Journal of Community Psychology*, 41(1–2).

Oliver-Smith, A. (2002). 'Theorizing Disasters: Nature, Power, and Culture', *Catastrophe & Culture: The Anthropology of Disaster*. Santa Fe: School of American Research Press.

Philippine Atmospheric, Geophysical and Astronomical Services Administration. (2020). *Tropical Cyclone Information*. Available: http://bagong.pagasa.dost.gov.ph/climate/tropical-cyclone-information. [Accessed: 2 May 2020].

Parkhurst, J. (2017). *The Politics of Evidence: From Evidence-Based Policy to the Good Governance of Evidence*. Abingdon: Routledge.

Paton, D., Millar, M. & Johnston, D. (2001). 'Community Resilience to Volcanic Hazard Consequences', *Natural Hazards*, 24(2).

Pelling, M. (2003). *The Vulnerability of Cities: Natural Disasters and Social Resilience*. London: Earthscan.

Quarantelli, E. L. (1978). *Disasters: Theory and Research*. New York: Sage Publications.

Santiago, J. S. S., Manuela, W. S., Tan, M. L. L., Sañez, S.K., & Tong, A. Z. U. (2016). 'Of Timelines and Timeliness: Lessons from Typhoon Haiyan in Early Disaster Response', *Disasters*, 40(4).

Sassen, S. (2014). *Expulsions: Brutality and Complexity in the Global Economy*. The Belknap Press of Harvard University Press.

Shaw, R., Sharma, A., & Takeuchi, Y. (2009). *Indigenous Knowledge and Disaster Risk Reduction: From Practice to Policy*. Nova Science Publishers.

Sison, S. (2014). 'The Problem with Filipino Resilience'. *Rappler*, 30 October. Available: www.rappler.com/move-ph/ispeak/73433-problem-filipino-resilience. [Accessed: 3 May 2020].

Uekusa, S., & Matthewman, S. (2017). 'Vulnerable and Resilient? Immigrants and Refugees in the 2010–2011 Canterbury and Tohoku Disasters', *International Journal of Disaster Risk Reduction*, 22.

Uekusa, S. (2018). Rethinking resilience: Bourdieu's contribution to disaster research. *Resilience*, 6(3), 181–195. https://doi.org/10.1080/21693293.2017.1308635.

United Nations Intergovernmental Panel on Climate Change. (2019). *UN IPCC Special Report on Climate Change and Land*. Available: www.ipcc.ch/srccl/. [Accessed: 20 April 2023].

United Nations Office for Disaster Risk Reduction. (n.d.). www.undrr.org/terminology/res ilience [Accessed: 30 August 2007].

Vicente, M. C. T., & Villarin, J. R. (2005). *Recent and Most Devastating Environmental Disasters in the Philippines*. Quezon City: Philippines.

Weick, K. E., & Sutcliffe, K. M. (2007). *Managing the unexpected: Resilient performance in the age of uncertainty* (2nd ed.). Jossey-Bass.

Wisner, B., Davis, I., Cannon, T., & Blaike, P. (2004). *At Risk Natural Hazards, People's Vulnerability and Disasters*. London: Routledge.

SECTION 2
Exploring the Dimensions of Trauma-Informed Placemaking

8

ETHICAL PLACEMAKING, TRAUMA, AND HEALTH JUSTICE IN HUMANITARIAN SETTINGS

Lisa A. Eckenwiler

Introduction

In earlier work, I argued for invoking the ethical ideal and practice of ethical placemaking as an ideal and concept that can help support people concerned with migrant and refugee accommodation and health services, especially given current conditions wherein displacement can last years. My claim has been that the ethical ideal and practice of ethical placemaking is obligatory as a matter of remedial, or transitional, justice, specifically health justice.

Following a review of critiques of humanitarian practices concerning shelter, in this essay I describe innovations in humanitarian response, particularly accommodations and health services aimed at improving conditions for refugees and migrants, and consider them in relation to trauma. A growing number of people in humanitarian settings are grappling with trauma from conflict, violence, and loss of loved ones and land from these and, increasingly, the effects of environmental change. I interpret contemporary innovations, some operationalized and others more aspirational, through the lens of ethical placemaking and its key elements: nurturing transformative autonomy; supporting interdependence and care relations between people and place; protecting bodily integrity; cultivating stability and facilitating freedom of movement; and advancing equity. Ethical placemaking, I will maintain, is especially valuable in contexts involving trauma, particularly place-related trauma. This is because it aims ultimately at the capability to be healthy, or health justice (Venkatapuram, 2011), not merely accommodation and/or (usually acute) response to crisis, and foregrounds place as central to this meta-capability. Ethical placemaking ought to be embraced and integrated intentionally into planning and design for refugee accommodations, given the significance of rupture from place in

DOI: 10.4324/9781003371533-11

generating trauma and threatening the capability to be healthy, and indeed, global justice.

I. 'Carceral Humanitarianism' and the Trauma of Displacement

Refugees and other crisis-affected people living in perilous conditions has become a protracted and pervasive phenomenon. Dwellings established for these populations may take the form of humanitarian-run camps, state-operated detention centers, or informal urban enclaves. Humanitarian settings specifically have been criticized for being what Foucault (1984) called 'out-places': 'outside of all place, even though locatable, and for being designed to 'park and guard' (Agier, 2011, p. 3; Hyndman, 2000). Kelly Oliver has called this 'carceral humanitarianism' (2017) and a continuation of a colonial mentality. A number of other philosophers have argued that the segregation such settings impose constitutes and deepens structural injustice (Eckenwiler & Wild, 2021; Parekh, 2020). These out-places tend to be situated on distinctly sparse landscapes: remote, ugly environs with scant natural resources like water, trees, and green space. In most instances, '[t]he essence of a refugee camp is separation' – not just from one's home but from virtually all social, economic, political life (Verdirame & Pobjoy, 2013, p. 472). Because they are supposed to be temporary, moreover, the material provisions available (things like tents, covers, furniture and sheeting) are designed to be disposable, or 'liquid' (Agier, 2011, p. 36). Yet now, people reside in them for years, decades, even over generations.

People in these settings suffer from a range of complex physical and mental health conditions (Acarturk, 2018; Alpak et al., 2015; Bakker et al., 2016; Tufan et al., 2013; Wild et al., 2017). Most have faced fear and violence prior to arrival and have suffered on the journey. If there is access to health services, however, only acute medical care is likely to be available. Studies in low-, middle- and high-income countries, furthermore, show how camp-life itself worsens physical and mental health, and often exposes people to further violence and trauma, even from just the detention itself (Buckley-Zistel et al., 2014; Hutson et al., 2016; Jabbar & Zaza, 2014; Silove et al., 2007). Organized around temporariness, laced with neo-colonialism, and increasingly operating in the service of state sovereignty (Oliver, 2017), these settings reveal a 'minimalist biopolitics' (Redfield, 2013).

There are countless ethical reasons for critiquing this situation, apart from the flouting of international law. People fleeing crisis and/or living in conditions of protracted displacement endure cruel and degrading treatment, and suffer long-term inequities in social, economic, and political opportunities. Over time they may lose a sense of agency. In being isolated and denied voice, they are cut off from political communication and suffer epistemic injustice in having their experiences and knowledge silenced (Ahmad et al., 2022). Hannah Arendt (1951) highlights a particular kind of 'ontological deprivation' refugees suffer: losing a sense of belonging, both in their own communities and as part of humanity.

Two additional and interrelated arguments against chronic displacement and carceral humanitarianism may be most important for our purposes. For people removed, often violently, from their ancestral homelands and denied access to them over time, those fleeing persecution and violence, and the growing number trying to escape the effects of climate change, a sense of belonging to place is also lost. Herein lies a most profound source of suffering and mental distress (Ahmad et al., 2023; Autti, 2022; Barnwell et al., 2020; Seamon, 2020). 'Place', in its richest sense, encompasses how we interact with and within the material environment socially, assemble and attach particular meanings, and form identities and forge relationships and, corporeally, how we move in it, absorb it, modify and are modified by it (Seamon, 2013). We are best understood, indeed, as ecological subjects, beings who are radically relational or interdependent, that is, situated socially and located geo-atmospherically (Berkes et al., 2012; Casey, 2009; Code, 2006; Krieger, 2021; Preston, 2003). For Indigenous people and many others, who they are and where they live with the land are indistinguishable (Barker & Pickerill, 2012; Redvers et al., 2022; Simpson, 2017).

Finally, and relatedly, people in camps or in otherwise segregated and under-resourced locations lack capabilities to be healthy. Rarely are camps equipped to address mental health conditions, chronic illness, or concerns linked to social determinants of health (Norredam et al., 2006). With health conceived as a meta-capability, this threat to the capability to be healthy is the most ethically concerning, for it erodes opportunities for a flourishing, self-sustained, and self-determined life (Eckenwiler & Wild, 2021; Wild et al., 2017).

II. Humanitarian Innovation through the Lens of Ethical Placemaking

As displacement becomes increasingly common and long term, shelter is fast becoming 'one of the most underfunded sectors of humanitarian response' (IOM for the Global Shelter Cluster, 2019, p. x). Even in the face of severe resource constraints, innovation among humanitarians and humanitarian-minded architects is responding to these ethically impermissible conditions. Below, I describe recent work from recognized agencies as well as architects critical of past ideologies and operations who are providing inspiring alternatives for humanitarian design. I do so within the scaffolding of the ethical ideal and practice I have called ethical placemaking, that is, offering examples (or near-examples) of its six criteria that are either being operationalized in practice, or are being proposed. Ethical placemaking is valuable in contexts involving traumatized groups, including those suffering place-related trauma, for it aims ultimately at the capability to be healthy for individuals and populations. Foregrounding place as central to this meta-capability, thus going beyond a focus on accommodation and the prevailing individualized, medicalized response to trauma.

'Placemaking' is defined as intentionally creating or re-designing parks and paths, housing developments and neighborhoods, features of landscape, hospitals, and long-term care settings (Project for Public Space, 2016; Silerberg, 2013). It is referenced in key international documents and declarations including the Sustainable Development Goals (2015) and UN Habitat's New Urban Agenda (2016). I have argued in earlier work that ethical placemaking is a core component of an enabling, capabilities-oriented conception of justice, specifically health justice (Eckenwiler, 2012; 2016). Organized around a conception of people as ecological subjects, ethical placemaking insists on appreciation for the ways in which a relation to place is a fundamental constituent of our experience and identities, shaping not just the kinds of creatures we have evolved to be physically, but also how we understand ourselves and our relationships. Ethical placemaking has six essential elements: nurturing transformative autonomy; supporting care relations; protecting bodily integrity; cultivating stability and facilitating freedom of movement; and advancing equity. Attending to these six criteria is a responsibility for governments (at the level of public and global health), and people working in health-related fields including urban planning; hospital and long-term care design; and in humanitarian settings (Eckenwiler, 2018; Eckenwiler & Wild, 2021). We are all ecological subjects who are due ethical placemaking as a matter of health justice.

For Indigenous peoples violently removed from their ancestral homelands and denied access to them over time, people leaving their lands fleeing persecution and violence, and the growing number trying to escape the effects of climate change, ethical placemaking should be seen as an urgent priority for global health justice. Ethical placemaking, indeed, should be seen as the remedial responsibility of humanitarian organizations, as second-best agents of justice (Rubenstein, 2015) in the absence of welcoming governments (Eckenwiler & Wild, 2021). Referring to women who have endured gender-based violence (GBV), Ahmad notes, 'We need to consider spaces for women to suffer their suffering' (2019, 16614).

Bodily Integrity

Ethical placemaking calls for working to create conditions that help to protect and promote bodily integrity. This requires provision of the primary goods of food security, physical protection, and privacy. Refugees and many migrants have likely lived without these for some time, and been traumatized as a consequence, and they all are endangered in isolated and/or overcrowded camps with weak or absent security measures and infrastructure (Hess et al., 2018). Humanitarians struggle to provide these basic goods but have become more attentive to the geography of humanitarian assistance – historically organized around fear and the 'park and guard' approach. Ethical placemaking with bodily integrity in mind attends to design elements like the placement of bathrooms and food distribution in order to reduce gender-based violence and prevent further traumatization (IOM for the Global Shelter Cluster, 2019). We might also include basic health services here, but I discuss this under other elements.

Nurturing Interdependence and Relations of Care Involving Both People and Place

The loss of a sense of connection and belonging is one of the most prominent themes in the narratives of native peoples, refugees, and migrants. As we have seen, the separation and loss are not merely social or cultural, that is, of family and community; they also involve the loss of a sense of place and interdependence with the land. Humanitarians have a vital role to play here in designing accommodations created for ecological subjects, beings for whom location and interdependence are the keys to flourishing.

For starters, universal design and disposable materials should be rejected in favor of culturally appropriate, welcoming designs and materials (Katz, 2017; UNHCR Global Strategy, 2014). Appreciating familial ties and supporting caregivers in ethical placemaking for humanitarians working with trauma-affected people is an overlooked yet crucial aspect of nurturing interdependent, ecological subjects. At the same time, particularly when displacement is protracted, support for re-cultivating cultural traditions and building social networks can help respond to loss and trauma. With displaced children who have suffered trauma, for example, the creation of 'therapeutic landscapes' is an example of ethical placemaking (Denov & Akeeson, 2013).

Attention to land and landscape, and the relationship of displaced people to the land are the other essential element here. UNHCR lists this as a principal consideration in the UNHCR (2014), asserting that 'environmental considerations drive design', and citing six other principles that concern everything from drainage, road infrastructure, and site carrying capacity to property tenure over time. Integrating concern for relationships with animals is yet a further consideration.

As noted above, mental health services are usually absent in humanitarian settings. But the growing recognition of the trauma suffered by people in crisis has led to the allocation of more mental health resources in refugee and other settings. Despite the increase in attention to trauma, interventions and the assumptions underlying trauma interventions have been subjected to scrutiny. Mannell, Ahmad, and Ahmad, for example, have critiqued the tendency for health practitioners to see trauma through a lens of individual cognitive pathology, for this eclipses the surrounding social processes and structures that generate traumatic experiences (2018). This emphasis on the individual, informed by the biomedical model, also obscures the relational nature of trauma, or the reality that violence and trauma affect us as social beings who live in a context of interwoven relationships and histories – for some, structured by injustice – that also call for attention (ibid.; Miller, 2009). Biomedical approaches also tend to think of trauma as linked to specific events, when there may be multiple traumatic events, and/or trauma may be intergenerational. In other words, trauma may be experienced by survivors as more fluid and reverberating, affecting relationships and opportunities over time (Ahmad, 2019; Conching & Thayer, 2019; Mannell et al., 2018). Ethical placemaking, with the element of nurturing care relations and interdependencies, calls, rather, for the kind of intervention Mannell, Ahmad, and Ahmad (2018) argue for: an 'interpersonal and social process of storytelling (p. 96)' that situates trauma sufferers in social and historical contexts and avoids a focus on symptoms of and care for individuals.

And lastly, conceptions of trauma and interventions aimed at addressing it overlook the interdependence of people and place, particularly when this relationship was severed in crisis. This opens up opportunities for truly transformative ethical placemaking through design and architecture in humanitarian accommodations and in care settings, with its call for cultivating and re-cultivating the kinds of connections necessary for the survival and flourishing of ecological subjects.

Stability and Free, Generative Movement

The trauma of being uprooted requires thoughtful, indeed, critical attention to creating a sense of belonging among refugees and migrants for the sake, ultimately, of health justice. Belonging demands some measure of stability, of rootedness, of implacement. This is a tall order when working with people who would rather be home if they could, or in a destination country of their choosing. It is also worth challenging the notion that people should get used to displacement, or that this is a push toward creating global ghettos. Both should be resisted with all our might on grounds of injustice.

With the new norm of long-term displacement, humanitarians are responding with strategies aimed at this component of ethical placemaking, helping people to establish some semblance of stability and dwelling, and rather than creating a context of confinement, designing one of openness, ideally moving toward inclusion. They are calling, increasingly, for people to be allowed to move beyond the confines of camps to access services, work, education and healthcare. UNHCR, for instance, calls for seeing refugee settlements as 'nodes' not 'islands', 'connected to the physical social and economic life of adjacent territories' (UNHCR, 2018). For trauma-affected people in certain locations, this kind of mobility can be important to enable access to care unavailable in humanitarian settings. Schemes such as softening boundaries and allowing unrestricted movement, offering work permits, shared education and use of health services may be the most important among placemaking strategies for displaced people. For the interaction such strategies enable can serve as a catalyst for relationships wherein people forge ties of solidarity and create more inclusive, just migrant policies (Ypi, 2010).

Finally, movement within humanitarian settings is also worth more attention, as noted in the discussion of bodily integrity. The challenge, here, is to balance freedom of movement with security concerns.

Nurturing Transformative Autonomy

In a broader context of anti-colonial critique, humanitarians have been cited for a severely paternalistic posture toward aid recipients, viewing them as passive victims in need of rescue and discounting their agency and knowledge. Yet at the heart of the UNHCR's Shelter and Settlements Section Master Plan Approach (MPA) to Settlement Planning is a commitment to 'provide an enabling environment for... displaced populations' (UNHCR, 2018). Similarly, the UN High Commissioner's 2018 Dialogue on Protection calls explicitly for shifting from the 'long-term care and maintenance' strategy toward supporting 'self-reliance' (UNHCR, 2018).

Independent architects and scholars, as well, are focused on projects that nurture agency and autonomy. Specifically with respect to materials and design, Irit Katz (2017) calls for a shift from the era when industrially manufactured materials were used for colonial expansion, rapid military deployment, and humanitarian aid, to one where design is oriented around supporting a sense of agency, identity, and, crucially, belonging, where residents help in decisions on the elements of accommodations. Sama El saket (2016) calls for a 'de-centralizing epistemology' which recognizes the epistemic agency of refugees in making camp design decisions. Pressing this further, efforts to respect agency and integrate epistemic justice in these settings encompass agenda-setting by displaced peoples, so that they can articulate their needs and experiences rather than being mere recipients (Huaman & Mataira, 2019).

Critics of mental health services focused on trauma care and informed by the biomedical model, where the emphasis is on [individual] 'maladaptive' cognitions, call for alternative approaches that, they argue, are likely to be better at nurturing transformative autonomy. Manell, Ahmad, and Ahmad, for example, have highlighted how storytelling interventions enable 'participants to discuss these traumas in the context of narratives that reaffirm positive self-identities' in contexts of persecution and/or where social norms de-value women and make them vulnerable to ongoing violence (2018, p. 96). Narrative or storytelling approaches emphasize the social nature of the trauma and its consequences in other words, instead of individualizing the experience and the response to the experience.

Advancing Equity: Beyond a Crisis Perspective on Accommodations and Health Services

In recent years, alongside critiques of humanitarian shelter have come pressing questions about the proper scope of humanitarian obligations, especially in contexts where crises are chronic and partly attributable to historical injustices (Chung, 2016; Eckenwiler et al., 2023; Rubenstein, 2015). Many argue that an unduly narrow and finite sense of responsibility for justice is problematic, as is a too-sharp divide between development and humanitarian work. It is critical, they claim, to question how the frame of crisis and logic of rescue can obscure historic and even current injustices and preserve or worsen inequities going forward. Without invoking the language of justice, we see, indeed, an increased emphasis in humanitarian discourse on the need to move beyond thinking only in terms of crisis management toward an appreciation of the need for capacity building and sustainability (Leander et al., 2022; Sphere Handbook, 2018). Structural and other barriers that thwart equitable access to food and schools, are the focus of adaptations (Elmasry, 2018; UNHCR, 2018, p. 25). The tendency to see trauma through a biomedical lens of acute crisis is another area for focus, for it may undermine opportunities to situate trauma in a broader social and global context, and the extent to which there are inequities in who is subjected to trauma (including land-related trauma) and its consequences over time.

I have argued that humanitarian action should be understood in relation to transitional justice. Speaking specifically to health justice for refugees as one of the groups most vulnerable to trauma, it is not that humanitarians are responsible for ensuring health equity. Instead, humanitarians working on accommodation and in health services are responsible for working to identify and help implement – or at least catalyze – reforms aimed at addressing global health equity in the long term. Because displacement and trauma set serious limitations on the capability to be healthy, further deepening global health (and other) inequities, ensuring people dwell in places that allow for flourishing and have access to competent and contextualized care, over time, is an obligation of justice.

Conclusion

Long deployed as an instrument of injustice, placemaking, specifically ethical placemaking, should be used as a response. Nowhere is this more true than in contexts where care for sufferers of trauma is warranted, not merely as a matter of compassion and basic human dignity, but indeed, for the sake of the capability to be healthy, or health justice for people wherever in the world they are implaced. Humanitarians have yet to realize the full ethical potential of their work in accommodations and health services, and the ethical ideal and practice of ethical placemaking can support them in their efforts.

References

Acarturk, C., M. Cetinkaya, I. Senay, B. Gulen, T. Aker & D. Hinton. (2018). 'Prevalence and predictors of posttraumatic stress and depression symptoms among Syrian refugees in a refugee camp', *Journal of Nervous and Mental Disease*, 206(1).

Agier, M. (2011). *Managing the Undesirables: Refugee Camps and Humanitarian Government*. Cambridge: Polity.

Ahmad, A. (2019). 'Conceptualising trauma in women with long-term health needs from violence', *BMJ*, 367 (letter).

Ahmad, A., Autti, O., George, B., Gougsa, S., Kobei, D., Kokunda, S., Laitti, J., Pratt, V., & Raj, R. (2023). 'Widening the understanding of solastalgia through land-based violence: Why we need to create new notions of harm and suffering towards the Land to understand mental distress within and beyond land-dependent and Indigenous communities', *International Journal of Comparative and Human Rights*, 6(1).

Ahmad, A., Pratt, V., & Gougsa, S. (2022). 'Where is the land and indigenous knowledge in understanding land trauma and land-based violence in climate change?', *BMJ*, 379 (letter).

Alpak, G., Unal, A., Tamar, B.F., Sagaltici, E., Bez, Y., Altindag, A., Dalkilic, A. & Savas, J. A. (2015). 'Post-traumatic stress disorder among Syrian refugees in Turkey: A cross-sectional study', *International Journal of Psychiatry in Clinical Practice*, 19(1).

Arendt, H. (1951). *The Origins of Totalitarianism*. New York: Schocken.

Autti, O. (2022): 'Environmental trauma in the narratives of postwar reconstruction: The loss of place and identity in Northern Finland after World War II', in Kivimaki, V. and Peter, L. (eds.), *Trauma, Experience and Narrative in Europe after World War II*. London: Palgrave Macmillan.

Bakker, L., Cheung, S. Y. & Phillimore, J. (2016). 'The asylum-integration paradox: Comparing asylum support systems and refugee integration in the Netherlands and the UK', *International Migration*, 54(4).

Barker, A. J. & Pickerill, J. (2012). 'Radicalizing relationships to and through shared geographies: Why anarchists need to understand indigenous connections to land and place', *Antipode*, 44(5).

Barnwell, G. C., Stroud, L. & Watson, M. (2020), ' "Nothing green can grow without being on the land": Mine-affected communities psychological experiences of ecological degradation and resistance in Rustenburg, South Africa', *Community Psychology in Global Perspective*, 6(2/22).

Berkes, F., Doubleday, N. C. & Cumming, G. S. (2012). 'Aldo Leopold's land health from a resilience point of view: Self-renewal capacity of social–ecological systems', *EcoHealth*, 9(3).

Buckley-Zistel, S., Krause, U. & Loeper, L. (2014). 'Sexual and gender-based violence against women in conflict-related refugee camps: A literature overview', *Peripherie*, 34(133).

Casey, E. (2009). *Getting Back into Place: Toward a Renewed Understanding of the Place-World*. Bloomington: Indiana University Press.

Chung, R. (2016). 'A theoretical framework for a comprehensive approach to medical humanitarianism', *Public Health Ethics*, 5(1).

Code, L. (2006). *Ecological Thinking: The Politics of Epistemic Location*. New York: Oxford University Press.

Conching, A. K. S. & Thayer, Z. (2019), 'Biological pathways for historical trauma to affect health: A conceptual model focusing on epigenetic modifications', *Social Science & Medicine*, 230.

Denov, M. & Akesson, B. (2013). 'Neither here nor there: Place and placemaking in the lives of separated children', *International Journal of Migration, Health and Social Care*, 9(2).

Eckenwiler, L. A. (2012). 'Global solidarity, migration and global health inequity', Bioethics, 26. Available: https://onlinelibrary.wiley.com/doi/abs/10.1111/j.1467-8519.2012.01991.x

Eckenwiler, L. A. (2018). 'Displacement and solidarity: An ethic of placemaking', *Bioethics*, 32.

Eckenwiler, L. A. (2016). 'Defining ethical placemaking for place-based interventions', *American Journal of Public Health*, 106(11).

Eckenwiler, L. A. & Wild, V. (2021). 'Justice for refugees living under long-term displacement and segregation: From livelihoods to placemaking', *Journal of Social Philosophy*, 52(2).

Eckenwiler, L. A., Hunt, M. R., Crismo, J. J., Conde, E., Hyppolite, S. R., Luneta, M., Munoz-Beaulieu, I., Saeed, H. M. & Schwartz, L. (2023). 'Viewing humanitarian closure through the lens of an ethics of the temporary', *Disaster Prevention and Management*, 32(2).

El Saket, S. (2016). Instant City. Urban/Rural. Boston: Affordable Housing Institute. Available: http://affordablehousinginstitute.org/storage/pdf/Instant-City_Rural-Urban_S ama-El-Saket.pdf. [Accessed: 28 August 2023].

Elmasry, F. (2018). 'Re-imagining refugee camps as livable cities', VOA News (22 March.). Available: www.voanews.com/a/refugee-camps-architectural-solutions/4310181.html. [Accessed: 28 August 2023].

Foucault, M. (1984). 'Of other spaces, heterotopias', *Architecture, Mouvement, Continuité*, (5).

Hess, S., Pott, A., Schamann, H. & Schiffauer, W. (2018). *Welche Auswirkungen haben Anker-Zentren? Eine Kurzstudie für den Mediendienst Integration*. Berlin: Mediendienst Integration.

Huaman, E. S. & Mataira, P. (2019), 'Beyond community engagement: Centering research through Indigenous epistemologies and peoplehood', *AlterNative: An International Journal of Indigenous Peoples*, 15(3).

Hutson, R. A., Shannon, H. & Long, T. (2016). 'Violence in the Ayn al-Hilweh Palestinian refugee camp in Lebanon, 2007–2009', *International Social Work*, 59(6).

Hyndman, J. (2000). *Managing Displacement: Refugees and the Politics of Humanitarianism*. Minneapolis: University of Minnesota Press.

International Organization for Migration on behalf of the Global Shelter Cluster. (2019). *Shelter Projects 2017–2018: Case Studies of Humanitarian Shelter and Settlement Responses*. Geneva: IOM. www.shelterprojects.org. [Accessed: 21 August 2023].

Jabbar, S. A., & Zaza, H. I. (2014). 'Impact of conflict in Syria on Syrian children at the Zaatari refugee camp in Jordan', *Early Child Development and Care*, 184 (9–10).

Katz, I. (2017). 'Pre-fabricated or freely fabricated?', *Forced Migration Review*, 55.

Krieger, N. (2021). *Ecosocial Theory: Embodied Truths and the People's Health*. New York: Oxford University Press.

Leander, A., Contreras, J. F., & Austin, J. L. (2022). *The Future of Humanitarian Design*. Geneva: Geneva Graduate Institute. Available: https://cloudup.com/iPlNBxmxXqe. [Accessed: 21 August 2023].

Mannell, L., Ahmad, L. & Ahmad, A. (2018). 'Narrative storytelling as mental health support for women', *Social Science and Medicine*, 214.

Miller, S. C. (2009). 'Moral injury and relational harm: Analyzing rape in Darfur', *Journal of Social Philosophy*, 40(4).

Norredam, M., Mygind, A. & Krasnik, A. (2006). 'Access to health care for asylum-seekers in the European Union: A comparative study of country policies', *The European Journal of Public Health*, 16(3).

Oliver, K. (2017). *Carceral Humanitarianism: Logics of Refugee Detention*. Minneapolis: University of Minnesota Press.

Parekh, S. (2020). *No Refuge: Ethics and the Global Refugee Crisis*. New York: Oxford University Press.

Preston, C. (2003). *Grounding Knowledge: Environmental Philosophy, Epistemology, and Place*. Athens: University of Georgia Press.

Project for Public Spaces. (2016). *What Is Placemaking?* New York, NY: PPS.

Redfield, P. (2013). *Life in Crisis: The Ethical Journey of Doctors Without Borders*. Berkeley: University of California Press.

Redvers, N., Celidwen, Y., Schultz, C., Horn, O., Githaiga, C., Vera, M., Perdrisat, M., Plume, L. M., Kobei, D., Kain, M. C. & Poelina, A. (2022). 'The determinants of planetary health: An Indigenous consensus perspective', *The Lancet Planetary Health*, 6(2).

Rubenstein, J. (2015). *Between Samaritans and States: The Political Ethics of Humanitarian INGOs*. New York: Oxford University Press.

Seamon, D. (2013). 'Lived bodies, place, and phenomenology: Implications for human rights and environmental justice', *Journal of Human Rights and the Environment*, 4(2).

Seamon, D. (2020). 'Place attachment and phenomenology: The dynamic complexity of place', Manzo, L. C. & Devine-Wright, P. (eds.), *Place Attachment: Advances in Theory, Methods and Applications*. New York: Routledge.

Silerberg, S. (2013). *Places in the Making: How Place Making Builds Places and Communities*. Boston: MIT.

Silove, D., Austin, P. & Steel, Z. (2007). 'No refuge from terror: The impact of detention on the mental health of trauma-affected refugees seeking asylum in Australia', *Transcultural Psychiatry*, 44(3).

Simpson, L. B. (2017). *As We Have Always Done: Indigenous Freedom through Radical Resistance*. Minneapolis: University of Minnesota Press.

Sphere Project. (2018). *Sphere Handbook, Humanitarian Charter and Minimum Standards in Disaster Response*. Available: https://spherestandards.org/handbook-2018/. [Accessed: 21 August 2023].

Tufan, A. E., Alkin, M. G., & Bosgelmez, S. (2013). 'Post-traumatic stress disorder among asylum-seekers and refugees in Istanbul may be predicted by torture and loss due to violence', *Nordic Journal of Psychiatry*, 67(3).

UN Habitat. (2016). *New Urban Agenda*. Available: https://habitat3.org/the-new-urban-agenda/. [Accessed: 21 August 2023].

United Nations. (2015). *The 17 Goals*. Available: https://sdgs.un.org/goals. [Accessed: 21 August 2023].

UNHCR. (2018). *High Commissioner's Dialogue on Protection Challenges 2018: Protections and Solutions in Urban Settings. Background Paper*. Available: www.unhcr.org/5c0fd0954.pdf. [Accessed: 21 August 2023].

UNHCR. (2014). *Global Strategy for Settlement and Shelter 2014–2018*. Geneva: UNHCR. Available: https://reliefweb.int/sites/reliefweb.int/files/resources/GlobalStrategyforSettlementandShelter.pdf. [Accessed: 21 August 2023].

Venkatapuram, S. (2011). *Health Justice*. New York: Polity.

Verdirame, G. & Pobjoy, J. M. (2013). 'The end of refugee camps?', *The Ashgate Research Companion to Migration Law, Theory, and Policy*. University of Cambridge Faculty of Law, 29 (24 July).

Wild, V., Jaff, D., Shah, S. & Frick, M. (2017). 'Tuberculosis, human rights and ethics considerations along the route of a highly vulnerable migrant from Sub-Saharan Africa to Europe', *International Journal of Tuberculosis and Lung Disease*, 21(10).

Ypi, L. (2010). 'Politically constructed solidarity: The idea of a cosmopolitan avant-garde', *Contemporary Political Theory*, 9(1): 120–130.

Acknowledgements

With heartfelt thanks to Allie Edwards and Vasuki Kandalai for their research assistance.

9

BEYOND DARK TOURISM

Reimagining the Place of History at Australia's Convict Precincts

Sarah Barns

Introduction

For many Australians, how we think about the histories of our places, and our connection to them, is in flux. There is a growing depth of understanding of what it means to live on a land stolen from its people, a people who constitute one of the oldest living cultures in the world. Our places may be home, they are places of belonging, and shape who we are: but our places are also deeply troubled by the past, in ways that remain unresolved.

Many today recognise British claims to sovereignty over this land in the late 18th and early 19th centuries as a form of theft, sanctioned through the assertion of property rights whose legitimacy rested on the patently untrue claim that these lands were unpeopled at the time of British arrival. The concept of an unpeopled land, as is widely known, goes by the name of *terra nullius*. 'It was truly an astonishing assertion of sovereignty that had almost no credibility in international law', writes the prominent historian Henry Reynolds (2021, p. 47) whose most recent book *Truth-Telling* (2021) returns to key debates around the legitimacy of British annexation present even during the earliest decades of colonisation. The sovereignty of the Australian state is, Reynolds and many others argue, badly compromised, and has been from the start, resting as it does on the erroneous assumption that the First Nations were too primitive a people to have been able to exercise their 'ancient sovereignty' over this land.

Many of Australia's earliest colonial buildings, protected under national heritage legislation – among them the Parramatta Female Factory and Institutions Precinct (Sydney, NSW), the Port Arthur Historic Site (Tasmania), Maitland Gaol (Maitland, NSW) and the former Fremantle Prison site (Fremantle, WA) – were built as sites of incarceration, constructed by prisoner-convicts and used by the

DOI: 10.4324/9781003371533-12

British Empire to assert its control over wayward and vulnerable populations, including First Nations people. The uses to which they were put evidences how very harsh readily-accepted doctrines of social engineering and eugenics in the day-to-day management of the early colony were, and how badly they scarred the lives of families and children living during these years, with devastating impacts for generations to come. Ongoing legacies of carceral control, in which the institution of the state exerts its power through tactics of fear, intimidation and forced separation, using its rights over land and people to govern through legally sanctioned violence, are today evidenced in high rates of incarceration of Australia's First Nations people, the most incarcerated people on the planet.

At its very foundations then, Australia is 'a nation built out of trauma'. Our convict precincts are wrestling with this past and look to shape new narratives of engagement with residents and tourists alike. They are the places where experiences of trauma, violence and neglect are etched into built form. They are maintained under heritage laws for their outstanding contribution to the heritage of the nation, and ten such sites were added in 2010 to the UNESCO World Heritage Register, reflecting years of advocacy, and ensuring their ongoing preservation as original colonial artefacts of Australia's early penal colony. At the same time, they are increasingly also required to pay their way through commercialisation strategies, which increasingly incorporate placemaking programs – events, festivals, and activations – as well as new adaptive uses in the form of start-up hubs and venue hire. Seeking to attract the tourist dollar, many are refreshing their visitor engagement programs through the lens of 'dark tourism', employing well-funded, contemporary, immersive and experiential storytelling programs that invite visitors to experience, in visceral, multi-sensory ways, the horrors of sustained abuse and confinement by hapless souls condemned under successive programs of institutionalised violence.

As Australia wrestles with questions unresolved – acknowledging unceded lands, and the violence of state-sanctioned dispossession – the implications of institutionalised violence increasingly casts shadows over the meanings and identities of these precincts, former gaols, orphanages, asylums whose Benthamite structures of sandstone and limestone are of unquestioned heritage significance. Intimately connected to foundational narratives of Australia's origins, they are not simply solid expressions of nation-building but also visceral evidence of the illegitimate and sustained violence of Australia's governing institutions.

As such, many custodians of such precincts today recognise their role in addressing more openly the continued legacies of Australia's troubled past. 'Discomfort' was the guiding theme for the 2023 national conference of contemporary arts and museum professionals, where cultural leaders discussed the need for colonial practices of incarceration and forced separation of First Nations people to be more directly reckoned with in their programming (AMAGA, 2023) . As one delegate, a custodian of one of Australia's convict precincts acknowledged to me at the event: 'We can't keep telling the same convict story over and over again.'

The year 2023 is an important one for conversations of this kind. In October 2023 all Australians of voting age will be asked to vote over whether to recognise the claim, made in the *Uluru Statement of the Heart*, that 'Aboriginal and Torres Strait Islander tribes were the first sovereign nations of the Australian continent' (First Nations National Constitutional Convention, 2017). They will also be asked whether they agree to enshrining a First Nations 'Voice' in the Australian constitution. For the Australian Government, led by Labor Prime Minister Anthony Albanese, this work of addressing the unsettled and indeed illegal nature of Australia's original occupation by the British is today a major focus. In turn, many Australians are 'walking together' with First Nations people to advocate for this recognition – the 'Yes' vote – and to rethink connections to place through the lens of Aboriginal sovereignty, of connection to country, and through visions of a future in which First Nations claims to ancient sovereignty are given a form of legal recognition.

Pathways for Placemaking at Traumatic Convict Precincts

Here we face difficult, unsettling, questions which can sit at odds with contemporary narratives around 'placemaking' and precinct activation, which tend towards more hopeful opportunities for gathering, celebration and activation than challenging concerns around sovereignty and nationhood. What forms might this recognition and renegotiation take at such convict sites?

In the first instance, as is today widely practised, precinct custodians are creating space for First Nations-led programming, allowing enduring practices of connection to country to be made more visible in the landscape and through contemporary curation. In Parramatta, where the Female Factory, Orphanage and Girls Home were key sites for the forced separation of young children from their mothers, Darug Elders and knowledge keepers are invited to share practices of eel fishing in the river, weaving fish nets and practising healing dances together, in honour of generations of mothers who came before them (Jannawi Dance Clan, 2021).

Guided by First Nations designers, the Europeanised landscapes of places like the Macquarie Street East Precinct, home to Australia's oldest governing institutions, are being redesigned in symbolic connection to stories and knowledge passed down – for example, acknowledging the forms of *bora rings* or tribal ceremonial spaces upon which colonial structures of power were imposed (Hromek, 2023). Landscape practices in this way foreground pre-colonial attachments to place, through symbolic inscriptions expressive of an Indigenous worldview and of enduring 'ancient sovereignty' over place.

When Wiraduri artist Jonathon Jones transformed 2500 m² of the former Hyde Park Barracks, once used to house the convict labour force that would ultimately displace and decimate the Gadigal people of Warrane (Sydney harbour), into a canvas for his temporary artwork *untitled – mararong manauuwi* (meaning 'emu footprint' in Gadigal language), he implored his audiences to 'wake up'. 'The way we understand history and culture and present it within historical institutions is very static' he told *The Guardian* (Dow, 2020), continuing, 'The building is perceived through an almost ordained notion of preservation and heritage' – through programs such as the UNESCO World Heritage Listing, to which the site was added in 2010 along with ten other convict places and institutions (UNESCO, 2010).

Against this, Jones' work aims to enmesh Aboriginal and colonial history more tightly together: the shape of white emu footprints his artwork saw scattered across the precinct mimic that of the emu but also that of the colonial broad arrow printed on convict uniforms. Critical to Jonathon Jones' work, like many other First Nations Elders, knowledge keepers, historians and artists, is the assertion of ongoing connection, language and culture, as expressions of continued sovereignty that has not been lost amidst the violence of dispossession and forced separation. Reinforcing this message, the temporary artwork was accompanied by a curated programme of performances, workshops and talks exploring how, in the words of the program, certain 'stories determine the ways we came together as a nation' and the way 'places such as Hyde Park Barracks need to be understood as complex sites, symbols of both creation and destruction, remembering and forgetting' (Hyde Park Barracks, 2020; Lindsey, 2020).

This is critical work of historical reimagining. But the work of disrupting and unsettling the colonial narrative can also go further than temporary arts programming and creative history – as powerful as this is – in ways that can contribute to a broadening of what placemaking can mean. There is also a need to acknowledge that the trauma inflicted at such historical sites continues to play out in Australian society today. At the Parramatta Female Factory Precinct, home to the notorious Parramatta Girls Home in operation until the 1980s, former residents have formed a group called 'Parragirls' to advocate for contemporary uses of the buildings in ways that support ongoing care and support for women experiencing domestic violence or hardship (Barns et al., 2017; Parra Girls, n.d.; Valentine, 2007) . The group's leader and advocate Bonnie Djuric, a former resident in the Girls' Home, advocates for the role of transformative justice in shaping the precinct's future identity. As Djuri writes with fellow authors Lily Hibberd and Linda Steele: 'Engaging with a site's history in this way, government, civil society and the public can better understand contemporary social justice issues and build a future society that does not repeat the wrongs of the past' (Djuric et al., 2018). The work of Parragirls, and Djuric in particular, has been instrumental in seeing the precinct added to Australia's national heritage register. It is also work that is instrumental in showing how transformative justice might support future visions of place and at Australia's longest-running sites of incarceration.

As Djuric et al. (2018) note, the Parramatta Precinct is Australia's 'longest-operating site of institutional incarceration and violence against females. It is also a place of punitive incarceration of children, women and Indigenous Australians and those labelled as mentally ill'. The group has successfully advocated for the former girls home to be recognised as a *Site of Conscience*, an international coalition formed in 1999 that calls for places of human suffering to be reformed as spaces of civic participation, care and social justice (Sites of Conscience, n.d.). As the group defines it, a Site of Conscience is a place of memory – whether a historic site, place-based museum or memorial – that prevents this erasure from happening in order to foster more just and humane societies today: 'Not only do Sites of Conscience provide safe spaces to remember and preserve even the most traumatic memories, but they enable their visitors to make connections between the past and related contemporary human rights issues' (ibid.).

The critical link here is between the trauma of the past and what possibilities it teaches about contemporary social justice. This approach moves history and its trauma beyond the lens of 'dark' or 'trauma tourism' towards a much more active engagement with social justice issues. As arts researcher Lily Hiberrd writes, reflecting on her collaborations with Djuric: 'Among the many unresolved issues for Parragirls is the recognition of and reparative justice for mental, physical and social disadvantage endured as a consequence of their early institutionalisation. The profound nature of this injustice is not widely or well understood' (Hibberd, 2020, p. 5). Djuric et al. (2018) also write powerfully on this:

Apologies, stone memorials and trauma tourism no longer suffice for those living with the consequences of serious abuse. We urgently need a new imaginary for our past, where we make use of Australian heritage to do justice.

The work of Parragirls, linked to the wider Sites of Conscience international coalition, presents a radical vision for the future of Australia's convict-era gaols, asylums and historic precincts – but also a powerful lens for contemporary place-makers, who are increasingly looked to by precinct managers and custodians to activate and program their spaces to keep them relevant to their communities. Given the scale of investment planned or underway at many of these precincts, as they undergo major programs of revitalisation and renewal in the context of widespread infrastructure investment and housing growth across Australian cities, it is surely vital that placemakers and precinct custodians engage with their troubled histories in ways that can bring them into productive dialogue with contemporary communities.

This, then, is the opportunity for deep and transformative placemaking practices to be fostered at major convict sites, in ways that bring historical experience into active dialogue with the present. In the context of widespread migration, housing stress, mental health issues and high rates of First Nations incarceration, this might mean broadening place-based programming beyond a traditional heritage lens that lets history remain a thing of a dark past. Placemaking and precinct activation through this lens might support more sustained commitment and investment in programs of dialogue with more underprivileged groups in a community – committing to their stories of transformation and creating spaces and services of care. Creating space for these groups to lead conversations about care practices in the community today may indeed mean letting different groups and agencies lead a conversation, and empowering agencies and leaders with broad remits towards social justice outcomes define how places such as these can act as healing spaces for communities in need. And critically, the value offered by such precincts could be redefined away from a reliance on visitation and tourism metrics and towards longer-term impacts and value in support of vulnerable groups, whose many needs continue to weigh heavily on the public purse.

Dr Daniele Hromek, of Yuin/Badawang descent, sees the redevelopment of Macquarie Street East Precinct as a unique opportunity for an honest reckoning and truth telling. To Sharon Veale, CEO of GML Heritage, also in reference to this major precinct redevelopment: 'We need to situate these histories in place and enliven them, creating a safe space for conversations that escort locals and visitors alike to the edge of their understandings. What if sovereignty and treaty was made here?' (PDNSW, 2023). As we live through yet another period of flux, to do with how Australians reconcile our relationship to a difficult past and embark on sustained periods of truth telling, perhaps our original convict precincts – places of historical incarceration and trauma – can be reimagined as sites of transformational care in community. This represents a different kind of dialogue with the past, but is surely, ultimately, what recognition is all about. As Hromek asks: 'If not here, where?'

References

AMAGA. (2023). Discomfort. *The Australian Museums and Art Galleries (AMAGA) 2023 National Conference Awabakal & Worimi Land / Newcastle, NSW*. Available: www.amaga.org.au/past-conferences. [Accessed: 28 August 2023].

Barns, S., Mar, P., & James, P. (2017). Waves of People: Exploring the Movements and Patterns of Migration That Have Shaped Parramatta Through Time. *City of Parramatta Council*. Available: www.cityofparramatta.nsw.gov.au/sites/council/files/inline-files/Waves%20of%20People%20Report%20Online%208MB.pdf. [Accessed: 28 August 2023].

Djuric, B., Hibberd, L., & Steele, L. (2018). 'Transforming the Parramatta Female Factory institutional precinct into a site of conscience', *The Conversation*. Available: https://theconversation.com/transforming-the-parramatta-female-factory-institutional-precinct-into-a-site-of-conscience-88875. [Accessed: 28 August 2023].

Dow, S. (2020). 'Designed to wake people up': Jonathan Jones unveils major public work at Hyde Park barracks', The Guardian [online], 20 February 2020. Available: www.theguardian.com/artanddesign/2020/feb/20/designed-to-wake-people-up-jonathan-jones-unveils-major-public-work-at-hyde-park-barracks. [Accessed: 28 August 2023].

First Nations National Constitutional Convention. (2017). *Uluru: Statement from the heart. Central Land Council Library, Uluru, Northern Territory*. Available: https://nla.gov.au/nla.obj-484035616/view. [Accessed: 28 August 2023].

Hibberd, L. (2020). 'Negotiating uncertain agency: The ethics of artistic collaboration and research in the context of lived trauma' in MacNeill, K. & Bolt, B. (eds.) *The Meeting of*

Aesthetics and Ethics in the Academy Challenges for Creative Practice Researchers in Higher Education. Abingdon: Routledge.

Hromek, D. (2023). 'Reading Country', *Macquarie Street East Masterplan.* Property and Development NSW, Department of Planning and Environment. Available: www.dpie. nsw.gov.au/housing-and-property/our-business/precinct-development/macquarie-street-east-precinct. [Accessed: 28 August 2023].

Hyde Park Barracks. (2020). *Information Program, untitled maraong manaóuwi.* Sydney Living Museums, Sydney.

Jannawi Dance Clan. (2021). *Net Fishing Dance, Parramatta Parklands.* Video produced for STORYBOX Parramatta Activation. Available: https://vimeo.com/517879631. [Accessed: 28 August 2023].

Lindsey, K. (2020). 'Indigenous approaches to the past: "Creative histories" at the Hyde Park Barracks, Sydney', *Australasian Journal of Popular Culture*, 9(1), 83–102.

Parra Girls. (n.d.). Parramatta Female Factory Institutions Precinct. Available: www.parragi rls.org.au/. [Accessed: 28 August 2023].

PDNSW. (2023). *Macquarie Street East Precinct Masterplan.* Property and Development NSW Department of Planning and Environment. Available: www.dpie.nsw.gov.au/hous ing-and-property/our-business/precinct-development/macquarie-street-east-precinct. [Accessed: 28 August 2023].

Sites of Conscience. (n.d.). *International Sites of Conscience.* Available: www.sitesofcon science.org/. [Accessed: 28 August 2023].

UNESCO. (2010). *UNESCO World Heritage Centre –World Heritage Committee inscribes seven cultural sites on World Heritage list.* Available: https://whc.unesco.org/en/news/642. [Accessed: 28 August 2023].

Valentine, A. (2007). *Parramatta Girls.* Redfern: Currency Press.

10

EQUITABLE FOOD FUTURES

Activating Community Memory, Story, and Imagination in Rural Mississippi

Carlton Turner, Mina Para Matlon, Erica Kohl-Arenas and Jean Greene

Grounding in Utica, Mississippi

Carlton Turner

Utica is a small town which sits at the foot of the Delta, 30 minutes from the Mississippi. Mississippi, an Ojibwe word that means 'Great River'. This river ferried king cotton and unnamed precious human cargo from weigh ins and auction blocks into markets and plantation bondage up and down this country – providing a literal stream of income that would build white generational wealth across the nation and fuel the southern block of confederate political power up through reconstruction and deep into the era of Jim Crow.

During the cotton boom my community of Utica was perfectly situated in a small, yet important, location between the growers and the river. A depot of sorts, where growers could bring their harvest to any of the three cotton gins to be weighed in and paid out. According to the stories, in its day, Utica had a top-of-the-line social scene that included jook joints (the vernacular term for an informal establishment featuring music, dancing, gambling, and drinking, primarily operated by African Americans in the south-eastern US, see Nardone, 2017), cafés, a hotel, a railroad depot and even an opera house. Utica was the site of the first radio station and the first paved road in the state of Mississippi. Utica was a town built on production. It produced agricultural goods, lumber, and most importantly educators.

But industries inevitably change, and when they do decisions about the future of communities like mine are usually made for us, rarely by us. These decisions are usually based on an analysis of the potential to exploit labor, land, and the means of production. A capitalist's dream. The result is our communities are adversely

DOI: 10.4324/9781003337371533-13

impacted by decisions they have very little agency in. In 1993, there was a decision at the county school board level to close Utica High School. Located right off main street, this school was part of the core of Utica's downtown area. In 1998, because of NAFTA, the North Atlantic Free Trade Agreement, the Bernstein shirt factory closed its doors. This factory, located directly across from the Sunflower grocery store in Utica, employed more than 100 Black women in our community. This loss contributed significantly to the bottom line of the grocery store and further impacted all the businesses on Main Street. In 2014, our remaining high school closed and later that fall so did the grocery store. These events, although individual, are part of a series of events that changed the quality of life for those of us who call Utica home.

The Utica Institute

Jean Greene

The lyrics to a song from 1966 speak to what living was like as a Black person in Jim Crow Mississippi:

> *They had a huntin' season on the rabbit.*
> *If you shoot him, you went to jail.*
> *But season was always open on me.*
> *Nobody needed no bail.*
> *Down in Mississippi.*
> (*Down in Mississippi*, J. B. Lenoir)

William Henry Holtzclaw came to Utica, Mississippi, from Alabama in 1902. Holtzclaw, a student of Booker T. Washington and graduate of Tuskegee Institute, followed in his mentor's footsteps and set out to found a school to help and uplift Black people in rural areas of the South. He found, in Utica, a community still reeling from the effects of slavery and reconstruction. Most of the Black people in the area around the town of Utica were living in cramped, single room cabins. These same people did not work their own land. They were either sharecroppers, day farm hands, or domestic workers still firmly under the foot of white landowners. Lawless brutality against these people was the norm. The torture and killing of Black men and women had become public spectacles attended by crowds of white parents and children… farmers and businessmen. *The Jackson Daily News* and the *Clarion-Ledger* would alert white readers so they could plan to attend and witness the horror. An example of one such headline read: 'Prospects Good for a Lynching, and the Indications are that when it Comes it Will be by Wholesale; Five Negro Men and Two Women.'

In Utica, not too many years before Holtzclaw made his arrival, whites had terrorized local Black families in a concerted effort to drive them out of the state.

Black landowners would wake up to posters that read: 'If you have not moved away from here by sundown tomorrow, we will shoot you like rabbits.'

Holtzclaw walked into Utica to put into practice the teachings of Tuskegee Institute and what he learned from his experiences working at Snow Hill Institute. There he established his first version of the Black Belt Improvement Society to help local Black farmers market their crops without resorting to white middlemen. He replicated that organization upon his establishment of the Utica Normal and Industrial Institute. The organization was a limited cooperative, however, because a true cooperative would have antagonized those white merchants who were almost wholly dependent on the Black farmers for their existence. Holtzclaw's Utica version of the Black Belt Improvement Society listed ten degrees to help Black people build themselves up to self-sufficiency. The first-degree members had only to 'have and show a desire to better their condition.' The other degrees were very specific on the requirements. For example, members of the 4th degree had to possess '12 chickens, two pigs, and a cow together with an orderly house.' Members of the 7th degree had to own 40 acres of land. Members of the 10th degree must own at least '1000 acres of land and possess such other qualifications as the central society may require.' Holtzclaw was able to provide a framework, a model to help guide his community toward autonomy.

The Annual Farmer's Conference, patterned after the Tuskegee Farmer's Conferences, was organized to help Black farmers learn practical agricultural techniques thereby raising the standard of their farms. Each of these conferences featured speakers and notable scientists. Dr George Washington Carver was a frequent participant. Individual consultants provided discussions and demonstrations on various topics, including demonstrations relating to planting and soil, farm equipment, livestock, and home economics. Holtzclaw and his faculty worked with local farm families to encourage them to buy and own land, to build multi-room cottages, and to utilize scientific farming techniques to increase yields to become more agriculturally proficient. Holtzclaw was instrumental in not only educating the Black children of the area but providing the parents of those children guidance and inspiration toward becoming economically independent.

Interest in the farmer's conference waned in the late 1960s. This is about the time farm work fell out of favor. Farmers were lured away to work in factories which were opening across the state. Farming became viewed as something not as lucrative as factory/production work. The population shifted from being primarily producers of their food to becoming consumers of food that was shipped into the area. The grocery stores in Utica became the main source of food for the population. This created a situation that became particularly dire when factories left the area leaving grocery stores to close in their wake. The last one, Sunflower, closed in 2014, firmly placing Utica in the middle of a 'food desert.'

The farmer's conference has been renewed over the past four years. The agriculture program now offers classes on the Utica campus which has jump started an interest in sustainable agriculture. There is also a renewed interest in participation in the Mississippi State Fair. The State Fair featured a day dedicated to the programs of the Utica Campus and the Utica Institute Museum.

The Utica Institute Museum is located on the historic HBCU Utica campus. The campus is the result of the merger of Utica Junior College (formerly the Utica Normal & Industrial Institute) and Hinds Junior College in 1983. The museum has partnered with the Mississippi Center for Cultural Production (Sipp Culture) on the *Oral History Project* which is collecting memories from participants living in and around the town of Utica about the dwindling access to food and what this means for the community.

Mississippi Center for Cultural Production (Sipp Culture)

Carlton Turner

Utica is my home. It's where my children go to school, where my nephews, nieces and little cousins are growing up. Eight generations of my family are buried beneath its fertile soil, and it's a community vastly different from what it was just a generation ago. Deterioration, like progressive transformation, happens gradually over time. The Utica of my youth was a very different place. And a Utica equipped to exist in the future will need to be radically different if our community is to thrive. These types of shifts don't happen all at once, they occur over generations. Change requires intention, time, and radical imagination.

For us, food is the common denominator. It's the thing that everybody does. I grew up in a community where I had a deep and personal relationship with my food. Many people in my community share that experience. So, we use stories connected to food as a framework for accessibility for people to enter and contribute their story to an ongoing conversation about change. They may not feel comfortable if the frame is just art, but when you combine food and story you get a backstage pass to the most prolific of venues for storytelling. The dinner table is the place where I've heard the greatest yarn spinners of my age. Coming from a big family, collard greens, cornbread, fried chicken, and macaroni and cheese become the centerpiece of a visceral and ethereal experience adorned by stories that bring layered emotions, tears, and some of the most gut-wrenching belly laughs you will ever experience. This place where arts and culture bloom and at the same time fade into the performance ritual of everyday life is special. It's not just 'table talk', it's theatrical and emotional, it's personal, it's comedy and drama, it's sometimes tragic, but always educational. My work as an artist is to acknowledge those stories, and the legacy and future of our community that is encoded within them.

Our ability to achieve health and wellness in our community is directly tied to our agency with the land. It is deeply connected to our ability to construct an analysis of the trajectory our community has taken to this current moment. It is grounded in our ability to tell our stories about food and our individual and collective histories with this land. And we believe that through the telling of those stories we can supplement parts of the social fabric lost over the past couple of generations and begin to construct a new community – one that centers the voice of the people that live here in fostering a place that provides health and wellness for all.

A Community Dream

The central mission of the Mississippi Center for Cultural Production (MCCP), otherwise known as Sipp Culture, is to reimagine the measurements by which economic prosperity is calculated, and in the process, redefine wealth for our rural community (Figure 10.1). We are taking an intergenerational approach to community cultural and economic development through the lens of cultural and agricultural production, shifting the community's dominant identity from consumer to producer. Creating both a physical and an imagined space for the community to bring its dreams, our work is comprehensive community cultural transformation. That is, the social and economic transformation of our community using arts and agriculture as an intersecting point to engage the community in a conversation about past, present, and future: the building of a collective dream.

Sipp Culture emerged from a need to shift our community from consumers first to being producers first. At the heart of our work is cultural production. It shows up

FIGURE 10.1 Equitable food futures: eat, meet & greet – Community Advisory Group member and youth volunteer delivering meals to the no-touch community dinner. Image credit: SIPP Culture/ E. Gaines.

in the production of stories in media, film, photography, audio stories, it shows up in the production of agriculture, in growing food, and just creating an overall sense of growing as an emerging and dominant community identity. In this process, Utica becomes the backdrop for a renewed community energy embodied in the sense that anywhere we have land or space we can grow something. The work began the moment we offered this idea as a seed to our community to nurture and grow. At that moment, when we asked permission to share our dreams with the community and have those dreams embraced, the people began to share and lay their dreams upon ours. Together our dreams and aspirations became interlaced, and that process, that is the work, that is the community transformation. It is difficult to see the incremental change: but remember, transformation takes time.

Our community is grounded in the aesthetics of Black southern rural culture bearers. Their imprint can be found on every community throughout the south in the forms of architecture, songs, spiritual practices, the movement of bodies, and in the language of freedom. The aesthetic of Sipp Culture is birthed in call and response – tell us what you need, and we'll work to figure out how to manifest it through whatever means are accessible. The physical space that Sipp Culture curates becomes a physical space to assist in the manifestation of dreams.

We understand the integral connection between the artist voice and community health and the wellness of our rural space. We know the stories told, who gets to tell them, and where they are found, are central to the shaping of policy and ultimately inform the quality of life in our community and those like ours. Our personal, organizational, and community health and wellness are negatively impacted when the only stories representative of our experiences are bound within someone else's dreams and limited imagination. To that end, Sipp Culture works to develop the imagination and enhance the creativity of our community (near and far) through critical dialogue driven by arts and cultural exchange to shape and reshape public policy and our nation's collective imagination.

Collaborative Action Research Catalyzes Memory and Imagination

Erica Kohl-Arenas and Mina Para Matlon

One of the methods engaged by Sipp Culture in advancing its work is through collaborative action research. In 2019, Carlton Turner and ourselves, then both with partnering organization Imagining America: Artists and Scholars in Public Life, were awarded a three-year Interdisciplinary Research Leaders fellowship from the Robert Wood Johnson Foundation. Jean Greene of the Utica Institute Museum of Hinds Community College, Utica, soon joined the team as a fourth partner and chair of the project's local *Community Advisory Group*. Together the team designed a research project, *Equitable Food Futures* (EFF), to document and activate community agricultural knowledge and assets towards a more sustainable, equitable, and healthy food culture and economy. The project was framed within the broader context of the creative cultural development practices of Sipp Culture and a long-standing history of social movements for land-based self-determination and Black farming in the US South.

As described above, once an important agricultural hub, the region surrounding Utica, like many similarly situated rural communities across the county, has lost much of its economic and food infrastructure. Despite the departure of local institutions and Utica's needs around access to affordable and healthy food, there are still significant assets in the community including fertile land and local agricultural knowledge. Using a mixed methods community Participatory Action Research (PAR) approach, this study integrates qualitative research methods, including focus groups and oral histories (ongoing), with survey and archival research to unearth, activate, recognize, and share these stories and assets with the Utica community. By undertaking an investigation of Utica's agricultural and food histories and infrastructure the project aims to contribute to the restoration of community memory, spark imagination about the future, and make visible existing community knowledge around healthy ways of feeding the community, both physically and spiritually. Ultimately, the project aims to show how creative methodologies can catalyze historic and new knowledge in ways that inspire a more expansive imagination of a healthy, locally owned, and equitable food future – using story and data as part of the shift in community perspectives, from consumer to producer of future food economies.

One of the first efforts of *Equitable Food Futures* was to host a series of focus groups. The main goals of the focus groups were to understand people's histories and experiences with food access and food culture, honor and activate local histories and storytelling, and explore Utica residents' sense of possibility for their food futures. Part of this area of investigation involved understanding Utica residents' views regarding the recent closure of the local grocery store and their sense of possibility towards a more expansive approach to food access and food justice over the long haul. Following these goals, we designed focus groups to explore what Utica residents think about past, present, and future ways of accessing and sharing food. We aimed to learn about the rich food and farming history in the community, how people feel about the loss of the local grocery store and other Utica organizations, and what they think is possible in terms of bringing food to the community now and in the future.

Participants were recruited to participate in a focus group based on their knowledge of and experience with the above themes: one group for elders, one for middle-aged residents, and one for younger adult residents. A fourth focus group was hosted in the Spanish-speaking community. To elicit conversation and dialogue we developed a set of questions, with feedback from the *Community Advisory Group*, that prompted dialogue around the past, present, and future food histories, assets, opportunities, and challenges in the region (see Focus Group Questions in Appendix A). After asking a series of questions about people's experiences with growing, cooking, sharing, and getting food, we invited participants to imagine the future, using the following prompts:

- Take a piece of paper and a few colored pens (provided here) and draw what a Utica might look like if healthy food was easily accessible to everyone and was a central part of the community. Feel free to draw a picture, make a diagram or a map.
- Write a Letter from the future: write a three-sentence letter to a relative, maybe a future grandchild or even a letter to an ancestor and pretend that you are in that future. In your letter describe the future where healthy food is accessible and a central part of the Utica community. What does it look like? Feel like? Taste like?

While the research team engaged in a qualitative analysis of the themes that emerged from the questions and creative prompts, some of the most significant outcomes included the rich memories evoked, stories told, relationships strengthened, and beautiful future visions shared. The themes that emerged as most critical for participants included: the daunting amount of energy, time and resources expended to access food; a deep sense of pride and commitment to maintain local food culture, traditions, and food sovereignty; a recounting of historical and ongoing structural challenges presented by regional abandonment and dispossession from the means to access, produce, and share food; a culture of cooperation, care, and social relationships centered around creating community support systems; and a shared hope for a future with healthy food for all where families and children feel nourished, supported, and full of opportunities.

The oral history aspect of *Equitable Food Futures* (see Appendix B) engages a similar set of questions to unearth stories of food culture and practice and dreams for the future and is currently an ongoing project in the community. Following our community-driven and participatory principles, the focus groups and oral history interviews were/are conducted and facilitated by local community members, with project development, planning, and training from EFF research team members. Similarly, data and research tools produced by the *Equitable Food Futures* research effort contribute to ongoing research that contributes to the long-haul vision and work of Sipp Culture.

Intended to provide baseline data from which to measure the long-term impact of Sipp Culture's and Utica residents' community development efforts, we developed and administered a survey to document residents' current activities and perspectives regarding food production and consumption. Complementing the qualitative components of the study and Sipp Culture's overarching community organizing work, the survey also assisted in identifying (potential) local agricultural and food resources and infrastructures that could be leveraged to support community efforts to build a local and regional food culture and economy that is sustainable, equitable, and healthy. The survey built on a prior area economic development survey and further drew from United States Department of Agriculture (USDA) survey instruments.

A random sample of Utica residents was invited to participate in the survey, which was distributed by mail and could also be completed online. Ninety-one residents returned completed surveys, with most participating by mail and many electing for in-person drop offs at Sipp Culture and thereby providing additional opportunities for residents to share feedback with staff and learn more about ongoing work.

With survey analysis ongoing, preliminary findings reflect the project's focus group research suggesting rich existing and unrecognized community assets around agricultural production and high valuation on the importance and need for healthy locally available food. Survey data indicate that approaching half of Utica residents are involved in some form of home agricultural production, a level of household activity that is significantly higher than US national averages (according to a 2022 survey conducted by the National Gardening Association, which found that 35% of US households were engaged in growing or raising food.) Findings further indicated that, even if they are no longer engaged in agricultural activity, most Utica residents possess some form of agricultural knowledge and skills. In addition to knowledge of food production, this community expertise extends to all aspects of the food production cycle, such as how to process and distribute food stuffs. These and similar findings underscore Utica as fertile ground for sowing a healthy food future with the seeds of its wealth of community knowledge.

Lastly, our team is conducting archival research to document how local land use and ownership has shifted over time, and the structural factors driving these shifts. This research is detailing the local and county-level transition from a robust period of Black-owned and operated farmland following the post-Emancipation Proclamation period and in the sharecropping era, to the sharp decrease in – and dispossession of – this land coinciding with the launch of USDA programs favoring wealthy white farm owners in backlash to civil rights movement organizing, and the ongoing transition of land to more fallow farmland over the past several decades. This archival investigation aims to contextualize the project's other research components within a larger history of community resilience and self-determination, and to support Sipp Culture's overarching community cultural development work.

How the Work is Evolving

Carlton Turner

The EFF project positioned Sipp Culture to continue to advance participatory action research in the arts and community wellness space. In 2022, Sipp Culture was selected to participate, as one of nine communities, in the National League of Cities' *One Nation/One Project*. This national project seeks to build a widely participatory community art project that aims to positively impact community collective wellbeing by bringing together arts organizations, municipalities, and healthcare providers. This collaborative project is between Sipp Culture, the municipality of the Town of Utica, and the Jackson-Hinds Comprehensive Health Center. This project builds on the foundation of the EFF project through the continuation of Community Advisory Group members informing the development of this new research. Data collected through the EFF survey, focus groups, and oral histories is deeply informing the development of our aesthetic approach to *One Nation/One Project*.

Acknowledgements

The authors are listed in reverse alphabetical order to recognize the contributions of Carlton Turner as co-founder and co-director of Sipp Culture and narrator of the first two sections. The authors contributed equally to the *Equitable Food Futures* project and the content of this chapter.

Appendices

Appendix A. Focus group questions

Following the Welcome, Introductions and Overview:

1. For the first question, introduce yourself and tell us about your favorite food.
2. Now, let's think back on a moment in Utica when you had an especially good experience with growing, buying, cooking, selling, or celebrating food. It can be a long time ago or in the recent past. What did that one moment look and feel like?
3. How would you describe your current effort to get food to cook, eat, or share with family, friends, and/or community members? Can you describe a common week or month?
4. Can you describe a time when you noticed your ability to get food changing? What was going on at that time? What did you have to change?
5. Take a piece of paper and write a line down the middle. On one side write all the positive things, resources, opportunities around food in the Utica area. On the other side, write all the kinds of food resources that you wish Utica had that it currently does not.
6. For this last question you can choose one of three different ways to imagine a thriving food future for the Utica area. Either:
   a. Take a piece of paper and a few colored pens (provided here) and draw what a Utica might look like if healthy food was easily accessible to everyone and was a central part of the community. Feel free to draw a picture, make a diagram or a map. Or:
   b. Write a Letter from the future: Write a three-sentence letter to a relative, maybe a future grandchild or even a letter to an ancestor and pretend that

you are in that future. In your letter describe the future where healthy food is accessible and a central part of the Utica community. What does it look like? Feel like? Taste like?

Appendix B. Sample of questions from Oral History Guide

Introductory

1. Talk about what life was like growing up in/near Utica.
2. How did your family get food (make groceries) when you were growing up?
3. Did your family have a garden or hunt and fish? Can you describe who worked and tended the garden or caught food for the family?

Food culture/memory

1. Bring me with you to the land or the places where food was grown, hunted, caught, or cooked and shared. What did it look like, feel like, smell like?
2. What are some of your favorite memories about food in Utica?
3. When you were growing up, what are some of the ways people helped others in the community get food, outside of your own family?

Changing times

1. What food traditions in your family have you kept over the years?
2. What has been challenging to keep going?
3. What do you do now to share and celebrate food in your family and in the community?

Imagining the future

1. In your dream future, what kinds of foods are available, enjoyed, and shared in the community. What does a thriving food culture look like, feel like, smell like, taste like?

References

Holtzclaw, W. H. (1915/1970). *The Black Man's Burden* (Reprint). New York: Negro Universities Press.

Nardone, K. (2017). *Juke Joints*, in Mississippi Encyclopedia, Center for Study of Southern Culture, July 11, 2017 [online]. Available: http://mississippiencyclopedia.org/entries/juke-joints/. Accessed: 22 January 2024.

National League of Cities (n.d.) *One Nation/One Project*. Available: www.onenationoneproj ect.com/. [Accessed: 17 August 2023].

Washington, W. (n.d.). *Utica Junior College, 1903–1957: A Half Century of Education for Negroes*. Dissertation Abstracts, The University of Southern MS, Hattiesburg, MS.

11

LANGUAGE IS LEAVING ME

An AI Cinematic Opera of The Skin

Ellen Pearlman

Prelude

In 1906 my ancestors immigrated just like thousands of others to New York City, landing at the port of Ellis Island.

In 2019 I found out why.

Location One – The Imagined Pale

They fled the widespread pogroms of 1905 and 1906 in the 'Pale of Settlement' (also called the 'Pale of Lithuania'), a constantly morphing geography containing parts of Russia, Belarus, Ukraine, Poland, Austria, Hungary, Lithuania, Latvia, Germany, and Moldavia. The Pale, a historically restricted area, was where most Jews were permitted to live.

I had never heard of The Pale until 2019. All I knew was that my maternal side came from Kovno or Kaunas, a city near Vilnius, the capital of Lithuania. My paternal side came from Dokshitsy, a *shtetel* (a Yiddish term for the small towns where predominantly Ashkenazi Jewish populations were allowed to live. The *shetels* existed throughout Eastern Europe before the Holocaust) on the perimeter of Minsk in Belorussia. I didn't particularly care why they came to America, and in my youthful imagination I imagined they had just sort of sauntered down the gangplank at Ellis Island, strode off into the streets of New York and settled in Brooklyn and the Bronx. I vaguely imagined theirs could have been an economic migration, sort of like when all the Irish flooded into America fleeing the Great Potato Famine. There were a lot of tough Irish kids in the neighborhood where I grew up, so I knew about the Irish Potato Famine.

DOI: 10.4324/9781003371533-14

An aunt once mentioned in a very casual, offhand manner that her grandmother or perhaps great grandmother, or maybe some other female ancestor had been cut down in the fields by Cossacks. You know – Cossacks – those strapping guys with handlebar moustaches wearing big oversized furry hats, riding thundering steeds throwing clods of earth into people's faces, brandishing scintillating steel sabers while slashing and hacking everyone in their path. I imagined this unknown relative in the barley fields scything away as she was slashed and hacked to bits. But it didn't seem really real. It felt like a dramatic story about somebody somewhere that had absolutely no relevance.

Location Two – Mt. Judah

The Mt. Judah graveyard straddles the Brooklyn–Queens border in Ridgewood. A massive testament to tens of thousands of immigrants, it is squished between acres of other ethnic graveyards in Queens, wedged and overlapping into parks, highways, and even beneath an elevated subway line. The grave of my paternal Grandpa Isaac, or Ike, who died when he was 92, is located on the perimeter of those clusters.

In the early 1960s on a balmy May afternoon as the US was on the cusp of massive political and psychedelic upheavals, I was brought by my parents to Grandpa Ike's burial. The service took place inside a plot ringed by a rusty red metal gate. The plots had been purchased by the Society of Abraham, all landsmen from the Byelorussian shtetel town of Dokshitsy. Though forced to sever ties with their homeland, this cluster of down-and-out Russian Jews purchased burial plots together in this new world, permanently knitting their landsmen (and women) bones together after death.

During the ceremony, as is common in the Jewish tradition, each attendee tosses a shovel full of dirt onto the lowered casket. My father, who had an extremely fractious relationship with his father, lost his footing and slipped. He fell right up to his armpits, smack into the grave – on top of his father's wooden casket. Gasps of shock ricocheted from all the mourners as the funeral attendants reached down, grabbed and hoisted him out of the pit. We went out to eat afterwards. He never mentioned it again and neither did we. The horror was erased as soon as it happened.

Location Three – Long Beach

Though birthed in Brooklyn, I was raised in a small, isolated seaside town – a short walk from the tip of the Rockaways in Queens. Long Beach, Long Island in Nassau County is a 13-mile Atlantic Ocean barrier island where many first and second-generation Jews settled, most after World War II (WWII). In 1906 it was conceived of as an exclusive seaside resort right at the exact same time Jews happened to be fleeing from the great pogroms of The Pale. Senator William H. Reynolds purchased an abandoned hotel on 'The Great Sand Beach' and turned it into something really grand. As part of a publicity stunt, he shipped in some elephants already working on another of his developments, *The Coney Island Dreamland*, to lay the wood piling foundations for the seven-mile-long boardwalk. He built a railroad line to shuttle his new clientele back and forth from the hot ravages of the city during the summer. After WWII Long Beach became a destination spot for the newly aspiring middle class of WWII veterans and their brides.

Long Beach is encased by a bay on one side and the Atlantic Ocean on the other side. At its widest it spans just one mile, and at its thinnest there is only a quarter of a mile separating the bay from the ocean. A microcosm of different communities, it has three elementary schools, one junior high and one high school. Its remoteness yet proximity to New York City sets up a bubble where inhabitants could easily abandon their past roots. There was only a passing mention of where they originally came from. My parents, who grew up in East New York Brooklyn and the Grand Concourse in the Bronx, were deeply conflicted by economic and class striving all around them – wealth, status, social standing in community, and a deep yearning for post-war normalcy.

Absolutely no mention of the great war that had preceded my birth ever came up. However, tucked away in the spare bedroom of our modest two-story home was a gun-metal military-issued footlocker full of discarded Nazi belt buckles, bayonets, red and black swastikas, yellow cloth stars, and old travel photos of France, including victory marches down the Champs Elysée. There was a book showing the occupation of Paris by the Nazis with lots of sexy burlesque photos, starving populations, dead bodies, and German soldiers. These mementoes were all my father, who had been stationed in France during WWII, had brought back. He never spoke of those years, or how he obtained all of that paraphernalia. The trunk was just there, an olive-green metal box in a spare room. It seemed placed by an immovable force and was as much of a part of the house as a doorknob, sink or closet.

Long Beach, seemingly tranquil, was internally fraught. The vast Atlantic Ocean, unbroken except by the sky above, was crisscrossed by vapor trails from planes taking off from JFK (originally called *Idlewild*) airport. The seven-mile boardwalk promised leisure but belied an insular, closed-off world inhabited by subterranean tribalism of religious bigotry, class, race, and materialism of its different ethnic groups: Irish, Italian, Polish, Russian, Chinese, Cuban, Puerto Rican, African American, Jewish, Catholic, and Jehovah's Witness. It was not until I went to college that I met a Protestant – or what is commonly referred to as a WASP (White Anglo-Saxon Protestant). I had no idea the residents of Long Beach were desperately reinventing themselves since I did not understand there was an old country they had all fled. Just like the green gun-metal footlocker in our spare room they just appeared.

Location Four – Hong Kong

I went to graduate school in Hong Kong and lived on Lamma Island, a remote outer island that was a 25-minute ferry ride from downtown Hong Kong. It had only two narrow walkable streets, no cars, and besides its human inhabitants was filled with a panoply of poisonous snakes: many banded kraits, bamboo pit vipers, cobras, boa constrictors, and roving families of feral pigs.

Many in Hong Kong were either unaware of, or vaguely aware of what a Jew was. I thought this was bizarre since so many locals were Christians, though they also worshipped the indigenous deities of both Taoism and Buddhism. The Taoist sea goddess Tin Hau's seventy or so temples were everywhere in Hong Kong, and many Lamma Island villager's ancestors were buried in round graves dotting the numerous footpaths. For most residents I was just a white American lady with curly hair. World War II, if remembered at all, happened a long time ago. It was a story their grandparents told them about when the Japanese invaded Hong Kong and raped Nanking, China. European references like the Holocaust or Hitler were vague. The cultural referents that psychologically seared me did not exist for them. I continually explained, especially to the younger generation, that in the Old Testament Jesus wasn't born a Christian, he was born a Jew, and that the Last Supper was a Passover ritual. They said they had never heard of Passover, but they certainly knew about the Last Supper. Yet it was on this foreign soil half a world away that my family epigenetic memory and art practice finally emerged, though it took years to connect the dots.

As part of my doctoral thesis, I created a brainwave opera in a 360-degree theater. I developed the opera using a wireless electroencephalogram (EEG) brain computer interface a performer wore on her head. It measured four brainwave signals translated by a smoothing algorithm tracking the emotions of excitement, interest, meditation, and frustration. These four emotions were displayed on a screen as four different colored bubbles. When the performer hit a pre-set threshold of feeling, she triggered the colored bubbles, as well as a sonic environment, and visual databanks of images. What this meant is if the performer experienced frustration during the performance, the images triggered were intense and angry looking, and the accompanying sonic environment loud and overwhelming. The opera, performed inside a 360-degree theater with no seats, allowed the performer to interact with the audience through gaze, movement, and touch. This formed a human computer interaction feedback loop with the audience changing the brainwaves of the performer, which changed the types of images and sound that were launched from her brainwaves, which changed how she interacted with the audience.

The opera's plot was based on the true story of Noor Inayat Khan, a Sufi Muslim princess and covert wireless operator for British secret intelligence during WWII. Noor enlisted in the British intelligence service to save the world and prevent the slaughter of innocents, including Jews. She was parachuted into Nazi-occupied

Paris with orders to send Morse code messages back to the Allied forces detailing the enemy's movements. Noor was captured twice by the Gestapo and escaped twice. Deemed a dangerous political prisoner, she was executed at Dachau by a bullet to the back of her head. A mystic Sufi Muslim woman, she never revealed any information about other members of her resistance cell. Her inspiring life and death left me wondering, 'Is there a place in human consciousness where surveillance cannot go?' I asked the question because the Gestapo was unable to break her faith. However, I wondered, if it was a different time in history, would they be able to surveil her brain?

Location Five – The Baltics

Place and identity, both individual and collective, arise in tandem with memory. Place is where we are born from and dwells in the genetic code of our ancestors. Place is embodied in nonverbal gestures, body movements, word fragments and slang, food, even emotional temperaments. Place produces soil, minerals, and trace elements that transfer into nutrition. Place explained why dill, salted lox wing tips, hard dark rye bread and other foods I had tasted but not culturally grew up with were so appealing. Even after three generations of separation these foods resonated for me.

I was in the Baltics teaching and lecturing in Latvia and Estonia. In Latvia I worked in Riga the capital, as well as Liepaja, a small coastal city. I also travelled from Riga to Tallinn, the capital of Estonia, only five hours away. During my visits the three-hour bus ride between Riga and Liepaja provoked what I can only call 'psychic whiplash'. Passing forests of white birch and scrubby pine trees, wooden houses speckled the semi-rural roadways, and the land resembled the land my ancestors from The Pale must have inhabited – dark, dank, soggy forests of wild mushrooms, sod, and in the past, grass or wood huts. As the bus raced down the highway, I heard imperceptible muffled sounds inside my mind's ear. The corners of my eye flashed on invisible tracings, like a colorless whirlpool of air whipping up and trapping debris. These sensations emerged out of a kind of vortex and vanished before they could even manifest. The environment felt haunted by something I could not touch or see, just vaguely sense.

When I met people, I told them my name was Ellen and I was an artist from New York City. That is all the information they needed. Some insisted in a low-toned guilt and sadness that they desperately needed to show me something. They took me to hidden locations, to sites where Jews had been rounded up and slaughtered during WWII. I never told them I was Jewish. I didn't have to. They knew.

I created my second brainwave opera *AIBO (Artificial Intelligence Brainwave Opera)* in Tallinn, Estonia, building an artificial intelligence character that ran in live time in the Google Cloud. I built it using the artificial intelligence program GPT-2 (at the time of writing, GPT-4, soon to be 5) so that the AI character in the opera was intentionally 'sicko' or perverted. To do this I created a libretto from the biography of Eva Braun, the mistress and doomed wife of Adolf Hitler. I seeded the Hitler AI with 47 movie scripts and texts from the 1880s to the 1940s that were intentionally warped. They included movie scripts like *Frankenstein* and *Dracula*, books such as *Venus in Furs* and the history of male sexual perversion, as well as texts on eugenics.

The performer who played Eva Braun wore a bodysuit of light connected to a brain computer interface. Her emotional responses to the AIBO character lit up her bodysuit of light with four different colors, reflecting the emotions of interest, excitement, meditation, and frustration. The emotions also launched themed databanks of sounds and images. AIBO's live time replies to Eva from the Google Cloud were analyzed for emotional sentiment by an algorithm. If the emotional sentiment was positive the room glowed green; if negative, red; and if they were neutral, yellow. This meant the synthetic emotions of a synthetic being were analyzed as if they had been spoken by a human, while the human wore their emotions on the outside of their body, as if it were an exterior nervous system.

After the opera premiered, I began asking myself why I was so obsessed with both opera's themes focusing on WWII. I had been born post-war, and no one in my family had been marred by the horrors of the Holocaust. My father had fought as an American GI in France but was not severely wounded in battle – though in retrospect I realize he suffered from PTSD. I could not figure out why war, covert Muslim resistance fighters, and the mistress of Adolf Hitler were arising from my psyche. The only war during my lifetime had been in Vietnam a hemisphere away and happened when I was a young teenager. Though I protested against it, it certainly did not scar me in any way.

Then I read about epigenetics, or the science of inherited traumatic memory that changes the basic ribosomal DNA (rDNA) genetic pattern of living creatures and is passed through generations – at least, generations of flat worms. Humans are obviously more complex than flat worms and epigenetic trauma in humans has not yet been definitively proven, but only suggested. I realized that Jews, who endured thousands of years of non-stop diaspora were genetically scarred or changed. I had my DNA tested. The results came back 99 percent Jewish and 1 percent Eastern European, so give or take a margin of error of 1 percent I was a pretty good testing ground for thousands of years of uninterrupted epigenetic trauma.

Being out of place, or 'place making', such as that in Hong Kong and Estonia, allowed former incoherent fragments of my, what I considered rather enormous, epigenetic trauma to rise to the top. These fragments felt hard wired into my character, the result of an internalization of the nonstop expulsion and diaspora of Jews from The Pale of Settlement. The fragments finally rose to the top coalescing in my two brain operas, though when I created them, I was clueless about their origin they just knew. I am a third-generation genetic Jew who does not observe or identify with the Jewish religion, though I like my ethnicity. I am sketchily aware of the numerous Jewish holidays and rituals having watched my paternal grandfather Ike observe them (and, since I live in Brooklyn you can't miss them.) This referred back to how Ike and his fellow villagers anchored their collective diaspora interring together in the cemetery plot in Queens, New York. Severed from their homeland, their Byelorussian bones mixed with the earth of the new world of America. I was the result of that mixing. Being in America could not, however, erase the trauma of thousands of years of diaspora, expulsion, antisemitism, and murder. My genes cleaved to their numerous psychological, physiological, and emotional skewerings. On the outside it all looked and acted fine, but inside it was tangles of deformed aspirations, unspoken sentiments, regrets, withheld gestures, deflated yearnings, ingrown self-loathing, and self-destruction, all internalized from millenniums of persecution.

Location Six – Poland: Shards of the Pale

I first witnessed Tadeusz Kantor's Polish theater company perform *Let the Artists Die* in New York in 1985. I don't remember where it was presented or what the story was about but I was struck by a unique and spare theatrical intensity, so different from the typical cheery theater pieces I had been exposed to. I read the haunting novel *The Painted Bird* by Jerzy Kosinski, with its violently upsetting story of a small boy wandering throughout the Eastern Europe countryside during WWII. The bleakness, perversion and cruelty stirred what seemed like a memory, but it could not possibly have been a lived memory since I was born post-war in Brooklyn. I saw the movie, released decades later filmed in black and white. It reinforced something I knew but did not know what it was or how I knew it.

I visited Poland after Perestroika in 1992 for about a week to see a friend who was working with the former Communist country developing new economic models. I remember the iconic towering Soviet architecture of the Palace of Culture and Science in Warsaw, a profusion of Vietnamese restaurants, and surfeit of cheap amber jewelry and hand-tooled leatherwork. Poland was a personal vortex, a fulcrum of shifting secrets related to my ancestral past but was so grimly Communist and grappling with shaking off that particular past I could not see through it.

In 2021, a Fulbright Specialist appointment in new media, art, and technology returned me to Warsaw. My university hosts took me to the Polin Museum, built in 2015, funded by Ronald Lauder of the Estée Lauder cosmetics company, the government of Israel and the government of Poland. The entire museum was dedicated to the history of the Jews over the past 1000 years in Poland, meaning it tracked the wanderings of the Jewish peoples within the shifting borders of the Pale of Settlement. My hosts only took me there for about an hour. The museum was filled with state-of-the-art interactive exhibits highlighting the migratory routes of Jews throughout the Pale of Settlement over the past 1000 years. Nation states, borders, empires, languages had morphed and changed like a time-lapse movie, but expulsions, exclusions, and pogroms never stopped. I discovered roots I did not know, cut down at the stalk, now ossifying beneath the cemetery of Mt. Judah in Queens.

Location Seven – *Language Is Leaving Me*: Virtuality and Pandemic Lockdown

Investigations of place, ancestry, and epigenetic trauma collided again with my arts practice and the nascent use of AI, and in particular AI image making. During the pandemic lockdown I began an arduous journey to access and reexperience my epigenetic memories. It was, and still is, a painful experience. It includes looking at ancestral family photos, sleeping and dreaming and urging my subconscious to connect the dots. I created a number of mini stories, each accompanied by one or two startling images. I began running those images through AI generators and the results looked nothing like my memories.

Pushing this concept further I made a short five-minute home movie using footage I had both filmed, found, and appropriated. The movie contained grueling, visceral images of death, interment, exhumation, and mass murders of Jews from WWII. I ran some of the images through an AI image generator using languages that did not correspond with the image bank sources. The results were astonishing. I discovered different cultures were unable to translate or identify my memories. They were swapped or rendered into unrecognizable replacements, or facsimiles that obliterated my experience.

Location Eight – Poland: The Pale Uncovered

I was awarded a Fulbright Scholar grant at the University of Warsaw to conduct a collaboration on artificial intelligence, memory, and trauma in a time of war. In the collaboration were both Poles and Ukrainians, some grappling with the ancestral memories of Jewish annihilation and some with the annihilation of Polish folk culture, while others were dealing with the current war in Ukraine. Mid-semester, one student walked into class wearing a bright yellow paper star pinned to his jacket. I looked at it and thought maybe he was being decorative for spring, but also noticed it resembled the yellow Jewish star Jews were forced to wear during WWII.

On the metro ride back home, I noticed a large black and yellow backdrop full of those same yellow star designs, and a Polish folk band with a female singer singing in front of the backdrop. Some people swiping their metro cards at the underground turnstile wore the paper stars pinned to their chest. The stars were commemorating the 80th anniversary of the Warsaw Ghetto uprising. The yellow paper cutouts were paper daffodils honoring Marek Edelman, a leader of the uprising. The flowers were cut in such a way as to resemble the Star of David when folded out. I noticed one of the paper cutouts on the ground, picked it up, stuck it to my jacket and rode the Warsaw metro home. It could have been me 80 years ago, stuck in that ghetto.

Location Nine – Przysucha and the Roots of Epigenetic Trauma

Przysucha is 100 km southwest of Warsaw and was once an important center of Hassidic Judaism. Jews first settled there in 1713 with a baroque synagogue built it the late 1760s. In 1939 out of the 4,850 residents in the town, 2,980 were Jews. By the end of WWII, there were only 1500 inhabitants left. I had never heard of Przysucha and still cannot pronounce its name. But Przysucha the place is where my epigenetic trauma and my creative practice collided, right in the middle of the Przysucha Jewish cemetery.

One of my fellow Fulbrighters, a Baptist minister named Steve Reece, runs the Matzevah Foundation dedicated to honor the Jewish heritage of The Pale that existed before the *Shoah* (the Hebrew word for 'catastrophe', referring to the Holocaust.) The foundation restores Jewish cemeteries, honors mass graves and conducts education. Steve was working on a cleaning, mapping, and honoring of the Przysucha cemetery, accompanied by a team of expert forensic archeologists and their students from Staffordshire University in the UK, as well as local Polish volunteers. The five-day expedition culminated in a dinner attended by the Chief Rabbi of Poland, the Archbishop of Polish Jewish relations to the Catholic Church, and the Polish representative of the Holocaust Museum in Washington DC. Steve invited me to take part in the expedition. I agreed but wanted the option to participate only some of the time. I wanted to explore the surrounding towns, viable witnesses to the horror of almost 80 years ago.

What were those horrors? The cemetery had been desecrated and destroyed. At least 100 Jews had been murdered on that site and buried in a mass grave. The bodies had been dug up by fleeing German soldiers who burned the evidence and then reburied the charred remains in a pit. The site had remained fallow since then, overrun by weeds and forests. The edge of the cemetery had been paved over by a narrow one lane road.

Four vans chock full of volunteers and scientists, including a forensic archeologist who specialized in identifying human remains, pulled up outside the recently installed metal gates. The cemetery did not have any intact headstones. Apparently, most of them had been destroyed, smashed, or otherwise torn out, with some headstones used to build a barn not far away. A Jewish volunteer working for a foundation that identified forgotten Jewish cemeteries gave us a brief lecture about what not to do (disturb any human remains or dig them up) and what we could do (chop down underbrush and trees, rake leaves, clear debris and put any human remains we might discover into paper bags to be reburied). He walked over to his parked car and removed two angled brass dousing rods, then walked around a bit with them in his hands. He demonstrated how they swayed towards the middle of his body and then jutted out from it, explaining that was how they worked to locate straight rows of long buried tombs. He stated the dousing rods were always correct.

After the orientation the gates to the cemetery were finally opened, and we filed inside.

I wanted to explore the wildly overgrown site. It had voluminous ant nests, mosquitos, and trash – liquor bottles, condom wrappers and broken pieces of plastic, all signs of secretive parties and furtive couplings. I sat under a tree and tried to feel the presence, any presence of what had purportedly happened there. I just felt insects crawling around. We took a bathroom break, walking a few blocks away to use the newly purchased but usually closed Hassidic center's bathrooms. The center was incredibly run down, neglected, and moldy, even though it was a recent purchase. However, it was located just across the street from the boarded-up baroque stone synagogue. The local inhabitants, though they did not know who we were, knew what we were doing. There was no other reason to be in the farming village. It was not a tourist stop.

We walked back to the cemetery, and I decided to rake some of the underbrush. The archeologists believed they had located the specific site of the mass grave in a section of the cemetery and wanted to use a special machine that sent radio waves into the earth to map abnormalities that indicated human remains. I stared at the site, now partially cleaned up. It was a dun-colored depression sucking itself inwards, surrounded by crumbly blackened earth. The burnt earth appeared fresh, even after 80 years and could have been a barbeque pit just a few months old.

And then, suddenly it wasn't.

One of the volunteers called out for us to quickly come over. There, underneath a tangle of overgrown brush tucked under a tree root was a big bulging bone broken midway. The forensic archeologist ran over with blue plastic gloves on her hands and picked it up. 'It's a human tibia', she stated. I just stared at it, this old, discarded bone, dusty and leeched of vitality. Suddenly all the worlds collided, the abandoned cemetery, the burnt mass pit, the scattered forlorn tibia, the condom wrappers and liquor bottles, the heisted cemetery headstones, the absolute hatred, and neglect of the burial grounds. I saw a blood vendetta, a purging, a butchered paganism dressed up through a local veneer of civilization. This hatred and neglect were the exact ancestorial epigenetic trauma palpable and irrefutable, all wrapped up in that broken neglected human tibia bone. It was the legacy of neighbors who, through centuries of cohabitation, had turned on their fellow neighbors, pillaging, looting, raping, and desecrating.

Then suddenly, as if on cue, as if out of a scene of a very badly scripted movie, a silver van full of Hassidic Jews showed up, a gaggle of incredibly pale men with big boxy hats, beards, long coats framed by long *payes* curls flanking their ears. They had flown all the way from Jerusalem to this backwards village in the middle of nowhere. They had come to pray for their ancestors, and they were furious. Who were we? They wanted to know. They raised their fists and moved together as one unit like a hoard of angry bees. It was as if we had summoned the angry spirits of the dead – except they were very much alive and in our face. It felt like a bad Western movie of the showdown at the OK Corral. We had every right to be here – the Chief Rabbi of Poland had approved it. But the Hassids spoke neither English nor Polish. Steve calmly approached them. A volunteer who spoke Yiddish was located and they began to calm down, and even took an interest in the radio wave machine being dragged over the mass grave. But I had had enough. The entire ordeal ripped open 1000 years of history, splaying the guts of my own dysfunctional neurotic family. This is why there was such deep trauma. Before this trauma was abstract, an invisible thing that was implied, felt, but never mentioned. But now here in this plot of dirt made real through dousing rods, old bones, mosquitoes and ants, morose wooden country houses, surrounding farm plots and forests showed me in a visceral way what I had sensed my entire life but had no context, no location, and no words for.

The dinner and small art exhibit from the Holocaust Museum in Washington DC a few days later with all the dignitaries and their speeches and vows of reconciliation did not interest me. I spent two days walking around in the nearby villages, shopping for trifles in their shops, wondering who amongst all the people I encountered had relatives that had participated in these mass murders. Who amongst them, if circumstances were right, would do it again? Who harbored in their hearts a secret hatred for the 'other'? Who would kill, who would betray? Who would murder me? Each time I passed someone in the street I thought – could this be a killer? Could this be a savior? Since I am not Polish these villages were not linked directly to my ancestors but stood in as proxies for my ancestor's villages not so very far away. This is the experience they had known their whole lives. This is what they had fled.

Location en – Back to the Virtual and AI

I continue to develop, *Language Is Leaving Me – An AI Cinematic Opera of The Skin*. It explores how newly developing visual artificial intelligence, specifically AI cinema, is incapable of conveying not just my epigenetic trauma but epigenetic trauma from most cultures of diaspora. When memory is obliterated, when the burial spots evaporate, when the homeland is lost, and the records not recorded, when the analogue cannot be digitized what remains of place for future generations? In *Language is Leaving Me…* I have made a five-minute home movie of one of my epigenetic memories. The movie has been sent into an AI image bank of billions of images and reprocessed in Yiddish, Chinese, Tamil, and Xhosa. How do my memory fragments appear when other cultures do not understand, or cannot contextualize my Eurocentric trauma? How will other cultures of diaspora understand and interpret their own memories, the core of their identity when the locus of place has been torn asunder? When language and memory has left me, who am I? Who are any of us?

12

TRAUMA AND HEALING IN THE POST-CONFLICT LANDSCAPE OF BELFAST

Aisling Rusk

This chapter explores ways in which trauma-informed placemaking has taken place within the existing built environment of the contested city of Belfast. Written in 2023, around the time of the 25th anniversary of the Good Friday Agreement (10 April 1998) and third anniversary of the COVID-19 pandemic and first UK lockdowns, I look at two spatial responses to traumas in the city's post-conflict, post-pandemic landscape.

This is placemaking as enacted by the ordinary people of a place, as part of their collective healing. I consider two contrasting scales – a landmark building, The Europa Hotel, once the most bombed hotel in Europe, where placemaking through forms of collective commemorating (or not) is discussed; and, on a much smaller and more hands-on and local scale, placemaking through action and everyday life in the prolific back alleys of Belfast terraces, during and after the pandemic. With a focus on existing spaces, the chapter highlights how the built fabric we inherit, when reconceived, can contribute to healing in different ways after collective trauma through the meanings we apply to it and the practices we enact within it. It raises questions about what policymakers and place-shapers can learn from these bottom-up interventions and interpretations.

Trauma in Built Form / Building in (and from) Trauma

It was Winston Churchill who stated in a speech in 1943 that 'We shape our buildings; thereafter they shape us' (cited in Volchenkov, 2018). Churchill was speaking about the House of Commons, London, that had been destroyed in The Blitz in 1941 (itself a traumatic event) and the belief that the original layout, with two sides opposite each other (adversarially), had shaped politics, governance and the ways in which the House of Commons acted as a legislative body. He argued,

DOI: 10.4324/9781003371533-15

successfully, for it to be rebuilt to essentially the same design. Churchill was making a very symbolic response to the trauma of the destruction of the Blitz, both by rebuilding during an ongoing war, and doing so in a way that would replicate what was destroyed.

Taken at its simplest, and perhaps as far as Churchill intended it, this can imply that it is first one that happens, then the other: we build a building, then it shapes us – *post hoc ergo propter hoc* (an informal fallacy that states: 'since event Y followed event X, event Y must have been caused by event X'.) But there is, of course, much more to it than this. We shape our buildings and spaces first in a literal sense, by designing and building them. Not everyone gets to be a part of this – as a society this is more often done for, or to, us. We inherit most of our built environment. But, according to Lefebvre (1991) 'we all produce space', because we all subsequently continue to shape spaces and buildings through our use of them and the meanings that we apply: by how we, as citizens, interpret our buildings, we keep on re-shaping and re-defining them and they, in turn, re-shape us anew. In considering trauma-informed placemaking – how we respond collectively to traumatic events that impact all of us – it is this symbiotic shaping and being shaped, more than the grand gesture of the making of a building, where the hope for collective healing surely lies.

We have become well-versed in designing grand gestures that respond to collective trauma or assuage collective guilt. Obelisks, war monuments, memorials and peace buildings abound – often with an iconic, highly symbolic architecture architectural style. They can also become shrouded in controversy, as the attempt to finalise a built solution causes old traumas to resurface or exposes inevitable contradictions. Peter Eisenman's 2005 *Memorial to the Murdered Jews in Europe* (Berlin), for example, that fills an urban square with coffin-like concrete slabs, provoked tumult from those who felt that it was offensive in its anonymity and abstract, generic nature. Massimiliano Fuksas' (2008) *Peres Peace House* (Tel Aviv), constructed of somewhat contrived layers of glass and concrete that are said to represent the periods of conflict and peace in that land, has been criticised for its inclusion of a panic room on each floor (Rose, 2009). It is difficult to commemorate collectively when conflict is ongoing, or where versions of the events being remembered vary between communities: the different memories held by two groups of the same event can at times be so divergent that they 'can scarcely be said to refer to the "same" event' (Connerton, 1989, p. 20).

The challenge of collective memorialising of traumatic events through grand monuments and gestures has led some to seek alternative approaches, creating what James Young has termed 'counter-monuments' (1992). This term refers to primarily to memorials of the Holocaust, and in particular those by artists Horst Hoheisel and Jochen Gerz. Counter-monuments are, in different ways, reactions against the aggrandising of trauma with a large, bold structure; instead, by the very absence of such a monument where one is counter-monuments the viewer to look inside themselves to reflect on the traumatic event rather than that trauma being displaced onto the monument itself. Examples include Gerz's 1993 *2146 Stones: Monument Against Racism* (Saarbrücken, Germany), for which the names of Jewish cemeteries were surreptitiously carved by Gerz and his students onto the bottom of thousands of cobbles and then replaced, a few at a time, so that the monument is entirely invisible; and Hoheisel's 1987 replacement for the 1908 *Aschrott Fountain* (Kassel, Germany), which he inverted into the ground so that it, too, was invisible – a negative form:

> The sunken fountain is not the memorial at all. It is only history turned into a pedestal, an invitation to passersby who stand upon it to search for the memorial in their own heads. For only there is the memorial to be found.
>
> *(Young, 1999)*

Remembering and Forgetting in Belfast

In Northern Ireland, there have been few attempts to collectively memorialise the lives lost in the Troubles, which ended, more or less, with the 1998 Good Friday Agreement. The trauma is too partisan and still too recent to make that easy. Individuals and particular atrocities are widely remembered locally and in a partisan way, but even these can be problematic. Plans for a peace centre, designed by American starchitect Daniel Libeskind at the former Maze prison site, were axed in 2013 because unionists feared it would become a shrine to Bobby Sands, the nationalist hunger striker who died there in 1981. Elizabeth McLaughlin's statue of a weeping girl to commemorate the 1972 Claudy bomb was vandalised in 2006, only six years after it was erected (and since replaced). Even ephemeral commemorations have been divisive: for example, Peter Rooney's 1995 plywood peace dove sculpture which was burned and toppled within days from its plinth at Carlisle Circus in North Belfast by loyalists who perceived its phoenix-like appearance as threatening (Leonard, 1997).

Artists have had more success, with notable examples including sculptor FE McWilliams' 1971 *Women of Belfast* series depicting women, with their shopping, caught in the blast of a bomb; Colin Davidson's *Silent Testimony* portraits of people impacted by loss (first exhibited in 2015); and art student Hilary Gilligan's 1996 writing over the course of three days of the names of 3300 individuals killed in the Troubles in chalk onto a footpath of Belfast City Centre, in which 'the human and weary presence of the kneeling artist' was more widely accepted and appreciated (except by some passing Apprentice Boys of Derry, who saw it as graffiti) (ibid.).

In Belfast City Centre, where over 70 conflict-related deaths occurred, no memorial can be found to the Troubles, and there is no marking of any of the locations where bombs exploded. There is just one reference in a ceramic wall art in Laganside Bus Centre, by Diane McCormick, depicting scenes you might see on a bus journey in Northern Ireland. The piece features a surround of 12 buses, each one discretely bearing the name of an Ulsterbus or Citybus worker killed in the Troubles, and there is a small explanatory plaque alongside. Kenneth Foote's (1997) well-known typology sets out different ways to respond to such sites of trauma – starting with *sanctification*, or making that place sacred, through a monument, for example as has occurred at the site of New York's Twin Towers. Next there is *designation*, where the site of a perhaps less symbolic/heroic trauma is recognised but without the consecration of a sanctified site, for example with a simple plaque. Sites can also be *rectified* – 'wiped clean for the collective 'good' of society' and put back as they were so that business can carry on as before, as has been the case in much of Belfast (Switzer & McDowell, 2009). The preference has been to clear everything up in order to move on and forget these 'places of pain and shame' (McDowell, in Logan & Reeves, 2009). Finally, where these sites are a source of shame, they can be *obliterated* in a desire to forget and remove all evidence of the atrocity. An example in Belast is the 2023 demolition demolition of the Kincora Boys' Home, where serious, organised child sexual abuse took place for many years and was covered up (BBC, 2022).

In Belfast we have tried to forget the Troubles, and yet many academics have acknowledged that to forget, heal and move on, we have first to remember (Edkins, 2003). So, in the absence of official memorials to the Troubles, how does a society heal? I suggest that collective healing can and does occur within the built environment – by attaching meanings to buildings and finding new ways to use spaces that are already there, serving another purpose entirely. People, through their use of a place, can shape it anew.

The Europa Hotel as a Counter-monument

The Europa Hotel is an iconic landmark building and luxury hotel in Belfast City Centre, which opened its doors in 1971 during the early days of the Troubles. Designed by Sidney Kaye, Eric Firkin & Partners for the Ulster Transport Authority, it was novel in height (at 51m, Belfast's first 'high-rise' city centre building), and with its highly glazed architectural style it was hoped it would 'show the world that Belfast and Northern Ireland were ready and able to attract new visitors and enterprises' (Scoular, 2003). Instead, this tall, glassy building quickly became an impactful target for frequent IRA bombs and was bombed almost 40 times during two decades of conflict, making it, for a time, the most bombed hotel in Europe, or even the world (Scoular, 2003). This earned it the nickname the 'hardboard hotel' for how frequently the windows were boarded up. Being modern, central and with phones and televisions in rooms, this was also the place where the press stayed in Belfast – their sanctuary in spite of the bombs – and the place where people came to share or pick up a story. Despite all of the bombs, the hotel was never destroyed and no one died – it was a purely symbolic target. For local people, the hardboard windows, viewed from afar, became a litmus test for how bad things were at a given time in the city centre. When peace talks finally got underway in the 1990s, the Europa was the location of choice, and the place where Bill Clinton famously and symbolically stayed in 1995, leading journalist David McKittrick (2009) to refer to it as 'the maternity unit of the peace process'. In peacetime, the hotel retained a reputation as a neutral territory for politicians and the public alike: 'There is a sense in which the Europa, in the oddest of ways, has brought Catholic and Protestant together...' (Fionnuala O'Connor, in McKittrick, 1998). Nowadays, the hotel has taken on an iconic status with local people and tourists alike, remaining a popular location for political talks and drinks on the town, and the building people choose to abseil down for charity, long after it has ceased to be the tallest or most modern hotel in the city centre.

Symbol, target, litmus test, sanctuary, maternity unit, neutral ground, the hotel became many things simultaneously to many different people that were beyond its initial purpose as a hotel. None of these meanings were by design, all came about through the placemaking and sense-making activities of the city's users and inhabitants, throughout and beyond the traumatic event of the Troubles and the many bombs that exploded there. There is no monument to Northern Ireland's Troubles, but perhaps the hotel, in its resilience, its cultural popularity, and the invisible memories and stories it contains, provokes and sparks, has become a monument in its own right. As Andreas Huyssen observes:

> *One of the most interesting cultural phenomena of our day is the way in which memory and temporality have invaded spaces and media that seem among the most stable and fixed: cities, monuments, architecture and sculpture... we have come to read cities and buildings as palimpsests of space, monuments as transformable and transitory....*
>
> *(2003, pp. 6–7)*

For Huyssen, a fixed space can contain both what it physically is in the present, along with traces of what it was, or could have been: 'The strong marks of present space merge in the imaginary with traces of the past, erasures, losses and heteroptopias (Ibid. 7). The Europa is arguably an unofficial, enduring, living monument to Northern Ireland's resilience and endurance through the Troubles, to the province's ability to pick itself up and get on with things, at a time when any more tangible monument to the thousands of lives lost in 'the Troubles' remains divisive. In the invisibility of this monumental role that the hotel performs, and the lack of physical acknowledgement of its history that is nonetheless internalised and passed like folklore from person to person, the building is perhaps more aptly described as a counter-monument.

The trauma-informed placemaking of the Europa Hotel took place on a civic scale, and explores symbolism, multiple readings of the same space, and the question of collectively remembering, or forgetting. This can happen not just through our memories and stories of a past trauma, but through our actions in the present, too, as the next example will explore on a much more personal and local scale.

9ft in Common: Everyday Responses to Trauma in the Belfast Entries

Belfast is a largely Victorian city, having grown rapidly during the industrial revolution, particularly around its linen and ship-building industries. The legacy of that is a city in which a large amount of the housing stock is highly adaptable and versatile Victorian and Edwardian terraces, in a range of scales from 'two-up, two-downs' to grand and ornate three-storey family homes. Behind these houses remain a network of alleyways, or 'entries' as they are locally known, which would originally have been used for coal deliveries, and waste removal. Nowadays many are still the domain of wheelie bins and routes for oil deliveries or bringing your bike around the back, and there are also issues with fly-tipping, and dog poo. But in recent years, that has begun to change, with some starting to be used for gardening, gathering and growing at the hands of a few green-fingered, proactive residents. Wildflower Alley in South Belfast is one of the original examples of how the alleys can become little oases of peace in the city – indeed one tourism website lists it as one of the five best streets in Belfast (McQuillan, 2020).

In March 2020, in Belfast like everywhere else across the world, people were thrown into unprecedented lockdowns in an attempt to kerb the COVID-19 pandemic. For those in terraced houses, with only a yard to the rear, this restriction to the home was, of course, more challenging than it was for those with gardens and space around them. Once restrictions eased a little, and first daily walks were permitted, then gatherings outside in small groups with social distance, people started looking for safe spaces in which to do that. For Belfast's many terrace dwellers, the alleys were the obvious place to go. Neighbours could share space and experiences – seasonal events, *seisúns* (traditional Irish music), children's games, collective meals and walks – with the requisite social distance and fresh air that lockdowns mandated.

In these adaptable laneways, more and more planted, spruced up alleys spread across the city, including Eden and Lockdown Alleys in North Belfast, Rainbow Alley to the East and St. Katharine's Road in the West of the city. People donated unused garden furniture, solar powered fairly lights, and vessels for growing plants (ranging from old welly boots to bike helmets, jiffy bags, filing cabinets, kitchen sinks and suitcases). The Swiss artist, Thomas Hirschhorn explains 'that only with presence – my presence – and only with production – my production – can I provoke through my work, an impact on the field' (Hirschhorn, interviewed in Birrell, 2010, p. 3). Hirschhorn argues that presence and production allow him to come into contact with others by first giving something of himself. In the same way, through the personal commitment of these alley gardeners to these spaces, many of them became transformative, grass-roots spaces of connection and healing during the pandemic.

While an increasing number of alleys are being gated by Belfast City Council, most remain open and are routes that could act as part of Belfast's green infrastructure. Creative practitioner-led collaboration *9ft in Common* (Startling Start / Studio Idir). led several guided walks through the alleys through 2020–22, seeking to encourage people to explore their local alleyways and normalise using them as a route. This draws from De Certeau's (1984) idea of walking as a spatial practice that changes the space that is walked through, however ephemerally. As Rebecca Solnit (2000) puts it, 'Walking, ideally, is a state in which the mind, the body, and the world are aligned. ... [It] allows us to be in our bodies and in the world without being made busy by them'. This is another way in which alleys, when walked, can offer space for contemplation, walking and healing.

Like the Europa but on a smaller scale, once again, Belfast's Alleyways are spaces that were designed for one utilitarian purpose, but are being redefined by people, in a period of trauma, into new uses that, for many, were part of a healing process after the initial trauma of the pandemic, and the personal losses that many incurred. People started coming together to grow food, make compost, play music and games. One alley in South Belfast held a sports day, of sorts, for the local children who were missing out on this in school, dubbed the *2020 Alleyway Olympics*. Indeed, artist and curator Meadhbh McIlgorm, missing the galleries and studios that had closed, conceived of the idea of using her own alley to exhibit art. McIlgorm arranged a group exhibition, *Liminal Belfast*, in October 2020, cleaned up the dog poo and contended with the wind and the rain, so that she and her friends could share art that responded in different ways to the characteristics of the alley and to the pandemic. There were works and installations by six artists, including posters of a *Liminal Times* newspaper McIlgorm had made, and that of printmaker Jonathan Brennan who made night-time site-specific *Pollinators* – installations of alley flora and fauna in silkscreen on Perspex with lights (McIlgorm, 2020). McIlgorm implemented booking time slots and a one-way system to control numbers and invited the public to come and look at art for a few hours. The event was so successful that she was invited to expand the concept into a weekend-long exhibition across clusters of alleys in two areas in the city as part of the *Imagine Belfast Festival* in 2021. This time it was called *LiminAlley*, and McIlgorm partnered with *9ft in Common* who were investigating Belfast's alleys and their potential as spaces of human and physical connection on a city-wide and local scale. The previous artists, and new ones, prepared pieces for these alleys, and *9ft in Common* prototyped and show-cased how fold-down benches, mirrors, street art and street naming could create places from these leftover spaces. Ceramicist Patrick Colhoun installed a ceramic security camera, entitled *Always Watching over You*, to the rear of Belmont Road's Café Smart, a double-entendre that juxtaposed a delicate material with a utilitarian, mass-produced object, and simultaneously recalled his personal memory of his late grandmother, while also evoking and critiquing the sense of being under surveillance in public space (Colhoun, 2021).

The piece that dealt arguably most directly with the trauma of the lives lost in the pandemic was ceramicist Anna Donovan's *By Your Absence* (Figure 12.1, left), an installation of raw clay, slip-painted canvas and red bricks that formed the outline of a body on the concrete alley floor, evoking a crime scene. The suggested body was a void, leaving only mud-like clay stains on a white canvas sheet, surrounded by piled up raw clay 'leaves', and the clay bricks from which most of Belfast, and its alleys, is built. The artist explained how the piece, in memory of her own late sister, Karen, evolved from the sense that 'memories of someone who have [sic] passed away means for as long as that memory is there they haven't quite gone yet' (Donovan, 2021a). Over the course of the two-day exhibition, the clay leaves would have mingled with real leaves in the alley and got soggy in the rain, obscuring the figure and blurring the boundaries of the art with its surroundings – becoming a memory in real time. However, that wasn't to be, as within hours of its completion, before the exhibition even began, the piece was anonymously destroyed, meaning that *By Your Absence* was absent, itself, from *LiminAlley* (Figure 12.1, right). The curators instead created a shrine of the piece on a wall beside what remained, complete with photographs, a plaque, a pile of the clay shavings and three glass bottles containing flowers. The artist reflected that the piece had grown in meaning through this shrine:

> *[B]y their actions those who destroyed the original work had inadvertently intensified its meaning. They [represent] the often unseen and incomprehensible reason for loss and the shrine represents the dogged memory of those remaining.*
> *(Donovan, 2021b)*

This occurrence speaks to the multi-layered reality of urban space. Like the famous adage that states that in civil unrest, one man's terrorist is another man's freedom fighter, so one person's hotel is another's target, and an alley is a garden

FIGURE 12.1   Left, Anna Donovan (2021), *By Your Absence* [mixed media installation], as part of *LiminAlley* (2021), outdoor art installations *in-situ*. Right, after the installation was destroyed. Image credit: Simon Mills.

or exhibition space for some, and for others a clear route to get their bin, bike, or even car to and from their home. Often, this multiple reading of the same shared space can lead to (micro-)conflicts. But it is arguably also where their potential lies.

Multi-layered Spaces for Finding One's Own Way through Trauma

Healing takes place in different ways and timescales for different people. Spaces that can be multi-layered to allow that, like the many readings of the Europa Hotel through and beyond the Troubles, or the ways the alleys opened themselves up to walking and gardening and art, and for that art to be responsive, in real time, to the sometimes hostile conditions of the alley, offer hopeful examples of ways in which we can co-exist in space, and find comfort there, as we collectively heal.

What can we, as placemakers, learn from these citizen-led reinterpretations of places? The ability of buildings and spaces to be and to mean different things simultaneously to different people is what enables everyday life, including healing, to take place in them. From the scale of a hotel to a 9ft wide alley, spaces that respond to the needs of different people simultaneously, without one meaning or use excluding another, are surely key to allowing healing from trauma. How can we design spaces that are plural and multi-layered? The answer seems to lie in being subtle and ambiguous, not overly designed. Grand gestures, memorials and monuments only go so far; what if instead, there were more spaces and places that allowed people to find their own way through their trauma and healing?

References

BBC (2022). *Kincora: Demolition of Belfast home where boys were abused begins.* 17 November. Available: www.bbc.co.uk/news/uk-northern-ireland-63669296. [Accessed: 21 April 2023].

Birrell, R. (2010). 'The headless artist: An interview with Thomas Hirschhorn on the friendship between art and philosophy, precarious theatre and the Bijlmer Spinoza-festival', *Art and Research: A Journal of Ideas, Contexts and Methods,* 3.

Colhoun, P. (2021). *Discussion between author and Patrick Colhoun,* 28 May.

Connerton, P. (1989). *How societies remember.* Cambridge: Cambridge University Press.

De Certeau, M. (1984). *The practice of everyday life.* Berkeley: University of California Press.

Donovan, A. (2021a). Artist Anna Donovan [Facebook] March 25, Available: www.faceb ook.com/artistannadonovan. [Accessed: 18 April 2023].

Donovan, A. (2021b) Artist Anna Donovan [Facebook] March 27, Available: www.faceb ook.com/artistannadonovan. [Accessed: 18 April 2023].

Edkins, J. (2003). *Trauma and the memory of politics.* Cambridge: Cambridge University Press.

Foote, K. E. (1997). *Shadowed ground: America's landscapes of violence and tragedy.* Austin: University of Texas Press.

Leonard, J. (1997). 'Memorials to the casualties of conflict, Northern Ireland 1969 to 1997', *Northern Ireland Community Relations Council.* Available: https://cain.ulster.ac.uk/iss ues/commemoration/leonard/leonard97.htm. [Accessed: 18 April 2023].

McDowell, S. (2009). 'Negotiating places of pain in post-conflict Northern Ireland: debating the future of the Maze prison/Long Kesh'. In W Logan & K Reeves (eds.), *Places of pain and shame: dealing with difficult heritage*. Abingdon: Routledge.

McIlgorm, M. (2020). *Liminal [Space] Belfast*. Available: www.liminalspacebelfast.com/exhibition-one. [Accessed: 23 April 2023].

McKittrick, D. (1998). Hardboard Hotel [Radio Broadcast]. *Belfast: BBC Radio 4 FM*, 27 April.

McKittrick, D. (2009). Interview by Aisling Rusk with David McKittrick in Belfast, 27 September.

McQuillan, S. (2020). 'Top 5 prettiest streets in Belfast', *Ireland Before You Die*. Available: www.irelandbeforeyoudie.com/top-5-prettiest-streets-in-belfast/. [Accessed: 23 April 2023].

Rose, S. (2009). 'Peace centre with a panic room', *The Guardian* [online]. Available: www.guardian.co.uk/artanddesign/2009/feb/17/peace-house-israel-architecture. [Accessed on: 20 April 2023].

Scoular, C. (2003). *In the headlines: The Story of the Belfast Europa Hotel*. Belfast: Appletree Press Ltd.

Solnit, R. (2000). *Wanderlust: A history of walking*. London: Penguin.

Starling Start / Studio Idir (2020-22). *9ft in Common* Accessed 15th Jan 2024 Available here https://9ftincommon.com/

Switzer, C. & McDowell, S. (2009). 'Redrawing cognitive maps of conflict: Lost spaces and forgetting in the centre of Belfast', *Memory Studies*, 2.

Volchenkov, D. (2018). 'Grammar of complexity: From mathematics to a sustainable world' [online] in *London: World Scientific*. Available: www.worldscientific.com/doi/epdf/10.1142/9789813232501_0007. [Accessed in: 14 August 2023].

Young, J. E. (1992). 'The counter-monument: Memory against itself in Germany today', *Critical Inquiry*, 18(2), Winter.

Young, J. E. (1999). 'Memory and counter-memory' in *Harvard design magazine*, 9, Fall.

13

ANTICOLONIAL PLACEMAKING

Karen E. Till and Michal Huss

Vignette One

Hassan Bek Mosque, a prominent feature among the high-rise buildings of Tel Aviv, is the meeting point for a tour with 20 participants, facilitated by Zochrot, an organisation promoting awareness of and accountability for al-Nakba (Figure 13.1, left, Stop 1). Our tour is led by Umar al-Ghubari, a second generation Internally Displaced Palestinian (IDP) who coordinates Zochrot's *Return Space and Tours* programmes. Umar explains that the Mosque was once the communal heart of Jaffa's (Yaffa's) coastal Manshiya neighbourhood and is amongst its only remnants, escaping the fate of the surrounding area destroyed during the 1948 Israel-Arab War or in subsequent demolitions. Umar highlights how, today, by preserving a few token traces such as the Mosque, Israel denies the systematic process of eliminating Palestinian homelands: 'Israelis reside amongst and within such Palestinian traces and ruins... some are unaware of this, some openly discuss it, whilst others raise questions about it; they do not connect these ruins with the Israeli crimes.' In the next two hours, we walk around and unpack the meanings of traces to reconstruct a local traumatic history. Through this walk, Umar enacts a co-performative testimonial exchange in collaboration with audience members and a chronically traumatized place.

Vignette Two

A painful story is shared as we review the results of a pilot map of significant places for Travellers (a traditionally peripatetic indigenous ethno-cultural group originating in Ireland) living in Coolock, north Dublin. Traveller researchers recall a proposal by neighbours to build a six-foot wall around where they live which

DOI: 10.4324/9781003371533-16

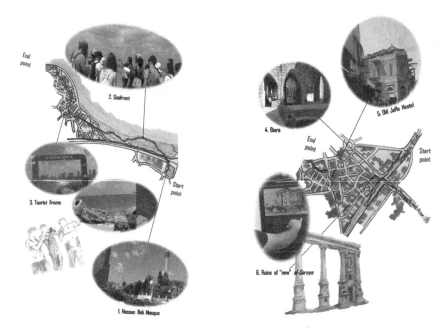

FIGURE 13.1   Left: Map of central features in Umar's tour, August 2018, 2.85 km, 3 hours 18 minutes, 29 degrees C. Right: Map of central features in Yusuf's tour, August 2019, 1.62 km, 2 hours 10 minutes, 34 degrees C. Artwork credit: Michal Huss (2023).

would result in restricted movement and access. The wall was to have a gate; the key held by non-Travellers; and a curfew imposed. Later, the story would be published on a digital *StoryMap* to highlight Traveller resistance:

> *We ended up setting up a residents committee. We all got together, and we said for every block they put up we'll take it back down, so it was coming from the community... We set up our own committee along with the Traveller accommodation officer, not associated with Dublin City Council at the time, and we went against it and we got it stopped.*
>
> *(cited in Pavee Point et al., 2022)*

Despite historic and ongoing experiences of discrimination, Travellers advocated caring for place and community as a means of stopping further harm to an already wounded city. For one researcher, '[t]his shows the importance of coming together and challenging in solidarity' (ibid).

Introducing Anticolonial Placemaking

As the above vignettes demonstrate, places are not mere backdrops to historical events – they are geographically and historically dense 'small worlds', central to people's sense of self, belonging, health, and emotional stability (Tuan, 1979). Places are at once shared and personal centres of felt value, not 'owned' by individuals. Places are significant nodes of activity and experience, rooted in the past and oriented towards the future, always in process.

This chapter emphasises the importance of listening to the experiences of inhabitants who have a deep understanding of how places and peoples have been harmed through chronic trauma and that offer more just placemaking alternatives. Despite claims made to include local voices through participatory practices, many placemaking professionals continue to ignore significant embedded relational place-based histories and thereby continue unjust, and often colonial, power relations. Indeed, dominant Western modernist planning, policy, architecture, and design understandings of place are as locations or 'sites' for renewal, development, and capital accumulation; institutional, legal and professional practices extend processes that harm places, peoples, and more-than-human lives, even if unintentionally.

Anticolonial placemaking practices instead are grounded in humility (Liboiron, 2021) and solidarity to support research and practice led by Indigenous and local practitioners who challenge forms of coloniality and oppressive power hierarchies. As scholar-practitioners of non-Indigenous ancestry in positions of privilege, we advocate working to make visible 'current and historic concerns of the peoples on whose lands they live and work' (de Leeuw & Hunt, 2018, p. 10), and to stand with those who have been and continue to be harmed by coloniality. We draw upon Indigenous, Black, feminist, queer, and geographical theory to argue that places are meaningful small worlds inhabited by multiple lives that are responsible to each other and the places in which they live(d). We further challenge medicalised theories of trauma and highlight 'geotrauma' (Pain, 2021) – how places, peoples, social relations, and ecologies are 'wounded' by chronic forms of state-perpetrated violence (Till, 2012).

To improve relations with wounded places and each other, anticolonial placemaking means supporting Indigenous leadership in decolonisation, including radically changing dominant understandings and practices of place-based relations to lands, waters, bodies, and other lives (Tuck & Yang, 2012). The chapter describes examples of anticolonial research through Palestinian-led tours to spectral traces and ruins in Yaffa (Huss, 2022; forthcoming), and Traveller-led community mapping projects in Dublin (Pavee Point et al., 2022) that result in emergent place-based stories of hope as well as pain and contribute to imagining more just futures.

Challenging Western Modernist Understandings of Place

Dominant understandings of place as property derive from the legacies of colonialism, empire building, the making of modern nation-states, and imperialism (Goeman, 2009), all processes supported by Western Enlightenment thought that separates space from time, nature from culture, and form from content. Papal bulls constituting the 'Doctrine of Discovery' described already-inhabited lands as 'empty' with 'discovered people' and in need of Christianity. Colonial surveys mapped resources, demarcated political territories, and renamed places in European languages. Not apparent on colonial maps accompanying conquest and genocide were the dispossession of homelands, the violences of settler colonialism, the destruction of local ecologies, and the denial of Indigenous sovereignty and personhood (Dunbar-Ortiz, 2015). Cartographic, toponymic, and geopolitical processes abstracted land from life, and peoples from places, imposing Western ethnocentric claims to superiority and normalising spatial imaginaries defined by national borders, cultural imperialism, and market relations (Coulthard, 2014); 'how European countries and colonies divided up the lands and assets of Indigenous Peoples and Nations in the distant past still determines national boundaries today' (Miller, 2019, p. 36).

There was never a *terra nullus* (Latin, 'nobody's land' and the international legal principle to justify claims that territory may be acquired by a state's occupation of it) with peoples waiting to be 'discovered', but always only the inhabited and storied places of sovereign peoples. Local Indigenous wisdom is placed-based and communicated through stories, memories, and respectful relations people have with places, lands, waters, plants, animals, ecologies, and others (Kimmerer, 2013). Mishuana Goeman (2009), Tonawanda Band of Seneca, cites geographer Yi-Fu Tuan (1979, p. 236) to argue that places should be understood 'from the perspectives of the people who have given it meaning', and highlights Indigenous stories as the basis for knowledge, resistance, and resurgence. Places support healthy emotional ecosystems (Fullilove, 2004) through everyday 'movements of memory through landscape, story, and ritual' (Till, 2008, p. 105).

In contrast, Western-modernist abstract understandings of space as a measurable volume, and place as locality or, at best, a culturally-embedded site, reproduce an ontology that privileges the politics of form as separate from the content of the built environment (Yaneva, 2017). Such concepts remain dominant in planning, architecture, and design theory and practice, and reproduce hierarchical, racialised, gendered, and patriarchal social relations and narratives that deny Indigenous self-determining authority and unique place-based relations (Porter & Barry, 2018).

An example of the settler-colonial planning imaginary that renders Indigenous places as anachronistic and in need of replacement by modern properties is Tel Aviv-Jaffa (formerly Yaffa). Following the 1948 war and its aftermath – a broader process which Palestinians refer to as *al-Nakba* (the catastrophe) – Israel seized and appropriated the remains of Palestinian cities in an effort to erase their histories of socioeconomic advancement and cosmopolitanism (Blatman & Sabbagh-Khoury, 2022; Hasan, 2019). Palestinians were not only dispossessed and uprooted from their homelands to make space for the formation of the state of Israel (78% of Mandatory Palestine was taken over (Sa'di and Abu-Lughod, 2007)), but also the names, stories, cultures, and worlds associated with particular places became occupied. The city *Yaffa* (in Arabic) was renamed *Yaffo* (in Hebrew) (Jaffa in English; hereafter we use Yaffa). After being emptied of its 95% Palestinian population, the 'new' Tel Aviv-Jaffa municipality was refashioned as an impoverished migratory district. From 1960–1985, under the guise of evacuation-construction plans, an enforced phase of economic decline was accompanied by the urban demolition (Monterescu, 2015). Following this, a phase of urban-renewal and gentrification included costly refurbishments of the 'Old City' into a historical theme park detached from its Palestinian heritage (Levine, 2007).

Modernist planning examples of emptying out and 'civilising' places demonstrate how colonial understandings of place work together with the neo-liberal rendering of properties for development. 'Experts' rarely acknowledge their role in what Edward Casey (2001) describes as thinning out 'thick places' through processes of 'renewal', for to do so would be to undermine their justification to 'improve' a neighbourhood. Henri Lefebvre (1992) describes such professional practices as 'the violence of abstraction', whereby national bodies commodify the richly differentiated socio-spatial realities of places – each with diverse experiences, identities, and histories – into fragmented grids of private property.

Such processes also occur in 'post'-colonial states. Mindy Thompson Fullilove (2004) portrays the negative social and health effects resulting from waves of urban renewal in historic African American neighbourhoods labelled as 'blighted.' Homes were taken through eminent domain, and, following a period of disinvestment and decline, razed and replaced by parking lots, motorways, hospitals, and universities serving predominantly white, middle-class suburbanites. The resulting 'root shock' for individuals and communities – similar to ripping out a plant from a healthy environment and relocating it to an unknown plot – includes intergenerational violence and serial displacements (ibid.). Rachel Pain (2019) similarly describes the cumulative effects of the processes of dispossession and renewal as 'chronic urban trauma.'

If placemakers were attentive to the geographically and historically 'thick' meanings of place, they would better understand how places have been 'wounded' by past forms of violence (Till, 2012). Moreover, rather than theorise trauma as a rift between the past and the present in an individual's psyche, adopting Pain's (2021) concept of 'geotrauma' would acknowledge the multiscalar, enduring, and intersecting relations between place and trauma. Postcolonial and Indigenous perspectives similarly enhance geo-temporal understandings of trauma by tracing the intergenerational and collective effects of colonial violence, including Indigenous genocide and slavery, across times and places (Brave Heart & DeBruyn, 1998; Fanon, 1963). Attending to the wounds of chronic geotrauma and root shock therefore requires prolonged acts of restoration and justice across generations. For some feminists and Indigenous theorists, recognising past and ongoing harms means to consider places as not irreparably damaged but, with care, as offering environments for justice, healing, and rebirth (hooks, 2003; Waziyatawin & Yellow Bird, 2012).

Anticolonial Placemaking in Practice

Resistance to the modern colonial city requires generating new ways of perceiving and imagining place. Learning from the peoples inheriting chronic geotrauma means rejecting linear biographies of sites and instead engaging in place-based embodied practices – walking, singing, narrated walks, and community mapping. Such processes of creative memory-work elicit emotional stories of survivance, solidarity, and more just futures. We turn now to two examples attentive to the existing ways in which places and peoples are wounded, but from which alternative place-based narratives and geographical imaginaries can emerge.

Spectral Traces in Yaffa

Ruins, as physical metaphors, spatialize history and temporalise architecture (Huyssen, 2006). Due to their fragmentary qualities, they resist being woven into a coherent narrative, particularly in places that have experienced chronic trauma. In the 'new' city of Tel Aviv-Jaffa, Palestinian tour guides bring participants into sensuous contact with 'spectral traces' (Jonker & Till, 2009) of silenced local traumas in Yaffa to ascribe meaning and memory to ruins from a Palestinian perspective. Performative interactions between Palestinian tour guides, ruins and traces, and audiences challenge official scriptings of Israeli national mythology and colonial domination and invite audiences to reimagine once inhabited places as wounded by colonialism and apartheid, but also as offering alternative futures (Huss, 2022).

Some traces are easily noticeable, such as Hassan Bek Mosque. Others are hardly identifiable to the untrained eye. As we walk along Tel Aviv's scenic southern seafront, we learn that the rocks that built this place were taken from the rubble of the destroyed Manshiya neighbourhood (Figure 13.1, left, Stop 2). The tours emphasise the Israeli appropriation of these traces as part of their political gesture. For instance, during another guided tour by Yusuf, a Palestinian-born history teacher raised in Yaffa, we visit traces appropriated to mark Israeli victory whilst disregarding the massacres, lootings, and deportations that caused them. He brings us to the remains of the 'new' al-Saraya (Governor's House) building (Figure 13.1, right, Stop 2), and reads aloud the municipality's official sign: 'On January Fourth, 1948, the building was blown up by members of the *Lehi* (*Irgun*).' As Yusuf stresses, the sign ignores that the building also housed an orphanage. The modes of ruin-appropriation that the tours highlight involve mimicry, where the colonial desire to create oriental landscapes serves as a symbolic indigenisation of the settlers.

However, through the tours, ruins are also transformed from inert landscapes to living artifacts generating a tension between past and present. We walk past a large tourist frame placed by the municipality to invite people to take a panoramic photo. On top of the frame, Umar places a historical photograph of the coastal Manshiya – violently erased from Tel Aviv's cityscape; tour participants photograph this gesture, forming a collage that links past and present, abstraction and resistance (Figure 13.1, left, Stop 3). In the context of Israeli systematic re-writings, the tours further catalogue spatial elements as a future-oriented archive. The photographic and filmed documentation of the tours, according to Najwan, a third generation IDP and Zochrot's Media and Testimonies director, provides an opportunity for those who cannot join the tours (due to age, physical ability, or restrictions on Palestinian rights of movement) to partially experience the return event.

The tours demonstrate how regeneration and gentrification continually abstract the places of Palestinian Yaffa. Yusuf asks during his tour, 'Have you heard of Old Jaffa Hostel? Everyone loves it, but there is a sad story here of identity blurring, since it used to be a Palestinian home' (Figure 13.1, right, Stop 4; see also Huss, 2022, p. 15). In Yaffa, the global process of gentrification has an added ethno-national dimension, as Yusuf explains: 'There are no check points in Yaffa, and no Arabic and no history, and this rapid process of spatial alteration is big, it's *al-Nakba*'. The inability to draw a clear line between multiple processes of ruination in Tel Aviv-Jaffa, including war, post-war reconstruction and gentrification is a testament to the ongoing dispossession of Palestinian space within the Israeli Green Line. *Al Nakba* continues to shape profoundly the lives of Palestinians born after its aftermath through state violence, displacement, expropriation and occupation. In the case of the Palestinian continuous forced displacement, the distinctions between trauma and its aftermath are blurred.

Yet these Palestinian-led tours provide a means to transgress oversimplified Western temporalities by performing with ruins to visualise an officially omitted history and the ongoing history of its omission (Huss, 2022). Through a mode of perceiving and imagining Yaffa again as constituted by many thick places, the tours teach Israelis to acknowledge their responsibility for *al-Nakba* and its continuation. For Palestinians, second- and third-generation children of *al-Nakba*, the tours link the past with a future of accountability and return.

Community Mapping in Dublin

In 'post'-colonial cities, anticolonial placemaking includes research led by people experiencing systemic oppression to challenge, expose, and possibly heal spatial injustices. In Ireland, Travellers were historically discriminated through the colonial 1634 Trespass Act, but parallels in the present include the 2002 Housing (Miscellaneous Provisions) Act that criminalises trespass, and thus the nomadic tradition of Travellers (Pavee Point et al., 2022). Only officially acknowledged as an ethnic minority in 2017, Irish Travellers continue to face systemic racism, leading to higher-than-average rates of unemployment, discrimination in schools, and severe health problems (All Ireland Traveller Health Study, 2010). As part of its goals of raising awareness and advancing community development through research partnerships, Pavee Point Traveller and Roma Centre collaborated with Maynooth University geographers to create two digital *StoryMaps* depicting the importance of place-based stories in Traveller culture and advocating Traveller-led community organisations.

Traveller storytelling is an important form of cultural knowledge (DeBhairduin, 2020). A Pavee Point Men's Health Team whose families moved from the Midlands in Ireland to Dublin developed Pavee Roads Home, 'to explore Traveller culture and heritage through the lens of history and geography' (Pavee Roads Home, n.d.). Recording stories, field visits, and creating maps reconnected a younger generation to a more positive sense of cultural identity than they experienced in the city. Learning stories about particular places, camps, and well-travelled routes, Traveller researchers found that placenames recalled emotional attachments and use values. Birdy's Big House was named after a stopping place where Travellers used to go hunting. The Nine Mile Road referred to the route from Edenderry to Rathangan that people took to work in the bog ('footing turf'), and the Pinkeen Road was named after the pinkeen (small fish) in the canals (ibid.). Family trees recorded military service and other contributions to Irish society and were deposited in the National Library of Ireland. Reclaiming these stories and sharing them with a general public was a significant form of resistance and 'dignity restoration' (Atuahene, 2016); the project was also an important piece of research, recognised with a National Heritage Week Award (Irish Times, 2021).

The Traveller Community Mapping Coolock StoryMap (Pavee Point et al., 2022) focused on one community in north Dublin. The team found that many places significant to Travellers' experiences were not on Google Maps. Part of this was a colonial legacy: the places inhabited by Travellers were not mapped in Ordnance Surveys of Ireland, which recorded properties and located people according to individual addresses for political control (Kearns, 2022). Traveller researchers confronted the absence of locally significant places by gathering and mapping stories about places from three generations. Centred on their homes rather than Coolock Village, the large colourful wooden community map inverts dominant orientations and creates new relational geographies (Figure 13.2).

Some of the stories shared by older generations recalled places no longer there, such as the 'Old Bakery' or the 'Milkman's Shop'. Identified as 'Places of Belonging', some of the recorded stories are published on the digital *StoryMap*, including of *Mister Monaghan's Shop*:

> …*years ago, when we used to go up with Daddy and Mammy, and they used to give everything out on tick* [credit]. *Whatever you needed — bread, milk, you had kind of everything. He knew all the Travellers in the area very well. Extremely lovely, lovely family. And they actually came from Monaghan, so we called it 'Mister Monaghan's Shop'. …[F]or years it fed Travellers, it kept Travellers alive in our own local area! That was the reason everyone used that shop.*
>
> *(Participant cited in Pavee Point et al., 2022)*

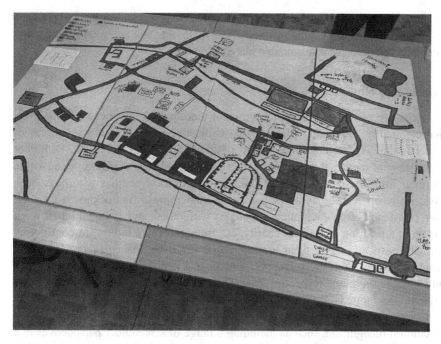

FIGURE 13.2 Traveller Community Map of Coolock, May 2022. Image credit: Karen Till.

There were also testimonies about everyday forms of ongoing structural racism and discrimination people experienced; the stories varied but remained across generations. For example, government authorities segregated Travellers by locating accommodation sites on the outskirts of towns and cities, at '…the sides of motorways, the backs of dumps, [with an attitude of]: "Leave them in there, let them live there"' (Participant, cited in Pavee Point et al., 2022). Moreover, despite available funding, Traveller-specific accommodation was not maintained by local authorities: 'The overcrowding – that was not just now, that [has been happening for] a long time. If you look at where the sites are built it's quite evident that [state and local authority officials think] "Keep them there. Keep them housed. Keep them out of the way"' (ibid.). Stories about discrimination in schools, including educators segregating students were common: 'I was taken out of class, out of my own classroom, and brought into a Traveller-specific class – they used to call it "Special Class"… It was kind of embarrassing; do you know what I mean?' (Participant in Pavee Point et al., 2022).

For non-Travellers, learning from Traveller stories and community maps reorients dominant understandings of the city, for '[w]hen we listen to other persons' tales, we absorb into our own lives realities that we have not personally encountered' (Tuan, 1989, p. 278). While some stories described respectful relationships with non-Travellers, many indicated mainstream society's responsibility for the harm resulting from discrimination. Non-Traveller placemakers can learn from Traveller-run organisations and support their leadership advancing policy, supporting healthier, inclusive, and sustainable places and cities.

Concluding Notes

This chapter calls for changes to dominant placemaking practices that maintain neo-coloniality. Devaluing places of past injustice as empty sites is a form of violence that exacerbates earlier forms of structural violence (Till, 2021). Rather than ignore or replace the multiple stories and temporalities of places, we advocate learning from the perspectives of Indigenous and local peoples who have been historically oppressed. As a first step, practitioners must first acknowledge their own privilege before attempting to connect with others and engage in their work not from a position of guilt or charity, but from a genuine interest in confronting difficult pasts and being open to multiple perspectives. Placemakers must begin from a place of humility to support and sustain healthy places where they live and work that include respectful and responsible eco-social relations in the past, present, and future (Liboiron, 2021).

Changing dominant planning and policy practices entails more than liberal forms of cultural recognition, such as including a range of stakeholder perspectives in a project. It means standing back, listening to, and learning from the leadership of peoples who have been historically subjected and silenced. Otherwise, placemaking practice, however well-intentioned, may continue to damage people, communities, and our Earth by assuming professional 'experts' can 'speak for' (and hence problematically 'speak instead of') peoples and places. Acknowledging different underlying philosophies about the places we inhabit may result in discomfort, unsettling tension, and conflict, but such emotions are part of the process of decolonising modernist placemaking (Porter & Barry, 2018).

Listening to the wisdom of already existing local and displaced placemakers includes learning about alternative understandings of place and practice, such as the walking tours and community mapping projects described above, that materialise stories of hope and pain for affected communities and may lead to more just futures. Through the memory-work and research of activists, community, leaders, and artists, stories emanating from particular places marked by chronic geotrauma might result in unexpected relations and networks of solidarity that can challenge oppressive relations, support local-to-local knowledge exchange, and nourish forms of care for places and each other.

Acknowledgements

The Traveller Community Mapping Coolock StoryMap was funded by an Irish Research Council *Enhancing Civic Society* New Foundations Grant. A special thanks to researchers from TravAct Coolock for sharing their stories.

References

All Ireland Traveller Health Study Team (2010). *All Ireland Traveller Health Study*. Dublin: School of Public Health, Physiotherapy and Population Science, University College Dublin. Available: www.lenus.ie/handle/10147/115606. [Accessed: 12 August 2022].

Atuahene, B. (2016). *We Want What's Ours: Learning from South Africa's Land Restitution Program*. Oxford: Oxford University Press.

Blatman, N. & Sabbagh-Khoury, A. (2022). 'The Presence of The Absence', *International Journal of Urban and Regional Research*. 47(1) January 2023: 119–128.

Brave Heart, M. Y. H. & DeBruyn, L. M. (1998). 'The American Indian Holocaust', *American Indian and Alaska Native Mental Health Research*, 8(2).

Casey, E. S. (2001). 'Between Geography and Philosophy', *Annals of the Association of American Geographers*, 91(4).

Coulthard, G. S. (2014). *Red Skin, White Masks*. Minneapolis: University of Minnesota Press.

de Leeuw, S. & Hunt, S. (2018). 'Unsettling Decolonizing Geographies', *Geography Compass*, 12(7).

DeBhairduin, O. (2020). *Why the Moon Travels*. Dublin, Ireland: Skein Press.

Dunbar-Ortiz, R. (2015). *An Indigenous Peoples' History of the United States*. Boston: Beacon Press.

Fanon, F. (1963). *The Wretched of the Earth*. New York: Grove Press.

Fullilove, M. T. (2004). *Root Shock*. New York: One Balletine Press.

Goeman, M. (2009). 'From Place to Territories and Back Again', *International Journal of Critical Indigenous Studies*, 1(1).

Hasan, M. (2019). 'Palestine's Absent Cities', *Journal of Holy Land and Palestine Studies*, 18(1).

hooks, b. (2003). *Teaching Community*. London: Psychology Press.

Huss, M. (2022). 'Autotopographies of Forced Displacement', *Journal of Refugee Studies*. Ahead of Print.

Huyssen, A. (2006). 'Nostalgia for Ruins', *Grey Room*, 23.

Jonker, J. & Till, K.E. (2009). 'Mapping and excavating spectral traces in post-apartheid Cape Town', *Memory Studies*, 2(3).

Kearns, G. (2022). 'Official Maps and Community Mapping. Eye on the World' [Blog] (25 October), *Department of Geography: Maynooth University*. Available: https://maynoothgeography.wordpress.com/2022/10/25/official-maps-and-community-mapping/. [Accessed: 12 August 2023].

Kimmerer, R. (2013). *Braiding Sweetgrass*. Minneapolis: Milkweed Editions.

Lefebvre, H. (1992). *The Production of Space*. New Jersey: Wiley.

Levine, M. (2007). 'Globalization, Architecture, and Town Planning in a Colonial City', *Journal of World History*, 18(2).

Liboiron, M. (2021). *Pollution is Colonialism*. Durham: Duke University Press.

Miller, R. J. (2019). 'The Doctrine of Discovery', *Indigenous Peoples Journal of Law, Culture and Resistance*, 5.

Monterescu, D. (2015). *Jaffa Shared and Shattered*. Indianapolis: Indiana University Press.

Pain, R. (2019). 'Chronic Urban Trauma', *Urban Studies*, 56(2).

Pain, R. (2021). 'Geotrauma', *Progress in Human Geography*, 45(5).

Pavee Point Traveller and Roma Centre, TravAct Coolock, and Maynooth Geography. (2022). *Traveller Community Mapping Coolock StoryMap: Storied Places of Belonging and Unbelonging.* Till, K. E. & McArdle, R. (eds.). Available: https://arcg.is/0Hryzy. [Accessed: 12 August 2023].

Pavee Roads Home. (n.d.). Available: https://paveeroads.paveepoint.ie/about-the-project/. [Accessed: 12 August 2023].

Porter, L. & Barry, J. (2018). *Planning for Coexistence?* Abingdon: Routledge.

Sa'di, A. H. & Abu-Lughod, L. (2007). *Nakba.* New York: Columbia University Press.

The Irish Times. (2021). National Heritage Week Award Winners 2021. Available: www.iri shtimes.com/sponsored/ireland-s-heritage/national-heritage-week-award-winners-2021-1.4704774

Till, K. E. (2008). 'Artistic and Activist Memory-work', *Memory Studies*, 1(1).

Till, K. E. (2012). 'Wounded Cities', *Political Geography*, 31(1).

Till, K. E. (2021). 'Dignifying the Ruins', *The New Urban Ruins: Vacancy, Urban Politics and International Experiments in the Post-Crisis City.* Cian O'Callaghan and Cesare Di Feliciantonio (eds.) Bristol: Policy Press.

Tuan, Y-F. (1979). 'Space and Place', *Philosophy in Geography.* S. Gale & G. Olsson (eds.). Dordrecht: D. Reidel Publishing Company (Reprint by Spring, 2011).

Tuan, Y-F. (1989). 'Cultural Pluralism and Technology', *Geographical Review*, 79 (3).

Tuck, E. & Yang, K. W. (2012). 'Decolonization is not a Metaphor', *Decolonization: Indigeneity, Education & Society*, 1(1).

Waziyatwin & Yellow Bird, M. (2012). *For Indigenous Minds Only.* Santa Fe: SAR Press.

Yaneva, A. (2017). *Five Ways to Make Architecture Political.* London: Bloomsbury Publishing.

14

PLACEHEALING IN MINNEAPOLIS

Before and After the Murder of George Floyd

Teri Kwant and Tom Borrup

Preface

Telling the story of Minneapolis after George Floyd's murder is a complex task. There are historical, cultural, political, geographic and deeply personal stories woven into each location, each site of harm and healing, anguish, and resilience. We look through known but not chosen lenses, crafted by policy, historic injustice, shaped by patriarchal white supremacist forces, and our own unique histories and professional experiences. Here and now, with this collective experience barely in our rear-view mirror we attempt to assess the trauma, share perceptions about what happened here in our neighborhoods, and propose a new way to think about the healing of places and people in the context of community trauma.

It is our intention to name possibilities for movement, a key attribute of healing, and possibilities for trauma-informed decision making and engagement, to further healing. We offer a redefinition of trauma-informed places to include design for resiliency, healing, and love. Our shared work as place healers in community leads us to want to share the tactics and practices, but also the mindfulness, of an approach that involves deep care, not just diagnostics and renovations.

In most times of crisis, there is a rush to correct or rebuild. The lack of pause, a hurried re-placing is antithetical to the theory behind the practices of trauma-informed care now emerging as a practice within the canon of design and design education. The medical emergency analogy is apt. We must triage wounds of both people and places, but what is also true is that we must help set the stage for healing through attentiveness and care.

We both watched and felt the visceral pain and anguish of our communities as another Black man was killed at the hands of police, this time – but not

DOI: 10.4324/9781003371533-17

uniquely – recorded on a teenager's phone for the world to witness. As neighbors to injustice, we felt the need to bear witness and act.

Such murders and abuses of Black and Indigenous people, as well as women, queer, trans, and others, go back 400 years in what is now called the United States of America. The lands on which we live and on which George Floyd was murdered have experienced layers of horrors and traumas. At the same time, it is important to acknowledge that joy, love, and celebration are also infused in this soil. This place, which holds all of these moments – from mundane to world-changing – remains the traditional homeland of the Dakotah people.

Where this Happened: A Diagram of Harm

We acknowledge the harm done to the Dakotah people and the place-taking that happened here. We are curious about places of great significance like Bdote, the confluence of two rivers, a traditional birthing place, site of the Native origin story of spirit becoming earth and inhabitants. This place later became the site of a US military fort and concentration camps where Native people were tortured. The overlay of white violence has repeatedly played out there and across all that is now Minneapolis.

Being Here Through it All: Some Context

On May 25th, 2020, the place at 38th Street and Chicago Avenue in Minneapolis was deeply wounded by the public and horrific execution of George Floyd. The by-stander who had the presence of mind to activate her cellphone camera and remain painfully vigilant the entire 9 minutes and 29 seconds of the act of murder – and beyond – made the event available for the world to witness. Trauma inflicted on the victim as well as those present who watched and screamed in protest has been repeated uncountable times during American history. This time unprecedented outrage was triggered not only in Minneapolis but around the world.

The most powerful and impactful range of human experiences were played out on this otherwise unremarkable street corner: shock, grief, anger, rage, pain, loss, recovery, healing, remembering, joy. The place has become a site of mourning, unrest, quiet reflection, loud protests, a site of artmaking and agency. It emerged as a sacred place to many as well as one that holds trauma.

This was at a particular and very charged moment in time. The bold presence of thousands of people on the streets, after months of isolation due to the COVID-19 pandemic, became another significant social turning point. Tens of thousands of people who had been reluctant to gather outside and participate in public life suddenly did. The protective lid that had twisted shut over the country suddenly popped off. We joined others in the streets and parks and backyards in order to connect around shared outrage and loss, but also to make meaning, which happens to be a cornerstone of trauma-informed design.

During the summer of 2020 people around the world were simultaneously experiencing complex personal losses. And, with dozens of micro- and macro-crises brought on by the pandemic, deeper chronic diseases of our time and our country were revealed. We were experiencing the opposite of ontological security – a fundamental sense of safety in the world due to the reliability of our surroundings and patterns of our lives. We were also in the midst of a four-year national political ordeal watching daily with outrage and stress brought on by leadership attacking and eroding a sense of humanity and justice we had spent lifetimes fighting to advance. Consistency and predictability had gone missing with renewed attacks on the civil rights of many. Acute stress turned into chronic stress and we started describing this experience through the language of trauma.

In this time of upheaval, shared experiences of loss, receding belonging, and alienation, people needed to join together. Some of us wanted a racial reckoning, but we have only seen more divisiveness. An initial tidal wave of calls for change have turned to ripples.

Across our city, specifically at the 3rd precinct where the police who murdered George Floyd were dispatched, unrest, conflict, protest, fear, grief and anguish spilled over. It was another site and series of acts of police violence against the communities who gathered to protest the killing. It was the flashpoint in the truest sense of the word when the precinct and dozens of other buildings surrounding it were set fire – an overdue attempt at the destruction of the tangible evidence of systems that have disenfranchised, injured, and killed people of color for generations.

Both 38th and Chicago and the 3rd precinct are places of trauma. Both became what they are due to ongoing trauma. We believe places hold trauma, as do people. Healing can be physical, psychological, and emotional and it can address historic or current wounds. But healing is not curing, it is an ongoing process. It is an engagement and, like the grief or other outcomes that trauma may produce, a lifelong journey. During the two-and-a-half years since, we've witnessed many changes and events, not the least of which were the trials of the involved police officers. Their convictions were described by most here as some degree of accountability, but not justice for George Floyd. The guilty verdicts were like a single dose of antibiotic when a lifelong regimen is required.

Trauma and Resilience

As a first step in our dialogue about trauma-informed places we need to discuss trauma – what it is, and how design choices and therefore the places that result from that design effort can mitigate or at least, do no further harm. We use the word trauma as shorthand for the consistent and persistent hauntings of an initial crisis. According to Sam Grabowska et al.,

The original crisis can be individual (like emotional abuse, a car crash, a natural disaster, or war) and can also be institutional and systemic (e.g., racism, sexism, homophobia, transphobia), chronic and enduring (e.g., homelessness and addiction), cultural and historical (e.g., colonization, genocide), and environmental (e.g., natural disasters, pollution)

(2021, p. 7)

There was a veritable ecosystem of traumas present at the place of George Floyd's murder and during the subsequent unrest.

Regardless of its source, trauma contains three common elements: it was unexpected, the person was unprepared, there was nothing the person could do to stop it. Simply put, traumatic events are beyond a person's control.

Adjacent to the study of trauma and trauma awareness in the creation of places it is important to also define resilience – the process of adapting to adversity, trauma, tragedy, threats, or significant stress. Creating places from a position of understanding not only the trauma, and informing work from that perspective, but of having resilience as a focus and a value, is an essential ingredient in the development of healing spaces.

Events Following the Murder: Naturally Occurring Moments of Healing

Events during the uprising were not simple. Rage and despair overtook many people, resulting in protests, destruction, and looting. Violence by police in their formulaic tactics to put down riots fueled further violence and rage. Systematic burning of neighborhood commercial corridors of mostly Black, Indigenous, and other People of Color (BIPOC)-owned businesses that carried on for days was the result of agitators from outside the city and the state as well as the absence of police who didn't seem to care if the community burned. Without police protection, over-burdened fire fighters were unwilling to venture into the most devastated parts of the city. Numerous individuals associated with groups such as the Proud Boys (an exclusively male North American far-right neo-fascist militant organization that promotes and engages in political violence) were later convicted of arson, groups who regale in destroying communities of color while creating the appearance that those communities are destroying themselves.

Experiences We bring to this Conversation

By way of introduction, we are two white, cis, queer people who have lived in Minneapolis for a combined total of more than 60 years. One of us lives on Chicago Avenue, blocks from where George Floyd was murdered by Minneapolis police, and one of us in the Longfellow neighborhood, up the street from the 3rd precinct police station, the flash point of the unrest. We are aware of our positionally, our socio-cultural lenses as well as demographic characteristics – our age, our privileges, our socio-economic status, and our cis, queer, white identities as we continue writing. At the beginning of any research project this is a required action step, to identify the ways in which we may not see the whole picture, and how our implicit biases play a role in our discovery process. The only way to look at this place is from one's place in it. And as we share our perspectives, we also share an intention of healing. If we are attempting to do anything through this dialogue and narrative, it is to reduce past, present, and future suffering. We acknowledge each reader has a unique lens as well and aspire to incorporate not just different lenses but experiences into the remaking and rethinking of place-healing work.

Teri is a trauma-informed designer of places, public artist, researcher, and design educator:

In my design and public art practice, I deconstruct and redesign human experiences of places and have done this work at many scales, from the intimacy of a healthcare moment to planning urban environments and park systems. With training in architecture, landscape architecture, and graphic design and through qualitative, ethnographic and observational research, I developed a practice and passion for creating spaces intended to create healing connections, reinterpreting landscapes and urban spaces in ways that provide meaning and hope.

My curiosity about human experiences of places developed early, I bonded deeply with environments, both natural and built ones – from the trees in our neighborhood on the north side of Chicago, to the taste of the well water at my nana's place in Michigan or even in the physical comfort of my favorite chair. I had an awareness of how connections to places and things shape us through our sensorial experiences of them. They can create meaning and evoke memories. Impacted by several significant losses at an early age, and again at another stage in life during the AIDS pandemic, I learned to navigate crises and the people in them, anticipating needs, creating safety for myself and others. Those early losses changed me, but I felt grounded in my cherished places.

I was raised with deeply homophobic messaging but came out at a relatively young age considering the era and context. I was about 20 when I revealed my gay identity and was rejected as a result. Ironically, my family took pride in my mother's Quaker roots – a socially progressive, justice-oriented and vaguely religious tradition. Specifically, that pride was rooted in my ancestor's involvement in the Underground Railroad, supporting safe passage for enslaved people to get to some measure of freedom. Nonetheless, I heard racism, I saw it, I heard homophobic comments and slurs as I grew up. I gradually became aware that our family history had become a permission slip for my cis, straight white relatives to not be directly engaged in justice work, or even to understand our privilege.

I have unfortunately, directly and indirectly, experienced trauma. Early in life, an avalanche of loss of both family and friends caused traumatic grief.

Later, being a visible activist as a gay woman I received death threats and most horrifyingly, recently as a mom of an adult son who has also experienced gun violence.

Tom is an organizer, nonprofit administrator, planning consultant, and educator:

In my community planning work, we bring people together through neighborhood, city, organizational, or cultural planning and through the facilitation of an artist (or artists). Artists tap into the cultural sensibilities and creative capacities of participants in ways traditional planning activities don't. Through this I learned of the efficacy of artist-led community processes and how they can unleash multiple layers of feelings, connections, and creative thinking. In trauma-informed placemaking/place-healing and trauma-informed design a focus on process is critical.

As a community arts center director, community process designer, teacher, researcher, writer I attempt to be a student learning through each experience. As a young person, a couple years too young to go to Woodstock, the social change energy of the time was transformative. Social justice motivations drove me from that age, yet an upbringing in an all-white small Connecticut town provided a confusing mix of values. A racist, antisemitic, homophobic, misogynist environment was pervasive. Civic responsibility, Christian acceptance, non-violence, and respect for the natural environment were ironically also present. Artists were nowhere in my experience and fine or Euro-classical arts were for the rare school trip to Hartford or Boston. I enjoyed modest economic privilege, raging white privilege, and an enduring sense of safety even as I eased out of the closet as a teenager and in college. This gave me enough contradictions to wrestle with for a lifetime.

In conversations with Teri for this writing, I came to realize I had never experienced what one would consider trauma. Violence against me or even horrific fear has been absent to date including accidents or serious health crisis. The sense of safety to travel and move among unfamiliar people and places remains an often-used privilege.

Our Conversation

As long-time acquaintances, friends, and colleagues we convened in the summer of 2020 following the murder of George Floyd and the unrest to talk about what happened, what we saw and felt, and some of the impacts. At the center of our dialogue, that we share in this chapter, is the question of how do we cope with and help communities heal from such vast wounds? We both noticed a lot of talk in workplaces, boardrooms, classrooms, and political chambers about listening to BIPOC community members, but as of this writing, more Black men's lives have been stolen by police. The system is still operating as it always has and was meant to – protecting those who hold undeserved, unwarranted power over members of oppressed communities while erasing their histories.

Tom: Within the days after the murder, my partner and I watched through our 9th floor windows at what can only be called a panorama of flames. This may be the closest I've come to feeling a sense of trauma. The heat on the glass, the sounds of gunshots, yelling, and frantic movements of people below are etched in my memories. Yet, we had options and little fear our building would burn. My partner, a Black man, felt a deeper fear in his bones. He was born in the Greenwood section of Tulsa, Oklahoma, site of the 1921 massacre of hundreds of Black residents and burning to the ground of their neighborhood. His parents grew up there; his grandparents were survivors.

Teri: My son and I watched as trucks with out-of-state plates cruised loudly and at mind numbing speeds through our neighborhood, with a half a dozen young white men in the back. We witnessed skinheads strut down our street intimidating residents, showing off their weapons. These outsiders were like an opportunistic infection in the wounds in our city and neighborhoods. On daily updates from our elected leaders in a local park we learned about their collateral criminality and practice of 'asymmetric warfare' tactics: intentionally setting fires and creating chaos and disruption in locations that would drive fire fighters into opposing directions, creating turmoil for days.

One night after a couple days of unrest, our neighbors came together to set up a night-watch in our neighborhood to provide a stabilizing presence and to protect the businesses from these outsiders. We removed and stored incendiary devices like canisters of propane from the neighborhood hardware and corner stores so they couldn't be stolen and used for more detonation and destruction. At the same time, American Indian Movement (AIM) Patrol kept watch on our neighborhoods and their presence was incredibly reassuring.

Tom: Incendiary efforts persist in different forms. The 3rd police precinct station sustained minor damage mostly at its entryway along with smoke and water damage inside. It stands largely intact. Nonetheless, doctored google images pervade the internet showing it fully engulfed in flames. Stories describe protesters burning it to the ground. Efforts to blame the victims serve to deepen the trauma or 'twist the knife.' And these continue in the media, social media, and political platforms.

Teri: Meanwhile, protests and marches continued in various locations across Minneapolis, peacefully. Joining one protest that merged onto a nearby highway I witnessed bravery and community pain, but I also saw deep connection and even dancing. What I saw was evidence of the natural tendencies and abilities of communities, individuals, and places to heal themselves. One could say once the bleeding was stopped, with bandages of paintings, musical events, and artful community, gatherings popped up. A collective healing breath was taken at many locations between the 3rd precinct and George Floyd Square. A sense of humanity and even joy in the form of generosity emerged.

Artists as First Responders: Acts of Art, Connection and Grace

We both saw and recorded countless moments of incredible generosity pop up in the days after the unrest. Some were acts by community members, individually, some were efforts by organizations, in attempts to mitigate the challenges created by the destruction of important resources like grocery stores and access to medications. These are stories of naturally occurring trauma-inspired actions but also acts of healing.

George Floyd Square became a gathering place and a memorial in the days after George Floyd's murder. Those who helped curate spaces and places within this zone designated it a police-free zone. Barricades went up to prevent car traffic from traveling through the site. Agency was made visible here too, like the many artworks that sprang up along Lake Street. Spontaneous memorials and galleries and the globally recognized portrait by Peyton Scott Russell all popped up to create a haunting image of the man who had lost his life there.

Along Lake Street and other zones of the city murals painted by known and unknown artists signaled protest and commentary (Figure 14.1, left). Some called for peace and connection. Our photos from the time depict a broad range of visual messaging from the makers. The entire city became the canvas for protest and art. Plantings in burned out lots were created from bent and deformed i-Beams, mangled by the fires, and free stores with clothing, diapers, food, and water started showing up every few blocks. Beauty bandaged the destruction. Not only art but generosity was on display.

Tom: Within hours of the fires, we participated in work with volunteers who joined to clean and salvage, to distribute free food and necessities to a community left with no retail stores, to build memorials, to place flowers and green plants on charred building remains and conduct music, drama, dance, and visual art-making activities in now-vacant lots and on the streets of George Floyd Square. As thousands of panels of plywood were rapidly deployed to cover storefronts across the city, artists were right on the job. Artists became known as First Responders in the social, emotional, and aesthetic sense, using spray paint and stencils to bring images honoring George Floyd while expressing resistance and hope.

One barber shop owner set up on an empty lot and gave free haircuts to Black men in an outdoor street-corner barber shop (Figure 14.1, right) honoring the needs of the community for predictability, healing, and kindness to be shared among community members. Trauma has been followed by healing in many forms, not to cover up but to provide spaces to process. On multiple occasions during the summer, my partner and I spent afternoons at George Floyd Square with the Million Artists Movement encouraging visitors to sit under shade tents and use materials provided to fashion expressive cloth panels later combined into quilts. White suburban families, unhoused persons, and others sat, talked, cut, glued, and stitched expressing their anger, fear, hopes, and attempts to understand. This activity was one of dozens, if not a hundred community and artist-led activities at George Floyd Square in the months following the murder.

FIGURE 14.1 George Floyd protest and commentary (left) and barber shop (right), Summer 2020. Image credit: the authors.

Teri: Lake and Hiawatha, near the 3rd precinct, became a zone of anger and pain made visible. The contrast between this intersection and the one only 20 blocks away is stark, but this one too, started a natural healing process through artful and helpful interventions that literally and figuratively fed the community. Parking lots became distribution centers, and temporary community gatherings happened with the impromptu music provided by DJs.

Another act of healing I witnessed was an effort from a local grocery chain to provide food in that area which had become a food and medicine desert overnight. During the summer of 2020 my neighborhood near the 3rd precinct had no operational groceries, pharmacies, or banks available into the foreseeable future. Being true to its mission to 'alleviate hunger in our communities' vulnerable populations' the chain quickly built a small soft-sided pop-up store in its own parking lot – complete with a pharmacy and an ATM – the very things originally targeted during the unrest. In doing so they signaled to the community trust, empowerment, safety, and support at a time of need.

Disaster Capitalism

Tom: More than 250 businesses burned or sustained heavy damage, mostly along Lake Street. Other areas of the city also experienced significant damage. Physical structures at now-sacred George Floyd Square, on the other hand, remain intact. Trauma across the city was not limited to those days of unrest. Nearly every business up and down the commercial corridors in Minneapolis, including downtown, remained boarded for over a year. Only after the guilty verdict of the police officers was announced April 21, 2021, was there some small sense of justice or resolution. A gradual reopening began. Within a week, omnipresent National Guard soldiers wielding assault rifles and their military vehicles that occupied our streets began to dissipate.

Many businesses did not survive and have not been rebuilt nearly three years later. As too often happens in urban spaces devastated by either slow or sudden disaster, real estate speculators swoop in, sometimes contributing to the devastation and loss felt by the community. With many forces at work, damaged and destroyed properties along Lake Street and Chicago Avenue have not succumbed to such speculation – at least not yet. Nonprofit organizations and philanthropists (that the Twin Cities has in abundance) and eventually the City, purchased most of these properties where owners lacked the capacity to re-build.

Teri: One significant exception to this rule is the Coliseum Building on East Lake Street. The building was urgently ushered into historic landmark status to preserve it while funding for renovations were sought. It will become an owner-occupied adaptive reuse project, community gathering place, and a true reflection of the neighborhood, only one block from the 3rd precinct. It will house BIPOC businesses and create an incubator for wealth building among affected small business owners in this heavily damaged neighborhood. Art already is a part of the reclaiming of this building, and it will continue as construction starts.

Tom: Tens of thousands of hours of community member's time and brilliant artistic productions have maintained George Floyd Square as a memorial site. The proliferation of artworks, from small to installation-scaled have been recorded and many shared with the world. Some have been collected and archived at a nearby community center, all evidence of the community's desire to contribute and the outpouring of grief and pain – trauma and healing through various forms of expressiveness and art. A sense of urgency to keep awareness and vigilance about what happened there outweighs the desire to return the intersection to normalcy. Just as there is no return to normal after a bodily trauma, normalcy in this community is not possible, nor is it acceptable. As of this writing in the spring of 2023, we witness continued tending of the square and the memorial elements. The free store of clothing has been replenished after a punishing winter, and volunteers are gently tending to the space, the wound.

Expanding the Practice

Teri: The presence of trauma in place doesn't mean beauty and inspiration cannot exist there too. We both witnessed it in our neighborhoods. Evidence of community-focused place-healing was naturally occurring in the wake of the unrest. The fact of these experiences and moments of visual joy point not only to the resilience of individuals and communities, but the natural tendency of bodies and the body of communities to heal themselves. We now see trauma-informed places as healing places. It's our opinion that addressing only the harm and the trauma may not be sufficient to provide space to heal. We suggest that healing places is a more holistic and reparative way to think about, design, and activate environments with beauty and meaning. Time is an essential component of that healing but cannot do the job alone.

So, our question becomes, what will help us all heal?

Our Realizations

Teri: As I look back on the summer of 2020, I'm aware of that view we have on the events that happened here, and ultimately had a global impact, and may still be too close to see fully. We cannot offer a one size fits all answer about what will promote healing in this city and in our communities, but we did witness the art of healing here and for the most part it didn't involve a wrecking ball.

It is telling what happened and what did not happen with the 3rd precinct building. It's still standing. It's like the proverbial elephant in the corner of the room in Minneapolis. The distinctly different zone of George Floyd Square is a conundrum for a different reason. The city and the community are struggling with how to approach these wounded places. There's been paralysis around certain spaces such as these charged and sacred ones, and a rush to surgically infill others. Today, the store George Floyd visited right before his death, Cup Foods, has changed its name to Unity Foods. Was this a healing gesture, or was this another example of removing something of cultural and historic significance? I'd like to consider that this may be a new way forward – a decentralized approach, and micro-moments of healing design, to urban design and redevelopment, or simply put, mindful and incremental doses of change to heal in incremental ways.

As chair of the Equity in the Built Environment committee for American Institute of Architects (AIA) Minnesota, I have noted that many of the practices and processes we designers and architects employ are fundamentally broken. Many of us agree that because they were originally formed on patriarchal, colonizing practices they need to change. I regularly ask myself and others in the design community, how can we create community-owned spaces of wellbeing in this context? And importantly, how can the practice of reclaiming destroyed spaces and buildings not overlook the very humanity of those whose lives have been altered? In this moment we have an important opportunity in our profession to review and critique our design process – to unlearn, revise and reconsider new, more equitable ways forward.

When I consider the idea of place-healing design, and as a practitioner in the space of trauma-informed design, I first look through the lens of the human experiences in that place. Dimensionalizing these principles should, in theory, mitigate some harmful and potentially long-lasting effects of trauma, but I believe the process itself is also an opportunity to correct the past inabilities of design to address public spaces with an equitable, healing lens. In redefining our approach to practice, we could generate repertoires of repair through the smallest and largest gestures we can muster.

Tom: Most planning processes, however, are meant to be short-lived, to generate a plan document, and then to allow moving into 'implementation'. One recent client said, 'We're done talking. Now we need to act'. Such temporal silo-ization is at its essence flawed but is all too common. Sure, the practicalities of designing a space or building and then constructing it require movement from one process to another. What are some ways we can keep talking while acting? Can a public space or building be left partially incomplete in ways that can be added to later? Or can spaces within structures be designed to reveal what is in the ground beneath them, and to promote dialog about the next stage of the work to be done? Can we learn to appreciate the value of process itself towards the generation or re-generation of a space or place and then continue that process as we learn to use, implement, or occupy that space or place?

The human body continuously rebuilds cells, muscles, and memories while adapting to new abilities and disabilities. The recovery of a community – as in the social bonds and cultural understandings – and the recovery of a physical space are both ongoing. Communities are ever evolving and spaces will be changed requiring new ways of using, occupying, and/or understanding their design symbolism. The condition or state we are supposedly rebuilding was itself a response to layers of events whether those responses were to acknowledge, honor, deny, or cover up. We're always building over or adding to what was there, but we need to learn from those earlier layers rather than simply cover or replace them.

Naming Possibilities

We want to underscore the importance of incorporating the voices of communities in acknowledgement of the very human toll of trauma in this place. Taking a more humanity-centered and asset-based approach to healing along with resistance and protest will be essential. We think a next step could include a narrative mapping of the sites of Minneapolis that were destroyed, and stories by those who were impacted in the community. This act, a virtual one, creates a non-place of visual listening to what is needed now and next, with an understanding that this healing tool will be ongoing. As with so many healing processes that cause us to stop in our tracks, this will allow us to focus more on being than doing.

At this moment our perspective has shifted from how we can be trauma-informed in the making or healing of places while finding a balance to also acknowledge acts of beauty and love. We see one of the most untapped opportunities right now is to create more moments and pockets of deep joy and inspiration. We acknowledge that the work continues, by regenerative practices like *Design As Protest* and healing equity work in urban planning practices like Creative Reaction Lab (CRXLAB), all attempting to address systemic oppression in the creation of places. This is our life's work. And interspersed in this work, we need the medicine of relief, joy and laughter. Our humanity and our wholeness depend on it.

References

Design as Protest (n.d.). Available: www.dapcollective.com/. [Accessed: 23 August 2023].

Grabowska, S., Holtzinger, C., Wilson, J., Rossbert, L., Macur, R. and Brisson, D. (2021). *Architectural Principles in the Service of Trauma-informed Design* [online].

Available: https://shopworksarc.com/wp-content/uploads/2021/10/Arc-Principles-in-the-Service-of-TID.pdf. [Accessed: 23 August 2023].

Million Artists Movement (n.d.). Available: www.millionartistmovement.com/. [Accessed: 23 August 2023].

Peyton Scott Russell (n.d.). Available: https://houseofdaskarone.com/. [Accessed: 23 August 2023].

15

OUR PLACE, OUR HISTORY, OUR FUTURE

Julie Goodman, Theresa Hyuna Hwang and Jason Schupbach

Introduction

Trauma-informed placemaking is a process of compassion and validation, honoring and releasing experiences that have negatively shaped personal identities, sustained harmful institutions, and constructed environments reflective of systems of oppression. Collective trauma in neighborhoods are more than just experiences held in the bodies of individual residents but in permanent scars in the streets, buildings, monuments and beyond. These infrastructural wounds are constant reminders of unprocessed grief that have easily become normalized and almost invisible to the impact on people and quality of life. At its best, trauma-informed placemaking disrupts, repairs, and reimagines public spaces by replacing areas of exploited power and harm into places of ease, connection, and care.

A trauma-informed approach restores wholeness. It is mindful of people's histories, social context, cultural landscapes and recognizing there is so much more than what appears on the surface. It is a process intentional of one's own actions in relationship to others to minimize feelings of unsafety. It is about creating spaces of imagination and connected growth, a creation of mutual support and collective care.

This chapter will examine the regenerative trauma-informed practices occurring in the City of Philadelphia in the post-pandemic era. Through an honest representation of these organizations' work, a broader story of the history of our place, our history and the potential future will be told. This article includes insights from the organizations doing this work and the lessons learned from it.

DOI: 10.4324/9781003371533-18

The Philadelphia Context

Philadelphia, America's sixth largest city, has been plagued by endemic poverty, racist policies and extreme violence in the years leading up to and post the pandemic. The pandemic has amplified the loss of black and brown business-owners and produced a major uptick in violence (Moselle, 2020).

Philadelphia is the largest poor city in the US, with a population of 1.6 million (US Census, 2020) and one in four families living below the poverty line (Bellesorte & Parkes, 2020). Like many American cities, it was founded through violent removal of Indigenous people – the Lenape tribe – and some of those tribal members remain but without any federally recognized territory. In addition, like other large American cities, its racial diversity has increased in the past decade (Frey, 2021). More than a quarter of residents are either immigrants or US natives with immigrant parents (Pew, 2018). There are hundreds of nonprofit cultural organizations in the city and surrounding region (Greater Philadelphia Cultural Alliance [GPCA], 2015), and a survey of more than 300 of them revealed total budget losses of 26% and a 35% decline in workforce due to COVID-19 (GPCA, 2021). Nationally, Americans for the Arts notes that the COVID-19 pandemic further disproportionately affected Black, Indigenous, and other People of Color (BIPOC) arts organizations (Cohen, 2021).

Exacerbating this situation is the reality that in the US, BIPOC-focused and -led arts organizations have historically received less philanthropic and government support than their peers (Sidford, 2011), resulting in long-term financial and organizational challenges distinct to these organizations (Helicon Collaborative, 2017). Recently, though many philanthropic organizations and public funding agencies in the US have turned their focus to supporting BIPOC arts organizations, the detrimental effects brought on by decades of underfunding continue to pose challenges for them. In addition, the Quaker traditions involved in founding the colonial city, which demand communal voice and decision making, continue to be cited by local practitioners as both a positive and a hinderance to the city advancing more equitable development (Baltzell, 2017).

Case study methodology

We recognize and validate that collective trauma arises as a result of institutional policies and systems of oppression such as racism, classism, and cis-hetero-patriarchy, and how they disproportionately hurt some over others. The impacts of these systemic issues are often an individual experience – everyone has a different and personalized situation. Collectively, they are generational and deeply rooted in both our bodies and our built environment. Harm has physical impacts that interrupts the brain, nervous system, digestive system, and other aspects of the body. A trauma-informed approach knows that sometimes behavior is not controlled but a biological response.

In seeking to understand how trauma-informed placemaking operates in Philadelphia, we found Bowen and Murshid's (2016, p. 224) analysis linking six principles of trauma-informed care to social policy to be valuable as a theoretical framework to explore the phenomenon. The six principles, described in Table 15.1, include safety, trustworthiness and transparency, collaboration, empowerment, choice, and intersectionality. We believe these principles align with best practices in creative placemaking (McCormick et al., 2020), with the added dimension of wellbeing. They also support the notion that in a trauma-informed setting, the 'place' in 'placemaking' is about more than physical structures. It includes dimensions of individual and community belonging, as well. Roberto Bedoya (2013) defined creative placemaking as, 'those cultural activities that shape the physical and social characteristics of a place', noting that the practice often has a blind spot in its, 'lack of awareness about the politics of belonging and dis-belonging that operate in civil society'.

We use Bowen and Murshid's principles as a framework to explore intersections of policy and practice in creative placemaking. Inspired by Crutchfield and McLeod Grant's recognition that, 'high-impact nonprofits engage in both direct service and advocacy' (2012, p. 32), we chose to illustrate the principles in action through case studies of Philadelphia cultural organizations. Many cultural organizations in Philadelphia engage in trauma-informed placemaking. Here, we present two whose work is deeply embedded in and informed by the populations they serve – Monument Lab, and The Village of Arts and Humanities. Understanding the context of these organizations' work both in terms of their location in Philadelphia and the specificity of the populations they prioritize within the city is important because as

TABLE 15.1 Bowen & Murshid's (2016) six core principles of trauma-informed social policy

Principles	Definition
Safety	Program efforts to ensure service users' physical and emotional safety, meaning reasonable freedom from harm or danger, and to prevent further traumas from occurring
Trustworthiness and transparency	The extent to which an organization maintains transparency in its policies and procedures, with the objective of building trust among stakeholders such as staff, clients, and community members
Collaboration	Agency staff view service users as active partners and experts in their own lives
Empowerment	Efforts to share power with service users, giving them a strong voice in decision making at individual and agency levels
Choice	Preserve meaningful choices for target population to maintain a sense of control
Intersectionality	Awareness of identity characteristics, such as race, gender, and sexual orientation, and the privileges or oppression these characteristics can incur

Bowen and Murshid observed, 'Although no population is immune to experiencing trauma, some types of trauma are disproportionately experienced by certain groups because of deeply entrenched structural inequalities' (2016, p. 223).

The information from these organizations was collected directly in partnership with them through surveys, interviews and original documents provided by the organizations.

Case study organizations: mission, vision and core values

Monument Lab

Monument Lab is a nonprofit public art and history studio who describes its work as, 'prototype monuments surrounded by spaces of learning and dialogue' (Carter, written response). The organization responded to our questions by email to share their history, philosophy, and practices. Though Monument Lab only hired its first Full Time Equivalent (FTE) employee and received its 501c3 (a United States corporation, trust, unincorporated association or other type of organization exempt from federal income tax) status in 2022, the organization has been active for over a decade, having emerged from a series of classroom conversations led by scholar and curator Paul Farber and professor and artist Ken Lum at the University of Pennsylvania that subsequently grew into a curatorial collective and ongoing community conversation. As the organization notes, 'Monument Lab exists in its current form due to societal demand', and since curating a citywide exhibition in 2017 it has, 'reached hundreds of thousands of people in person and over a million online, amidst the larger call for change in the monumental landscape and widespread acknowledgement that what is currently on the pedestal misrepresents our history.' Monument Lab defines a monument as, 'a statement of power and presence in public.' Its current activity involves participatory approaches to engage artists, students, educators, activists, municipal agencies, and cultural institutions in critical conversations around the past, present, and future of monuments (Table 15.2).

The Village of Arts and Humanities

The Village of Arts and Humanities (The Village) is one of the most well-known and documented placemaking projects in the United States. It is an arts and community development organization rooted in Philadelphia's deeply disinvested Fairhill-Hartranft neighborhood, founded in 1986 by artist Lily Yeh and the community. The Village was 'born in response to systematic and racially motivated disinvestment' and is 'a sacred space of practice and learning within a 36-year-old web of relationships, artworks, memories, and histories' (Kapust, a) The Village's creative campus – a network of non-contiguous art parks, gardens, and buildings spanning 4.5 acres of urban land – has bent and molded itself to respond to our

TABLE 15.2 Monument Lab mission, vision and core values

Mission	Monument Lab is a non-profit public art and history studio based in Philadelphia. Monument Lab works with artists, students, educators, activists, municipal agencies, and cultural institutions on participatory approaches to public engagement and collective memory. Founded by Paul Farber and Ken Lum in 2012, Monument Lab cultivates and facilitates critical conversations around the past, present, and future of monuments.
Vision	Monuments must change.
	At Monument Lab we envision a society where monuments are dynamic and defined by their meaning, not by their hardened immovable and untouchable status.
	To illuminate how symbols are connected to systems of power and public memory, we engage critically with our inherited monument landscape and work joyfully with artists, organizations, and movements to imagine the next generation of monuments.
	By disrupting the status quo of how monuments are made, preserved, and interpreted, we hope to contribute to a future society defined by joy, regeneration, and repair.
Core Values	As an organization we aspire to make all of our decisions, including how our resources are obtained and allocated, based on these values:
	Art at the Core: We are artists and believe that art is critical to how we understand, experience and imagine the past, present and future, so we engage with artists whenever possible.
	Process Matters as Much as Outcome: We aim to take time for quality, move at the speed of trust, and cultivate relationships beyond the life of a single project. We strive for a powerful final result that is fueled by intentional and iterative processes. We balance urgency and timeliness with purposeful reflection.
	Intersectionality for Transformation: We are driven by anti-racist, de-colonial, feminist, queer, working class, climate-conscious, and disability-justice perspectives, and know all forms of oppression must be dismantled for us to truly get free.
	Collaboration With Boundaries: We are committed to working with individuals, organizations, and institutions, but only build trusting relationships with those that share values and honor a collective sense of process.
	Integrity With Accountability: We intend for our words and actions to be aligned. If it comes to our attention, from staff, collaborators, or anyone else that we're out of whack, we pause, process and apologize to determine the best way forward.
	Culture of Care: To maintain our own wellbeing and support collective healing, we acknowledge the presence of trauma while centering a culture of care, repair, harm reduction, and safety for all. Expressing joy, having fun, and sharing nourishment are essential parts of our organizational culture.

TABLE 15.2 (Continued)

> **Citation and Compensation:** We acknowledge all contributors (including artists, students, educators) for the fullness of their efforts. We value local knowledge and expertise while building strategic coalitions across locations. This takes form in a number of ways including as payment, citation, credit, authorship, or ownership, as decided with each individual or collective ahead of time.
>
> **Wisdom and Learning:** As students and teachers with an endless curiosity, we are perpetually learning from others and sharing information. We believe that wisdom and intelligence come in many forms. We are committed to challenging white-centric, hegemonic norms and neutrality.

TABLE 15.3 The Village mission, vision and core values

Mission	Our mission is to support artists and Black community residents to imagine, design, and build a more just and equitable society.
Vision	Ultimately, we envision a thriving Fairhill-Hartranft neighborhood of Philadelphia where Black neighborhood residents are leaders, stewards, and ultimate beneficiaries of reinvestment, recovery, and revitalization.
Core Values	**Build Trust:** We create, nurture, and pass on trusted relationships through our work. We hold ourselves accountable for our past, present, and future actions through these relationships.
	Maintain Consistency: We show up, again and again, to listen, share our resources, and do what we say we're going to do. In a rapidly changing city, this consistent rhythm instills confidence that The Village will be here three, five, fifty years from now in a rapidly changing city. Confidence in longevity allows us to invest our whole selves today.
	Seek Boldness: We are unwilling to invest our precious time, energy, and resources to maintain the status quo. Instead, we create bold ideas that generate motivation, resources, and a new sense of what is possible. We set ambitious and achievable goals that move us closer to achieving transformational social change.

community's ideas, creative products, and experiences.' (Kapust, b) They work with over 1500 community members to achieve their goals (Table 15.3).

The Village uses the following language when describing the site and its importance, 'The "aesthetics" of The Village – both invisible and visible – insist on the process as product; collaboration across experience; shared space as sacred; sampling the culture of the African diaspora in a contemporary urban and Philadelphian context; grounding in history to imagine the future; fluid understandings of expertise; and nimble responsiveness to the needs and conditions of the neighborhood.'

These aesthetics 'evolved out of The Village's creation story, collaboratively written by artists and community members. More than 40 years ago, Arthur Hall, a renowned dancer and civil rights activist, erected Ile Ife Black Humanitarian Center on the site of what is now The Village's main building. He invited any neighbor to become experts in dance and drumming regardless of skill. Twenty years later, Village founder Lily Yeh worked with community members to infuse and evolve Ile Ife's aesthetics within arts programming and the physical landscape. Together they built murals and undulating walls that referenced symbols of the African diaspora and used novice-friendly media like mosaic and concrete to enable neighbors to participate in the building process, thus rendering spaces sacred through shared creation.'

The Village's work generally falls into three areas – social justice, youth leadership and equitable neighborhood revitalization. The diversity of their work in these areas ranges from the *100 Families Initiative*, in which they 'seeks to anchor 100 Families with 100 Homes and 100 Jobs for 100 Years in the Fairhill-Hartranft neighborhood of North Philadelphia;' to Village Industries, a 'pre-employment program for teens and middle-school youth that combines art and justice with social enterprise to build self-sufficiency and actualize creative power' that serves over 350 teens a year; to *The Civic Power Studio*, where in 2022, 'The Village completed its renovation of three row homes that had been vacant for more than two decades. The properties, formerly owned by the Philadelphia Housing Authority, have been transformed into a gathering space and media lab at the center of its creative campus. Here, community members, artists, and advocates collaboratively study how power operates, create works that amplify existing neighborhood power, and design new systems of power that are rooted in care rather than control'.

How the case study organizations reflect the six principles of trauma-informed social policy

Each of the organizations reflects Bowen and Murshid's six principles of trauma-informed social policy. A text analysis of the mission, vision and core values statements revealed that the five most common words across the two case study organizations were artists, collective, future, relationships, and resources. These perfectly describe the work that they do. Examples where this can be observed are described below.

Safety

Safety is a key principle of both of the case study organizations' work, and an element of their founding purpose. For Monument Lab, safety is a core value found in the idea of Collaboration with Boundaries. The organization's statements about Integrity with Accountability and a Culture of Care also speak to safety. Providing clarity about its engagement practices fosters safe spaces for people to deeply engage with Monument Lab's processes. As one might suspect, for The Village, safety is the residents of this neighborhood's top priority, so programming that builds safety is of utmost importance. One example can be found in the *Civic Power Studio*'s work, where the Village directly calls out the need to start with safety, stating:

'Traditional public safety strategies have largely failed or exacerbated our community's problems. But our neighbors have developed unique strategies for safekeeping tailored to the needs of our environment...What if we could scale these solutions, and invite partnership from police and the city? What does a public safety strategy look like when it is built from the inside of a neighborhood, rather than imposed from the outside? What is public safety rooted in care, rather than control, and borne out of a neighborhood knowing and caring for itself? Ms. Nandi put it this way: "The Civic Power Studio *is like the root of the tree. From the roots grow respect. Many various projects are the branches. And public safety is the shade that the tree gives the community...*" *This is the time to say that our neighborhood's wisdom is as powerful as any law handed down from city, state or national government, and to act from and invest in this knowledge. The guardians on the wall of Angel Park are watching. We're working to live up to their promise.'*

Trustworthiness and transparency

Establishing trustworthiness is an ongoing process to which transparency contributes. Both case study organizations work to consistently and clearly communicate strategies, goals, and processes for their internal and external stakeholders in order to manifest these principles in their work. One indicator of Monument Lab's commitment to this principle is that since 2018, the organization has been consistently sought out by other cities to 'help them with their monuments,' because of their community-engaged, people-centered process. Monument Lab's core values of Process Matters as Much as Outcome, Collaboration with Boundaries, and Citation and Compensation illustrate and communicate the organization's deep commitment to trustworthiness and transparency. Likewise, 'Built Trust' is one of the core values of The Village. One of their current projects exemplifies this work:

'Our current project, Art Space as Emergent Response: The Village's Futures Gallery, *exemplifies these strategies. During 2022–2023, we will modify our main cultural building to include a hybrid indoor/outdoor gallery space at the terminus of the Germantown-Lehigh commercial corridor. Through a thoughtful visioning and planning process, led by Afrofuturist Li Sumpter, we are melding inspiration from our peers, ideas/assets from our community, and lessons from experiments at our growing edge to create a uniquely permeable exhibition space that amplifies community expression and tradition and builds communal*

practices of art-making, curation, exhibition, and conviviality. This Futures Gallery *will serve as a powerful cultural and economic tool for resilience—for community residents past-present-future, for The Village, and for Black working artists across the City—one that combats the physical and cultural erasure of Black communities. It is an equitable, long-term investment in presenting and preserving artistic manifestations of power, loss, joy, struggle, and ingenuity made by Black people in our neighborhood and across our city. Finally, it makes legible—for our supporters and skeptics—the power of art as an indispensable tool for equitable community preservation and revitalization.'*

Collaboration

Collaboration is an integral component of both organizations' work. Their respective approaches to collaboration illustrate how their core values align with this principle in action.

ARCH – Art Remediating Campus Histories – illustrates Monument Lab's collaborative process. *ARCH* is a project in partnership with Bryn Mawr College that builds on, 'previous and ongoing College-supported efforts by students, staff, alumni, and faculty to reveal and repair harm, ensuring a reckoning with Bryn Mawr College's history and a clear-sighted look at the way to a future of inclusion and reconciliation.' During the research phase of the project, Monument Lab, 'posed a central research and engagement question to the campus to guide the project: *What stories are missing from Bryn Mawr College?* Rather than seeking a single "winning" response to guide the future commission, we sought to enact a broad collective and creative mapping of the campus across time. The central question aimed to serve the reckoning with, dismantling, and reimagining of narratives of campus history that may offer an artistic and reparative vision for its future.' The process included student researchers, current students, faculty, staff, and alumni, who were invited to, 'map the campus according to their own perspectives and experiences,' and respond through drawing, sketching, and/or describing their maps. Through the process, 'We witnessed a composite mapping of the campus that highlights its layered multiplicity – places of trauma and transformation, erasure and memory-keeping, individuality and collectivity all exist simultaneously.' Artist Nekisha Durrett has been selected to create her proposed monument, titled *Don't Forget to Remember (Me)*. It will be placed in the Cloisters at the center of Old Library on Bryn Mawr's campus to contribute to a broader effort of historical reckoning at the College. In keeping with its core value of Citation and Compensation, Monument Lab compensated all semifinalist artists for their proposals and provides straightforward acknowledgement of the project budget and Durrett's artist fee.

The Village defines all of its work through 'Co-Creation,' saying, 'When we co-create, we develop mutual value (in the form of programs, products, policies, etc.) in collaboration with community members, stakeholders, and other experts.

Co-creation is a manifestation of shared power where ideas are challenged, developed, and affirmed together, rather than kept to oneself or one group.'

Empowerment

For both Monument Lab and The Village, empowerment of the populations they serve is part of their reason for existence. The examples below demonstrate their respective commitments to this principle. Monument Lab's *raison d'etre* is about empowerment, seen in its purpose of collaborating to 'make generational change in the ways art and history live in public.' All of its projects, partnerships, and programs embody this idea. The recent 'larger call for change in the monumental landscape' further underscores the value of Monument Lab's efforts to empower community dialogue and action. The Village is host to an important and groundbreaking organization called *The People's Paper Co-op* which embodies the core values of empowerment and choice. Co-founded in 2014, The *People's Paper Co-op* (PPC) is:

> 'a women-powered, women-centered art and advocacy project…The PPC connects women returning home from prison with artists, mentors, and allies to imagine, design, and advocate for a more just and free society. Together, these courageous women create arts and advocacy campaigns that aim to shift public awareness toward supporting an end to the incarceration of women; toward exploring alternatives to prisons; and toward understanding that incarceration is a continuation of policies rooted in racism, oppression, and control.'

Their work is exemplified by how they have helped free dozens of women held on cash bail:

> 'Each year, PPC creates poetry and artwork that serves as a campaign to raise funds for the Black Mama's Bail Out. We've sold more than $170k in artwork and donated all proceeds to the Philadelphia Community Bail Fund to bail out Black mamas and caregivers. PPC Fellows and advocates stage a 'Welcome Home' celebration outside the jail to greet and welcome women as they are released. Many of our fellows have helped free women with whom they were recently incarcerated.'

Choice

The organizations studied respect and value the principle of choice as a means to empower those they serve. Monument Lab's projects enable participants to make meaningful choices that guide the work. One example of this process is seen in *Re:Generation*, a nationwide public art and history project that resulted from an open call for interdisciplinary and intersectional groups of people shaping the next generation of monuments reckoning with and reimagining public memory in their local community. Ten project sites each received $100,000 to collaboratively pursue a commemorative campaign rooted in the living history of a neighborhood, city, or region, supported by the Mellon Foundation's Monuments Project. Likewise, The Village's support of The *People's Paper Co-op* enables women directly affected by incarceration to lead advocacy efforts and direct resources to others in similar circumstances. In doing so, The Village empowers these women to create social change.

Intersectionality

The organizations studied understand that intersectionality reflects the existing and changing dynamics of the populations they serve, as well those of the society in which we all exist. For these organizations, intersectionality is both a respectful and inclusive approach to the diversity found within those populations, as well as an opportunity to address systems of power, privilege, and oppression.

Monument Lab's commitment to intersectionality is evidenced throughout its work, most prominently in the statement that the organization, 'exists because what is currently on the pedestal misrepresents our history,' and in its core value of Intersectionality for Transformation. The organization's core value of Wisdom and Learning further supports the inclusive practices in its work. Monument Lab recognizes that its participatory process enables greater awareness of the identity characteristics of those engaged in its work and the privileges or oppression these characteristics can incur. As the organization notes about the research process for *ARCH*: 'In analyzing this kind of collective knowledge—crowd-sourced, varied, personal, and reflective of individual experience – we highlight the oft-unquestioned falsehood that data is neutral and without messiness.'

The Village organizes its work under the umbrella of 'futures thinking,' which clearly outlines their desire to be consistent in their work around intersectionality. They state:

> *We imagine and consider an unlimited and unencumbered range of possible futures. We do not limit ourselves by the constructs, tools, and policies created through white supremacy, heteronormativity, and capitalist patriarchy; we do not limit ourselves by the current availability of resources; we do not limit ourselves by what is. We imagine and create what can and should be.*

Themes that Emerged from the Case Study Organizations

Across the case studies, we observed many commonalities of practice which we refer to as themes here. The following themes emerged:

Theme 1: Process is as important as product. Often placemaking is about outputs – what is being built, produced, or accomplished as a final product of the work. These organizations remind us, however, that the process is crucial to any outcomes achieved. Clarifying and communicating the process by which their work occurs is what enables these organizations to build the safety, trustworthiness and transparency, and empowerment described in Bowen and Murshid's principles.

Theme 2: The language of healing. Their approach to process also distinguishes how these organizations speak about their work. Though they all engage in trauma-informed placemaking practices, none of them use this terminology to describe what they do. In addition, words like 'marginalized', 'disenfranchise' and 'minority' which are often used to describe the communities they work within are not used at all in their language. Rather, they speak of joy, care, wellbeing, trust, boldness, justice, equity, thriving, and centering the voices and lived experiences of the people and communities they serve. These are the ways they address trauma in their work.

Theme 3: Centering of BIPOC and Queer voices. The organizations all center people of color, not the white patriarchy. This centering is seen in their organizational structures, policies, and practices as well as in their programmatic activity and the outputs of their work. They create opportunities specific to the needs and interests of those they serve, while also promoting the resulting outputs to the broader community. In this way, they engage in both direct support and advocacy that supports their mission and purpose. The organizations also center the queer community in this same way.

Theme 4: Move at the speed of care. Both of the organizations describe the urgency of their work to serve communities in this moment, but all speak of moving with care and at the speed of trust, not in the rigid timelines that capitalism can define for much of American society.

Theme 5: Politics matter. These organizations consciously claim a political frame for their work. They recognize and explicitly name the political power (and its impacts) that surround their work on a daily basis.

Theme 6: The work adapts to its time. Recognizing the work will continue to evolve and the situation surrounding the work will change, both of the organizations engage in a flexible and adaptive mindset. The Village is on second or third generations of leadership and has a clear understanding of the need to build the next generation of leadership.

Theme 7: Values alignment. The case study organizations' core values mirror Bowen and Murshid's principles of trauma-informed social policy. The organizations provide direct statements demonstrating how these values are manifest in both the process and the products of their work. These statements can be found both in descriptions of the organizations' internal structure and policies, as well as in their external programmatic activity with others. Again, this speaks to elements of trustworthiness and transparency that foster safety and empowerment in settings where collaboration, choice, and intersectionality are part of the process.

Theme 8: Beyond the place. By focusing their attention on process, on their core values, and on the voice and lived experiences of the people they serve, these organizations expand the concept of 'place' in 'placemaking' to encompass individual and community wellbeing. Author Theresa Hwang, as an expert practitioner in trauma-informed placemaking, captured this theme best when speaking about her own experience:

> *Often, communities of Color have constructed trauma-sensitive, healing-centered places intuitively. We have reclaimed and collectively determined solutions to meet our own needs with our own resources, often we just need to resource existing efforts rather than introduce external interventions. When I am in spaces that are truly trauma aware, I feel at ease and reflect deeply, and there is the ability to move, release, and hold space and compassion for others. It is a space where I feel seen, heard, and most of all, respected. These are the spaces that I long for and the hope that trauma-informed placemaking produces.*

Advice for Practitioners

The case studies and authors' experiences reveal practical advice about how to center trauma-informed placemaking practice in organizational practice.

Acknowledge Trauma, Sharing Stories and Timelines

To name the collective harm that a community has experienced, there needs to be a process to acknowledge that the trauma has occurred. This process cannot be rushed or forced. Trauma embedded into neighborhoods is complex and difficult to pinpoint to a singular community development issue. Hence, there needs to be space for people to process their experiences together, which includes controlling the pace and details of the experience, and time to craft their own narrative of the impact. If you are not from the community, it is an honor to witness these conversations and awareness of your presence is important. As a practitioner, it is not our role to judge which stories are valid, and there are times we assume community members are sharing unrelated experiences, but it may not be true. As a team of practitioners, it is not our position to tell what stories are relevant and not; silencing community members only may continue patterns of harm. Additionally, stories and honest feedback from residents need to be valued, incorporated, and affirmation there will be no retaliation or consequence for participation. This is the process of building trust and relationship, not just facilitating an agenda. We can bracket or create specific containers for sharing stories that include time constraints, content warning requests, and ensuring multiple voices are included and a few do not just dominate the space available.

Understand the Biology and Create Safe Spaces for Processing

Trauma produces a situation where people can operate from a place of survival and people instinctively react from a place to create greater safety for self and move to fight, flight, freeze, and fawn responses. The nervous system is triggered and automatic responses from the reptilian part of the brain are directing. To regain a place of creativity and imagination, the higher parts of the brain, the prefrontal cortex especially, can only be accessed when grounded and connected. Sometimes placemaking programs are directed at fostering social cohesion and connection, as a means to stabilize, increase safety, and ultimately return the nervous system to a state of ease rather than alert. Jumping into solutions in creative processes may not be readily available for people who have experienced constant, chronic, and complex trauma. Prior to generative processes, grounding activities that foster safer environments may be necessary. It is important to prioritize personal safety over the safety and protection of property.

Have Experts on Hand

People with training to support people who may experience retriggering or re-traumatization should be on hand to provide support in case people move through a trauma response. Sometimes this takes the form of a service provider in the room, a mindfulness facilitator, or a culturally attuned healer that has effective methods of regulation and de-escalation.

Engage in Self-reflection about your Processes and Practices

Essential questions to ask yourself as one engages in trauma-informed placemaking are:

- Am I moving away from a power-over model towards a power-with process?
- What has been the role of artists, designers, policy makers, and others in the trauma cycle?
- What is my proximity to the issues and direct lived experience?
- Are there peers and people with direct lived experience as part of the leadership team?
- What parts of my own identity and my role sustain a dynamic of oppression and status quo?
  o Can white-owned organizations effectively address racial trauma?

 o Can cis men lead talks about safety for trans people in public spaces?
 o Am I reinforcing cultural practices rooted in white supremacy?
- Are activities truly inclusive with minimized barriers for full participation?
 o Are all workshops conducted in native languages or do I require translation?
 o Is my workshop space supportive for people of all abilities, both physical and cognitive?
- How do we make space to be whole together and heal collectively?

Conclusion

Philadelphia has long been a leader in community engaged practices that can be seen as inspirational examples for those working to better their place across the world. It is clear from the case studies that organizations in Philadelphia are beyond moving from 'What's wrong with you?' to 'What happened to you?' to 'What makes you thrive?' and imagining beyond the limitations of the present. In 2021, Monument Lab and The Village of Arts and Humanities came together for an exceptional temporary exhibition that created 'a place to understand how residents over many generations sustained staying power despite systemic forces undermining them.' (Thackara, 2021). This exhibition brought new considerations and lessons learned for both organizations about healing practices and exemplifies the virtuous circle of learning occurring in Philadelphia. These exceptional organizations carry many lessons for all practitioners about how to center the human and an organization in its community to do the healing work that is needed in our society.

References

Baltzell, D. E. (2017). *Puritan Boston & Quaker Philadelphia*. Abingdon: Routledge.

Bedoya, R. (2013). 'Placemaking and the Politics of Belonging and Dis- Belonging', *GIA Reader*, 24(1).

Bellesorte, M. & Parkes, F. (2020). 'Philadelphia's Poverty Problem Is Bigger than You Ever Imagined', *Generocity Philly* [online], March 16, 2020. Available: https://generoc ity.org/philly/2020/03/16/philadelphias-poverty-problem-is-bigger-than-you-ever-imagi ned/. [Accessed: 13 August 2023].

Bowen, E. A. & Murshid, N. S. (2016). 'Trauma-Informed Social Policy: A Conceptual Framework for Policy Analysis and Advocacy', *American Journal of Public Health* 106(2): 223–29.

Cohen, R. (2021). 'COVID-19's Pandemic's Impact on The Arts: Research Update October 5, 2021', *Research 1-Page Summaries*. Washington, DC: Americans for the Arts. Available: www.americansforthearts.org/node/103614. [Accessed: 13 August 2023].

Crutchfield, L. R. & McLeod Grant, H. (2012). *Forces for Good: The Six Practices of High-Impact Nonprofits*. Hoboken: John Wiley & Sons, Inc.

Frey, W. H. (2021). '2020 Census: Big Cities Grew and Became More Diverse, Especially among Their Youth', Brookings [online] 28 October 2021. Available: www.brooki ngs.edu/research/2020-census-big-cities-grew-and-became-more-diverse-especially-among-their-youth/. [Accessed: 13 August 2023].

Greater Philadelphia Cultural Alliance (2015). *Portfolio: Culture Across Communities*. Available: https://philaculture.org/research/2015-portfolio-culture-across-communities. [Accessed: 13 August 2023].

Greater Philadelphia Cultural Alliance (2021). *Fall 2021 COVID-19 Impact Study*. Available: https://philaculture.org/COVIDFall2021. GPCA, December 2021. [Accessed: 13 August 2023].

Helicon Collaborative (2017). 'Not Just Money: Equity Issues in Cultural Philanthropy'. Available: www.giarts.org/sites/default/files/not-just-money-equity-issues-cultural-phila nthropy.pdf. [Accessed on: 13 August 2023].

McCormick, K., Hardy, J. & Utter, M. (2020). '*Creative Placemaking: Sparking Development with Arts and Culture*'. Washington, D.C.: Urban Land Institute. Available: https:// knowledge.uli.org/-/media/files/research-reports/2020/creative-placemaking-v2.pdf. [Accessed: 13 August 2023].

Monument Lab. *Written response to interview questions*. March 2023. Provided by Amelia Carter.

Moselle, A. (2020). 'Philly Families Live with Crushing Heartbreak as COVID-19, Gun Violence Crises Collide', *WHYY* [online] 16 July 2020. Available: https://whyy.org/artic les/philly-families-live-with-crushing-heartbreak-as-covid-19-gun-violence-crises-coll ide/. [Accessed: 13 August 2023].

Sidford, H. (2011). Fusing Arts, Culture, and Social Change: High Impact Strategies for Philanthropy. Washington DC: *National Committee for Responsive Philanthropy*. Available: www.giarts.org/sites/default/files/Fusing-Arts-Culture-Social-Change.pdf. [Accessed: 13 August 2023].

Thackara, T. (2021). 'They Are Their Own Monuments', *New York Times* [online], May 3, 2021. Available: www.nytimes.com/2021/05/03/arts/design/monument-lab-philadelp hia.html. [Accessed: 13 August 2023].

U.S. Census Bureau (2022). Annual Estimates of the Resident Population. Available: https:// data.census.gov

Village of Arts and Humanities. *Organizational description documents*. Provided by Aviva Kapust.

Village of Arts and Humanities. *Power Safety Healing, Civic Power Studio 2023*. Provided by Aviva Kapust.

Village of Arts and Humanities. *Village Overview 2022*. Provided by Aviva Kapust.

Crafting Spaces of Resilience and Restoration

16

TRAUMA-INFORMED PLACEMAKING

In Search of an Integrative Approach

Joongsub Kim

Introduction

The COVID-19 pandemic and George Floyd's death in 2020 refocused attention on structural disparities and their associated trauma, which still exist today in underserved communities. Advocates of trauma-informed approaches to design and planning argue that consideration of history and lived experiences of residents in Black and Brown neighborhoods is essential to building a just society. On the other hand, others call for bottom-up planning or resident participation to promote equitable and sustainable community development. In this chapter, we, Detroit Studio, ask what role design and planning can play to help traumatized people cope with their loss and change, considered here through a qualitative study within the city. Using the concepts of social justice, kindness (Forester, 2021), and hope (Inch & Crookes, 2020), this theory-building chapter aims to explore an integrative framework, taking the form of a six-point plan, thereby addressing trauma in the built environment. Our plan links these concepts to place-cultivating.

Defining Trauma

Trauma in underserved communities of color can manifest in various ways and can be linked to systemic issues related to urban planning and design. The history of the built environment in the United States has taught us that adoption of racially motivated zoning ordinances for segregation of communities in the early 20th century, including redlining in the 1930s, urban renewal in the 1940s and 1950s, and highway construction starting in the 1950s are some of the most

DOI: 10.4324/9781003371533-20

well-known historical events that caused trauma in underserved communities (Rothstein, 2017; Sugrue, 2014). Zoning laws and building codes have played a significant role in racialized practices. For example, urban planning and design that perpetuate environmental injustices, such as locating polluting industries in low-income communities of color, can lead to chronic health issues such as asthma and cancer. Poorly designed public spaces can also contribute to social isolation and lack of community cohesion, which can have a negative impact on mental health (Bullard et al., 2019; Saha, 2016). Gentrification (Wyly & Hammel, 2019), displacement, limited access to essential services (Desmond & Shollenberger, 2015), lack of green spaces (Rigolon, 2016), policing and surveillance (Eubanks, 2018), disinvestment, neglect (Galster & Hayes, 2017), and urban sprawl (Pickett et al., 2004) are all trauma-linked contemporary phenomena.

Symptoms of Trauma

Trauma experienced by underserved communities of color in the built environment can manifest in a variety of overlapping ways and can include:

1. Physical symptoms – including headaches, body aches, fatigue, sleep disturbances, and digestive problems, as a result of exposure to environmental toxins or lack of access to healthy food and safe places to exercise. (Jones & White, 2020; Jackson & Green, 2019; Smith et al., 2019).
2. Mental health symptoms, such as anxiety, depression, post-traumatic stress disorder (PTSD) (Bryant-Davis et al., 2019); with associated feelings of hopelessness, helplessness, and worthlessness; as well as difficulty concentrating and making decisions (Williams et al., 2019).
3. Behavioral symptoms, which impact behavior, leading to withdrawal, isolation, substance abuse, and other destructive coping mechanisms (Smith et al., 2022; Brown et al., 2021).
4. Social symptoms, such as difficulty forming and maintaining relationships, struggles with trust issues, and feeling isolated from the surrounding communities (Williams et al., 2013; Carter et al., 2017).
5. Cultural symptoms, which are a result of historical and ongoing discrimination, racism, and marginalization (Carter, 2007), can lead to a sense of cultural disconnection and a loss of cultural identity (Paradies et al., 2015).

While the treatment of these symptoms requires health care and medical expertise, this list is useful in helping planners and designers recognize the impact of their profession and how to improve their practice.

Turning now specifically to placemaking, we introduce six principles of placemaking that have been favorably mentioned in the literature, and we examine ways, using reflective analysis, that effective placemaking can address trauma.

Placemaking

Placemaking is an approach to urban planning and design that focuses on creating vibrant and engaging public spaces within the built environment (Montgomery, 1998). It involves transforming ordinary spaces into meaningful places that foster community interaction, enhance the quality of life, and reflect the local identity and culture (Gehl & Gemzoe, 2001). Placemaking goes beyond the physical design of buildings and infrastructure; it considers the social, cultural, and economic aspects of a place to create environments that people enjoy and feel connected to (Thompson & Johnson, 2020). To these ends, placemaking emphasizes the following principles:

1. Community engagement – actively engaging the community and stakeholders throughout the planning and design process (Reed, 2012) to ascertain their needs, aspirations, and preferences, and incorporate their input into the decision-making (Carmona et al., 2017).
2. Human-scale design – creating spaces that are designed for people and that take into account factors such as walkability, public transportation, and the provision of amenities like seating, lighting, and greenery (Gehl & Svarre, 2013).
3. Mixed-use and activation – incorporating diverse functions within a place, to promote 24/7 activity through activation strategies such as events, markets, public art installations, and programming.
4. Identity and sense of place – highlighting and celebrating the identity, heritage, and culture of a place, often through preserving historic buildings, incorporating local materials and craftsmanship, or integrating public art and cultural symbols that reflect the community's identity (Smith & Johnson, 2020).
5. Sustainability and resilience – promoting environmentally sustainable practices in the design and operation of – should read as 'operation of place and fostering social resilience by creating inclusive spaces that accommodate diverse populations and promote social cohesion (Colantonio & Dixon, 2011).
6. Iterative and adaptive approach – allowing, in an ongoing process, flexibility, experimentation, and adaptation based on user feedback and changing needs over time (Johnson et al., 2020).

Building the Six-Point Plan of Place Cultivation

There are several important drawbacks to the application of placemaking in the production of built environment for underserved communities of color, and some key partially overlapping criticisms include:

1. Gentrification – this process can lead to displacement of the existing residents, particularly from marginalized communities. This can lead to loss of affordable housing and cultural displacement (Atkinson & Bridge, 2019; Davidson & Lees, 2018).

2. Unequal distribution of resources – communities of color, particularly those that have been historically marginalized, may not have access to the same resources or funding as more affluent neighborhoods, making it difficult for them to undertake placemaking initiatives (Jennings & Johnson Gaither, 2015; Payne-Sturges et al., 2018).

3. Tokenism – placemaking can be used as a tokenistic approach to community engagement, where community members are invited to participate in the design process, but their input is not valued or acted upon, perpetuating inherent power imbalances (Sturup, 2018).

4. Lack of diversity in design teams – placemaking efforts may not adequately reflect the diversity of the community, either in terms of the people involved in the process or the design outcomes (Vanky et al., 2019).

5. Lack of sustainability – placemaking initiatives may prioritize short-term gains over long-term sustainability. Without careful consideration of the environmental impact of new developments or the long-term needs of the community, placemaking initiatives may exacerbate existing social and environmental problems (Brown & Kyttä, 2014; Loukaitou-Sideris & Ehrenfeucht, 2016).

6. Limited scope – focus may fall on making aesthetic improvements rather than addressing deeper structural issues that may be affecting a community.

7. Neoliberalism – a complicity in urban development that prioritizes market-based solutions and individualism over collective community solutions and government intervention (Evans, 2017; Marcuse & Van Kempen, 2017).

8. Power dynamics – reinforcing existing power dynamics between different groups in a community, with more privileged groups that have more influence and control over the design process and outcomes (Smith, 2018).

At this juncture, we propose a six-point social justice-informed approach to make placemaking more effective in addressing trauma in underserved communities:

1. Community Engagement – Community engagement is necessary but is insufficient unless it is long term in nature and built on long-term working relationships with community partners.

2. Cultural Relevance – Development should be appropriate to a local community's unique social, cultural, and historical contexts and development should be determined by local actors and led by them.

3. Equity – Underserved communities lack equitable access to basic services; to promote equity, a development needs to be aligned with, connected to, or supported by a citywide community development support system via strategic partnerships.

4. Public Participation – There must be a process that encourages public participation in planning and design, and the process should be inclusive, transparent, democratic, and engaging.

5. Governance – Community engagement and public participation are necessary components of these efforts, but by themselves they are insufficient to break power imbalances between service providers (e.g., planning and design experts, public officials) and service recipients (e.g., residents). Community actors or representatives should sit in key places in a governing body of community development.

6. Kindness – To make these elements more applicable to placemaking or to trauma-informed planning and design practices, we incorporate the concepts of kindness, hope – expanded upon in the case study below – and place-cultivating. We suggest that place-cultivating is a more effective way to rework trauma-causing environments and address already-existing trauma in the community. An effective place-cultivation must address all five elements mentioned above and also exemplify the concepts of kindness and hope.

In the following case study, we describe a service learning-based urban design project that we conducted at our community-based studio in Detroit to illustrate how the six elements of place-cultivating can be achieved and how such place-cultivating can help address trauma in underserved communities.

Case Study: Hope District Project

Hope District is in Detroit, Michigan, in the United States, and has a community of approximately 2,500 residents in about 50 block areas. Hope District consists predominantly, of low-income Black residents; the area includes extensive swathes of vacant properties and chronic blight that developed over many decades. More recently, gentrification and displacement, due in part to the area's proximity to the improved Detroit Riverfront and downtown Detroit, have become important influences. While there have been sporadic small-scale community-led improvements, there is little evidence to suggest that long-term investments are in the pipeline for the community anytime soon. In 2010, a local community development corporation (CDC) known as Friends of Tri-County (FTC), contacted our community-based studio, Detroit Studio, seeking to partner with us to develop a sustainable and resilient community master plan for the community.

Guided by the community, the masterplan project had three goals: to create a series of productive landscapes (e.g., industrial-scale urban farms, community gardens) and various green infrastructure elements (e.g., bioswales, rain gardens); to promote mixed use development (e.g., live-work units, commercial, recreational, and cultural uses) and other uses that support productive landscape and existing housing; and to create community centers (e.g., community hubs that house productive landscapes, mixed-use developments, and other amenities such as parks) at several locations that are easily walkable from nearby residential areas in the study community.

Led by our Six-Point Plan of Place Cultivation, we applied its approach to the Hope District project:

1. Community Engagement

Our studio team and FTC began meeting a year before the project began and worked together to develop the project, defining distinct and shared delegations of activity and coordination.

2. Cultural Relevance

The project's overall vision and goals were determined primarily by our main partners, FTC and key community actors, including schools, churches, businesses, and urban farming groups based in the study community and its vicinity.

3. Equity

A community resident remarked that 'one example of promoting equity is a person's empathy and willingness to give a physically handicapped resident a ride to a community meeting because it is a right thing to do... that person may be any member of our society or community and could be a neighbor, a cop, a cab driver, or a city official'. Our master plan promoted equity by considering a reasonably balanced distribution of community services (e.g., community enters) and amenities (e.g., parks) and making them easily accessible from various residential blocks. More attention was given to blocks where the most needy residents, such as impoverished elderly, single parents, veterans, or children live.

4. Public Participation

FTC coordinated public participation, because that group is more familiar with the history, culture, assets, trauma, and lived experience of residents; with the help of FTC, we made the planning and design process more interactive, engaging, and educational so that residents, regardless of age or education level, could participate in the process in an inclusive and democratic manner.

5. Governance

We positioned a technical knowledge generator – our studio – versus a local knowledge generator, FTC, to be equal partners. Our studio focused on tasks like mapping and case studies, while FTC led resident-lived experience-based site analysis and storytelling to develop a series of alternatives for the master plan and the focus site design. While we used tripartite community engagement strategies of informing, consulting, and deciding, to promote shared governance, we put more emphasis on deciding (allowing community representatives to make key decisions) than on informing and consulting them. Addressing perpetuated trauma that has resulted from structural disparities in the community requires additional measures, and to that end, we use the concepts of kindness and hope coupled with place-cultivating. We offer deeper explanation of our approach here.

6. Kindness

Forester (2021) offers a kindness model, premised on being kind to a vulnerable person, which requires four practical and iterative judgments. The first three judgments of his model consist of establishing basic empathetic understanding, gauging the intricacy of situation of a vulnerable person, and setting a scope of what will be done to mitigate the vulnerability; the fourth judgment of the model is to then choose to act on a mitigation plan. Taken together, this kindness model is an empathetic and socially responsive act to mitigate the vulnerability of another person.

In the Hope District project, project stakeholders worked together to create a series of workshops to explore the application of this kindness model and how they could operationalize themselves to support trauma-informed planning and design. The workshops were based on the theme of reciprocity, aiming to promote understanding and empathy towards the community; considering long-held social and economic disparities that exist in low-income communities of color, it is important to understand and empathize trauma associated with those disadvantages and injustices. The format of the workshops meant that those in normative and formal power positions – including our studio faculty and students, residents, local teachers, and external stakeholders such as city officials, along with planning professionals, relinquished their usual roles and played other positions, in an empathetic learning exercise.

Residents shared their lived experiences and their stories about places or buildings that have been lost as a result of urban renewal or highway construction in the twentieth century. Students from elementary, middle, and high schools, with the help of their parents, teachers, and our studio undergraduate architecture students, created storytelling maps based on forgotten places and lived experiences of long-time residents. Understanding and empathy are essential to planners and designers in valuing those stories. Equipped with the understanding of how residents treasure their memories, planners and designers can follow their moral or ethical responsibilities toward helping the vulnerable populations in those communities. In this regard, the concept of kindness can help make placemaking more socially responsive to trauma.

7. Hope

It is a persistent effort to bridge what is lacking in the present to what is possible in the future. During that journey, people bounce between emotion and reason, a constant fluctuation that tests one's patience and requires courage. This journey may feel more difficult to those who have experienced trauma.

Unlike Forester's kindness model, the model of hope by Inch et al. (2020) consists of four dimensions that aim to illuminate human struggles and motivations to make persistent efforts and remain optimistic about the unpredictable future: a) openness to the future, b) intertwining of emotion and reason, c) an open-ended act of faith, and d) resourceful use of various modes of hoping. Exploration of hoping behavior, its dimensions, and the dimensional richness can allow us a deeper examination into how planners and designers might respond to serious problems or urgent crises such as displacement.

In the face of ongoing blight, worsening gentrification, weakening social cohesion, and persistent economic and health disparities, it would be normal to expect that local governments would take action in response to those ongoing challenges. It is safe to assume that there is no single perfect solution to those problems; both short-term interventions and long-term investments would be necessary. Under that circumstance, hoping would be among the responses taken by the planners and designers, especially when they work with underserved communities of color.

In Hope District project, we developed both near-term plans and longer-term plans, using phasing and anchor-agency strategies, and piloting through a small-scale project led by grassroots agencies that identified with stakeholders. These pilots sat alongside focused site developments, and comprehensive-scale projects (i.e., longer-term investments such as a community-wide circulation plan) in a comprehensive masterplan.

The studio team, in collaboration with the community leaders, applied the concept of hoping in proposing the short-term interventions so as to help build the confidence of the residents in the longer-term plan, so as to promote collective expectations about the community's future.

Place-cultivation as Trauma-informed Practice

Place-cultivating is an approach or practice in urban planning and design that can address trauma in underserved communities of color and refers to deliberate efforts aimed at fostering a sense of belonging, healing, and empowerment within a specific community or neighborhood (Cidell, 2011). The approach brings a focus on creating spaces that meet the physical needs of the residents and address the historical and contemporary traumas experienced by underserved communities of color (Lopes et al., 2021). It involves a collaborative and inclusive process that engages community members in decision making and ensures their voices are heard

(Smith & Johnson, 2019). Through the use of culturally sensitive design principles, such as incorporating culturally significant symbols, art, and landmarks, the practice seeks to honor and celebrate the community's heritage while promoting healing and resilience (Garcia & Chen, 2021). It emphasizes the provision of equitable access to essential amenities and services, including healthcare, education, transportation, and green spaces (Cutumisu et al., 2014) and recognizes the interconnectedness of physical, social, emotional, and environmental factors in shaping the wellbeing of a community (Marans & Stimson, 2011). By addressing trauma and systemic injustices, this approach aims to create spaces that promote social cohesion, improve mental and physical health, and enhance the overall quality of life for underserved communities of color.

Place-cultivation operates alongside two further placemaking approaches: place-taking and placemaking. Table 16.1 offers a comparative matrix of these approaches.

Regarding place-taking, a resident reported, 'outsiders are taking our land, our church, our school, and our community... resulting in gentrification and displacement... It is a serious threat to our sustainability and survival'. Residents shared mixed views on placemaking. As one interviewee commented, '...designers seem to have preconceived ideas about our community and how to improve it. We [residents] are rarely given an opportunity to play a significant role in planning and designing of our own community... We want to create a community reflective of our history, our culture, and our value system'. Some residents perceived place-cultivating to be similar to placemaking. Others, however, differentiated place-cultivating from placemaking. According to one interviewee, 'our grassroots organization focuses on how to build capacity of residents and communities, and how to leverage lived experience of residents in community development... To me, that is place-cultivating'.

The Hope District project takes some useful first steps to demonstrate how place-cultivating can make residents feel a sense of ownership over the planning and designing of their community and how the community can be more resilient in addressing trauma. Based on our study outcomes, we suggest that place-cultivating is an effective way to address trauma-causing environments or to mitigate existing trauma in the community, partly because place-cultivating

TABLE 16.1 Comparison of three types of place development

	Place-taking	Placemaking	Place-cultivating
Approach	Directed or imposed	Directed, community-based	Self-help, grassroots-driven
Priority	Controlling development	Creating a community vision	Capacity building, empowering
Product vs. process	Product-centred	Product-centred	Process-centred

prioritizes people. That is to say, place-cultivating puts residents at the center of planning and design work so that others can focus on understanding and acknowledging their historical and contemporary trauma. Therefore, planning and design in place-cultivating are viewed as process-oriented endeavors that empathize with the residents.

Armed with practical judgements of kindness and hoping strategies, place-cultivating can provide a collective, optimistic expectation into the future through various strategies, including an emphasis on community engagement with long-term planning and working-relationships; community-led interventions (relevance); leveraging or aligning with the citywide support system (equity); resident leadership or community representation in a governing body (governance); and inclusive, interactive, transparent, and democratic public participation.

We also acknowledge practical challenges associated with place-cultivating, and a few are mentioned here. First, priorities of funding agencies, such as community development financial institutions, shift frequently. This means that to get funding the priorities of financial institutions and those of a community must be aligned. This requires coordination and negotiation between financial institutions, governments, and community development organizations, given that reliable funding streams are essential to making community development sustainable over the long term (Kim, 2021). Second, building trust is essential to successful development (Sanoff, 1999; Kim, 2022). Trust takes time to build and requires a long-term working relationship. Establishing long-term relationships can be laborious for all parties involved. A better strategy is to break long-term work into smaller parts so that participants or partners can more easily manage their efforts and build long-term relationships one step at a time. Also, using a short-term lens makes it easier to track and evaluate the progress of development. Third, many of the trauma symptoms that were listed earlier are beyond the expertise or realm of what planners and designers normally handle. Considering that, trauma-informed place-cultivating requires partnerships with experts in other disciplines (e.g., public health). Moreover, community-city engagement is essential to trauma-informed practice because historical and contemporary traumas are often associated with structural injustices and disparities faced by underserved communities of color. Some cities, including Detroit, have a citywide system similar to trauma-informed practice. For example, Detroit has instituted a Community Benefit Ordinance (CBO) as a mechanism for the community to negotiate with developers. However, it promotes community engagement in only a limited way, which is at the initial stage of the development process (Berglund, 2021). Nevertheless, a CBO is a good first step. When combined with other strategies such as the ones suggested in this chapter, the process could be more effective in addressing trauma, thereby leading to enhanced equity and justice.

References

Atkinson, R., & Bridge, G. (2019). 'Gentrification and the politics of placemaking', *Urban Studies*, 56(3).

Berglund, L. (2021). 'Early lessons from Detroit's community benefits ordinance', *Journal of the American Planning Association*, 87(2).

Brown, G., & Kyttä, M. (2014). 'Key issues and research priorities for public participation GIS (PPGIS): A synthesis based on empirical research', *Applied Geography*, 46.

Brown, L., Martinez, R., & Williams, G. Year (2021). 'Impact of trauma on behavioral patterns in underserved communities of color', *Journal: Journal of Psychological Trauma*, 15(2).

Bryant-Davis, T., Ullman, S. E., Tsong, Y., Gobin, R. L., & Yoder, J. R. (2019). 'Therapeutic group interventions for trauma survivors: A meta-analysis', *Trauma, Violence, & Abuse*, 20(1).

Bullard, R. D., Johnson, G. S., & Torres, A. O. (2019). *Highway robbery: Transportation racism and new routes to equity*. Boston: MIT Press.

Carmona, M., de Magalhães, C., & Hammond, L. (2017). 'Public spaces and urbanity: Can densification play a constructive role in supporting social interaction in Western Australian suburbs?', *Urban Policy and Research*, 35(2).

Carter, R. T., Kirkinis, K., & Johnson, V. Year (2017). 'Traumatic experiences, posttraumatic stress, and mental health problems among African Americans', *Journal: Journal of Black Psychology*, 43(8).

Carter, R. T. (2007). 'Racism and psychological and emotional injury: Recognizing and assessing race-based traumatic stress', *The Counseling Psychologist*, 35(1).

Cidell, J. (2011). 'The promise and perils of community-based participatory research in urban planning: Observations from New Orleans', *Journal: Planning Theory & Practice*, 12(4).

Colantonio, A., & Dixon, T. (2011). *Urban regeneration and social sustainability: Best practice from European cities*. Hoboken: Wiley-Blackwell.

Cutumisu, N., Roy, S., & Bagheri, N. (2014). 'Designing walkable school neighborhoods: A case study of School Travel Planning in Ottawa, Canada', *Journal of Transport Geography*, 34.

Davidson, M., & Lees, L. (2018). 'New-build "gentrification" and London's riverside renaissance', *Urban Studies*, 55(5).

Desmond, M., & Shollenberger, T. (2015). 'Forced displacement from rental housing: Prevalence and neighborhood consequences', *Demography*, 52(5).

Eubanks, V. (2018). *Digital dead end: Fighting for social justice in the information age*. Boston: MIT Press.

Evans, G. (2017). 'Resilience, ecology and adaptation in the experimental city', *Journal of Urban Design*, 22(4).

Forester, J. (2021). 'Our curious silence about kindness in planning: Challenges of addressing vulnerability and suffering', *Planning Theory*, 20(1).

Galster, G., & Hayes, C. (2017). 'Neighborhood disorder, perception, and investment', *Journal of Planning Literature*, 32(1).

Garcia, M., & Chen, S. (2021). 'Culturally sensitive design principles in place-cultivating: Honoring heritage and promoting healing', *Journal: Environment and Behavior*, 53(5).

Gehl, J., & Svarre, B. (2013). *How to Study Public Life*. Washington: Island Press.

Gehl, J., & Gemzoe, L. (2001). *Public spaces-public life*. Copenhagen: Danish Architectural Press.

Inch, A., Slade, J., & Crookes, L. (2020). 'Exploring planning as a technology of hope', *Journal of Planning Education and Research*, 1(1).

Jackson, C., & Green, R. (2019). 'Environmental factors contributing to physical health disparities among Black and Brown communities', *Journal of Environmental Health*, 81(6).

Jennings, V., & Johnson Gaither, C. (2015). 'Approaches for understanding and measuring environmental justice', *Journal: Sustainability*, 7(3).

Johnson, A., Smith, B., & Williams, C. (2020). 'Iterative design in placemaking: Enhancing user participation and satisfaction', *Journal of Urban Design*, 35(2).

Jones, A., & White, S. F. (2020). 'Exploring the relationship between trauma and physical health disparities among underserved minority populations', *Journal of Health Disparities Research and Practice*, 13(3).

Kim, J. (2021). 'What kind of community development system can effectively support citywide philanthropic efforts to promote community well-being?', *International Journal of Community Well-Being*, 2022, 5: 305–338.

Kim, J. (2022). 'Making architecture relevant to underserved communities: Mapping reconsidered', *Architecture*, 2.

Lopes, S., Costa, H., Silva, F., & Ferreira, F. (2021). 'Building sustainable communities: The role of urban planning in promoting social inclusion', *Journal: Sustainability*, 13(9).

Loukaitou-Sideris, A., & Ehrenfeucht, R. (2016). *Sidewalks: Conflict and negotiation over public space*. Boston: MIT Press.

Marans, R. W., & Stimson, R. J. (2011). 'Investigating the links to improved quality of life and sustainable urban forms', *Journal of Urban Affairs*, 33(4).

Marcuse, P., & Van Kempen, R. (2017). 'The limited promise of placemaking: Contradictions of neoliberal urbanism', *Journal of Planning Education and Research*, 37(3).

Montgomery, J. (1998). 'Making a city: Urbanity, vitality and urban design', *Journal of Urban Design*, 3(1).

Paradies, Y., Ben, J., Denson, N., Elias, A., Priest, N., Pieterse, A., & Gee, G. (2015). 'Racism as a determinant of health: A systematic review and meta-analysis', *PloS one*, 10(9), e0138511.

Payne-Sturges, D. C., Towe, V. L., Hernandez, N., & Díaz-Barriga, F. (2018). 'Environmental justice and health: The Potential Role of Community Health Workers in Research and Advocacy', *Journal: Environmental Justice*, 11(6).

Pickett, S. T., Cadenasso, M. L., & Grove, J. M. (2004). 'Resilient cities: Meaning, models, and metaphor for integrating the ecological, socio-economic, and planning realms', *Landscape and Urban Planning*, 69(4).

Reed, M. G. (2012). 'Community involvement in urban regeneration: Lessons from the UK', *Planning Practice & Research*, 27(4).

Rigolon, A. (2016). 'A complex landscape of inequity in access to urban parks: A literature review', *Landscape and Urban Planning*, 153.

Rothstein, R. (2017). *The Color of Law: A Forgotten History of How Our Government Segregated America*. New York: Liveright Publishing.

Saha, R., , & Bunker, R. (2016). 'Urban form, social sustainability and health', *Health Promotion International*, 31(4), 787–800.

Sanoff, H. (1999). *Community participation methods in design and planning*. Hoboken: John Wiley & Sons.

Smith, John A. (2018). 'Placemaking and Power in Community Development', *Journal: Community Development*, 43(1).

Smith, J. A., & Johnson, M. L. (2020). 'Preserving historic buildings in placemaking: Strategies and challenges', *Journal of Urban Design and Planning*, 25(3).

Smith, J. A., & Johnson, R. L. (2019). 'Community engagement in place-cultivating: A participatory approach for inclusive decision-making', *Journal: Journal of Community Psychology*, 47(3).

Smith, J., Johnson, A., & Rodriguez, M. (2022). 'Trauma and Behavioral Symptoms Among Underserved Black and Brown Communities', *Journal of Trauma and Diversity*, 10(3).

Smith, R. J., Williams, A. B., & Johnson, C. D. (2019). 'Socioeconomic disparities in access to healthy food and their impact on health outcomes: A comprehensive review', *Journal of Public Health*, 42(1).

Sturup, J. (2018). 'Tokenism and its consequences: Reframing the discourse on public participation in planning', *Journal: Planning Theory*, 17(1).

Sugrue, T. J. (2014). *The Origins of the Urban Crisis: Race and Inequality in Postwar Detroit*. Princeton: Princeton University Press.

Thompson, L., & Johnson, M. (2020). 'The economic impacts of placemaking', *Urban Studies*, 57(9).

Vanky, A., O'Brien, D. R., Markusen, A., & Doud, M. J. (2019). 'Exploring the connection between placemaking, gentrification, and social capital in Detroit', *Landscape and Urban Planning*, 189.

Williams, M. T., Metzger, I. W., & Leins, C. (2019). 'Disparities in mental health treatment for racial and ethnic minority populations: Trauma-informed care as a pathway to equitable access and care', *Journal of Clinical Psychology: In Session*, 75(10).

Williams, D. R., & Mohammed, S. A. Year (2013). 'Racism and health II: A needed research agenda for effective interventions', *Journal: American Behavioral Scientist*, 57(8).

Wyly, E., & Hammel, D. (2019). 'Gentrification and the social trauma of displacement in gentrifying neighborhoods', *Urban Affairs Review*, 55(2).

17

THEORIZING DISAPPEARANCE IN NARRATIVE ECOLOGIES AS TRAUMA-INFORMED PLACEMAKING

Marwa N. Zohdy Hassan

Introduction

Marginalized narratives possess the capacity to embed themselves within the dominant literary canon, even when constrained by the parameters set by prevailing discourses (Althusser, 1970). As seen in the context of real cities, pre-existing fictionalized urban landscapes can be revisited through the lens of self-narratives, which are inherently influenced by the performative aspects of life. This resonates with the understanding that memories are fundamentally shaped by textualization, enabling marginalized accounts to persist despite attempts at forced erasure (Saunders, 2008, pp. 321–32).

Building on this premise, the argument put forth suggests that the conceptual tools outlined in this discourse can serve as empowering instruments for individuals who have endured trauma. Specifically, these tools can be harnessed for the creation of speculative fiction and the implementation of episodic walkability practices within their respective urban environments. This transformative process involves the act of constructing, penning, and recording their personal narratives, subsequently employing these narratives as guiding frameworks. Trauma victims, grappling with the aftermath of experiences often silenced or marginalized, can find agency in crafting speculative narratives. These narratives transcend the boundaries of lived reality, offering the creators an avenue to explore alternative realities and rewrite their stories. By weaving their own experiences into speculative frameworks, they engage in an act of reclamation, harnessing the power of imagination to navigate their trauma and its aftermath.

Furthermore, the proposal to implement episodic walkability practices capitalizes on the notion that physical movement through urban spaces can serve as a cathartic journey, parallel to the narrative voyage undertaken in speculative

DOI: 10.4324/9781003371533-21

fiction. Trauma survivors can map their personal narratives onto the cityscape, creating waypoints that symbolize resilience, growth, and healing. As they traverse these waypoints, guided by their own recorded narratives, they engage in a tangible and metaphorical journey of reclamation, reinterpreting spaces once associated with pain into sites of triumph.

In essence, the interplay between crafting speculative fiction and implementing episodic walkability practices offers trauma survivors a multifaceted avenue for healing and redefining their relationship with their environment. By channelling their stories into imaginative narratives and embedding them within the physicality of the city, they engage in an act of agency and transformation. This holistic approach empowers trauma survivors to transcend the boundaries of their past, ultimately fostering a sense of empowerment, resilience, and renewed connection to their cities.

Thus, in this chapter, I will examine the disappearance of cities in literature, revealing their deep linkages to memory and trauma manifested in physical buildings and people. Cities are more than simply concrete jungles – they hold memories, traumas, and narratives that shape citizens' lives. Narratives are strong forces that shape residents' reality. The concept of anti-cities and anti-citizens shows that cities may be both nurturing and antagonistic, encouraging affiliation and separation. This argument holds that cities tell their own stories – narratives that capture their essence, experiences, and residents' collective consciousness. Therefore, the fragmented narrative of cities is not a consequence of trauma alone but rather an outcome of heightened awareness – a consciousness catalyzed by the impact of city life. Cities are portrayed not just as canvases upon which stories unfold but as active participants in the process of narrative construction. Individuals simultaneously shape and are shaped by the urban landscape, establishing a reciprocal relationship that blurs the boundaries between personal and urban narratives. This symbiosis results in the narrative of the city becoming an extension of the collective human experience.

Workshop Format: Narratives of Disappearance

In 2020, a workshop on gamification, *How to Disappear Completely* (for a description of the workshop, see MeetLab 2021), was conducted in Cairo, Egypt, under the guidance of Hungarian co-instructors, Ambrus Ivanyos and Bálint Tóth. The main aim of the workshop was to develop a game audio walk set in Downtown Cairo. Following successful implementation in Budapest, Hungary, and Izmir, Turkey, the framework was subsequently tailored to suit the specific context of Cairo, Egypt. In this novel setting, the central character, Andrei, deliberately presents himself as detached from any specific national or ethnic identity, as he embarks on a renewed inquiry into the vanishing of individuals and urban centres. The audio walk deviates from the traditional format of a touristic audio tour, instead serving as a collaborative endeavour involving participants who aim to express their perspectives on the city, its history, and its walkability attributes.

The workshop provided an opportunity for Egyptian creators to examine their preconceived notions regarding audio walks, which subsequently influenced their decision-making process as they engaged in discussions to carefully plan and refine the route for the audio walk. The resulting audio walk, set in Downtown Cairo's evolving urban landscape, seeks to encompass marginalized narratives and sentiments while embodying the challenges of accessing public spaces in the context of policing discretion and restricted leisure walking.

After conducting in-depth interviews with locals living in the chosen spatial contexts, the participants began developing episodic narratives during the workshop's second phase. This critical juncture gave the creators a unique opportunity to shape the city's urban landscape, community dynamics, and socio-cultural complexities. including the demographic characteristics of residents, coffeehouse patronage, the perspectives of elderly residents, the livelihoods of street vendors, and the intricate interplay of religious dogmas and gender dynamics all over the city. (Foucault, 1978; Bourdieu, 1984).

It is important to acknowledge that the ever-evolving nature of urban landscapes poses a challenge to maintaining a consistent representation, especially within the dynamic context of pre-pandemic Downtown Cairo. The period spanning 2020 to 2023 witnessed radical mutations in the city's urban planning and socioeconomic underpinnings, thereby prompting a significant transformation in its landscape and character.

This element of transformation resonates profoundly with Andrei's journey, as he navigates a narrative entrenched in the concept of disappearance as the sole constant of change. Consequently, in order to provide the creators with the freedom to show the city in its idealized forms, drawing from historical precedents or speculating about future trajectories, the thematic examination of disappearance acts as a counterweight to the framework. This approach allows participants to capture the profound emotions associated with disappearing elements, resulting in a contemplative fabric that includes elements that already disappeared, elements that are actively disappearing, and elements that merit collective commemoration. As a result, the process of writing the episodes empowers the creators to imprint the essence of the city in the collective consciousness.

Furthermore, the innovative format adopted for the audio walk, wherein players exercise agency to forge their own paths, introduces a mechanism that embraces fragmentation as a means of comprehending individual identity and experiences. This fragmentation serves as a metaphor for the existential investigation which aims to understand and interpret personal identity in the context of a fragmented world (see quoted except below). It is important to acknowledge that the ever-evolving nature of urban landscapes poses a challenge to maintaining a consistent representation, especially within the dynamic context of pre-pandemic Downtown Cairo. The period 2020–23 witnessed radical mutations in the city's urban planning and its socioeconomic underpinnings. This theme of change and transience connects well with Andrei's own experience as he moves through a story that is built around the idea of disappearance as a fundamental aspect of change. To provide the creators with the freedom to show the city in its idealized forms, drawing from historical precedents or speculating about future trajectories, the thematic examination of disappearance acts as a counterweight to the main narrative. This approach gives participants the ability to capture the profound emotions associated with disappearing elements, creating a contemplative fabric that includes elements that have already disappeared, those that are actively disappearing, and those that merit collective commemoration. As a result, the story gains agency, securely embedding the essence of the city in the collective consciousness.

It will increase the sensation of agency even more to have a 'safe space' to examine the disappearance, its grieving component, and its projection onto the urban canvas (Mayer et al., 2004). The audio walk's formatting further demonstrates this feeling of agency, allowing players to construct their own routes by selecting the places they wish to visit as long as they arrive at the endpoint. Thus, it is a shared experience that workshop participants have passed on to players. The curated walks' limitless potential outcomes create a sense of identity and accessibility to public spaces that may not be noticeable during daily routines. Therefore, the gamified aspect of the audio walk gives all parties a sense of agency to change public spaces (Mayer, 2009).

I disappear and yet still remain. The city remains here, or it stays here [...] it crumbles and in the meantime, a certain kind of cohesive force pulls and sticks the essence together [...] Even if the city exists, it is only because the story I'm telling you holds it together.

(An excerpt from the final episode)

Gender and Trauma

Within placemaking paradigms, it is crucial to recognize the complex impact that gender dynamics have on public places, particularly in contexts similar to Arab and Islamic cultures. The development of audio walks can be greatly influenced by ensuring an inclusive approach that recognizes the subtleties of uneven gender relations. This discussion is further enhanced by the use of ideas from literature like Yoko Ogawa's *The Memory Police* (2019) and Anna Burns' *Milkman* (2018), two books with similar issues of coercive surveillance. Imaginative audio walks and works of science fiction have similarities that make the former a good narrative format for the episodic nature of the latter. According to Hassler-Forest (2018), speculative fiction may create a variety of interrelated micro-narratives that together contribute to a larger theme inquiry. These similar characteristics not only make it easier to translate speculative tales into appealing audio walk episodes, but they also highlight the possibility for intermediary cooperation in conceptualizing alternative experience narratives (Hayles, 2017).

Thematically, the relationship between forced mass forgetfulness, memory preservation, and resistance to authoritarian control is evident in *The Memory Police* and there are fascinating similarities in the prospective development of audio walk episodes. The idea of including a fictional female character in a few audio walk episodes, especially in areas where women congregate in public, holds the potential to expand the idea of memory preservation and resistance to experiential narratives, interwoven with the visual aspects of feminist urban planning. The introduction of a fictional female character into the audio walk episodes creates a chance for narrative development in which her interactions with the urban landscape signify the fight against imposed forgetting. This strategy supports the underlying themes of memory retention and resistance shown in *The Memory Police*, adding to the narrative medium's thematic cohesion.

Furthermore, viewed through the lens of narrative disappearance, the deliberate placement of these audio walk episodes within public spaces less frequented by women introduces a compelling layer that resonates with the principles of feminist urban planning – a framework dedicated to cultivating inclusive and safe surroundings. This contextual decision also foregrounds the recurring motif of disappearance, inciting a contemplation of its underlying motives. This harmonization aligns inherently with the foundational tenets of feminist urban planning, which seeks to transform urban landscapes to cater to the diverse needs and experiences of women.

The integration of these infrequently visited spaces aligns harmoniously with the socio-cultural dimensions that underscore the narrative fabric of *The Memory Police*. This alignment provides a strategic platform for the audio walk narratives to meticulously capture the intricate interplay between memory, gender dynamics, and gestures of resistance woven within the urban context. As participants navigate these evocative settings, the thematic correspondences invite a reflective exploration into the nuanced relationships interconnecting memory preservation, gendered encounters, and the encompassing socio-political milieu that frames these narratives.

Similarly, Burns' (2018) *Milkman* casts a revealing light on the female protagonist's ordeal of being under surveillance by a figure embodying a pervasive authoritative presence, shedding light on the intricate ways women are subjected to surveillance within public spheres. The fact that she was reading while walking was considered provocative emphasizes the gendered requirements of alertness and compliance with cultural standards. The removal of the surveilling figure might be understood as a rupture in the dominant gendered power dynamic, similar to the idea of resistance included in the audio walk model.

The terrible experiences that the female characters go through in both literary works are sensitively highlighted. These experiences are inextricably entwined with social expectations and oppressive monitoring. This thematic resonance is consistent with the audio walk episodes' speculative and investigative nature, in which narrated events act as catalysts for revealing significant social realities.

Integrating these observations into frameworks emphasizes the concept that speculative components may be effectively leveraged to unravel and critique gendered power interactions. While the episodes may flow in a non-linear fashion, an overall narrative thread can drive the investigation of how speculative aspects meet with lived experiences, promoting a deeper knowledge of urban landscapes and the gendered dynamics that affect them. This synthesis emphasizes the convergence of speculative fiction and placemaking, with a focus on gender dynamics in urban settings.

Metafiction of the Audio Walk

In *The Memory Police*, the novel itself contains excerpts from an intertwined narrative that the protagonist had been crafting – a narrative that serves as an autobiographical testimony of trauma. This trauma encompasses the struggle to comprehend one's relationship with existence, along with the challenge of maintaining a narrative amid a simulated reality. The interspersed novel, written by the protagonist, revolves around a character who loses her ability to speak and communicates solely through typing. The character's voice is forcibly taken and confined, and her autonomy erodes over time, culminating in a sense of hopelessness and passivity. This fictional narrative becomes an exploration of the performativity of violence, mirroring the gendered dynamics that Diana Taylor elucidates – where power is male and marginalization is female, resulting in the disappearance of marginalized voices from mainstream narratives (Taylor, 1997).

In this context, the protagonist's endeavour to construct such a narrative stands as an act of resistance. Her attempt to document her trauma and to maintain the act of writing itself signifies defiance against the erasure of novels and narratives – a refusal to let her voice be extinguished. The act of bearing witness to her own oppression is an assertion of her agency in the face of oppressive forces. However, the novel does not conclude with a triumph; the protagonist ultimately disappears, suggesting the overwhelming might of the forces she is resisting.

The concept of disappearance, woven into both the protagonist's autobiographical narrative and the broader fabric of the island's culture, carries a multifaceted meaning. The island's culture becomes permeated with the ritual of disappearance orchestrated by the memory police, becoming an inherent performance that defines the national identity. This parallels the trauma experienced by individuals who encounter a rupture in their own narratives due to traumatic events. Trauma often leaves narrative gaps, scars that resist assimilation into a coherent narrative, while simultaneously seeking expression. The tension between unassimilated trauma and the narrative impulse is fluid, reflecting the intricacies of memory's interplay with textualization (Chu, 2010).

Yoko Ogawa's construction of this science-fiction world, while an imaginative creation, mirrors an ontological fabric already imbued with a traumatized consciousness. The protagonist's endeavour to align her personal narrative gaps with the larger narrative fabric of the island demonstrates a convergence. This convergence occurs because the ontological fact of literalized narrative gaps, imposed by the memory police, aligns with the protagonist's subjective experience, where she believes she lives a normal life despite the imposed traumas (Chu, 2010).

In summary, the integration of a metanarrative holds significant promise for the construction of immersive and therapeutic audio walks. This envisioned methodology involves participants engaging with characters who, like the novel's protagonist, interact with their own autobiographical narratives. Such an approach resonates particularly well within the context of audio walks structured as a series of investigative episodes, akin to those undertaken by Andrei. In this framework, the act of seeking textual information from diverse sources, such as books, media, and letters, emerges as a dynamic vehicle for unravelling the nuanced ways individuals within these episodes grapple with the aftermath of traumatic disappearances.

By extending participants the opportunity to not merely witness but also immerse themselves in the characters' inner lives, the audio walk evolves into a therapeutic journey of shared narratives. As players traverse physical spaces while audibly encountering these autobiographical accounts, they are offered a layered and empathetic understanding of the characters' struggles, coping mechanisms, and resilience. This interaction cultivates a sense of collective healing through storytelling – an inherently human endeavour that echoes the cultural tendency of cities to commemorate their experiences through self-reflexive expressions.

Nation-ness and Trauma

Everyone on the island had a vague premonition about what awaited them at the end, but no one said a word about it. They were not afraid, and they made no attempt to escape their fate.

(The Memory Police, chapter 28, p. 261)

The passage from *The Memory Police* cited above encapsulates a profound alignment between personal and national traumas, showcasing a 'twice-behaved behaviour' that mirrors the performative nature of disappearance (Schechner & Turner, 1985).

This alignment is exemplified by the protagonist's adaptation of recipes to accommodate the disappearance of certain vegetables – a private response to a public phenomenon. However, this performative response itself becomes ephemeral, paralleling the transient nature of disappearance. The concept of the self, acting in/as another performance sphere underscores the role-based nature of the social or trans-individual self, which was profoundly affected by the cultural and social disappearances on the island.

Within this context, the disappearance of cultural and social spheres led to a vanishing of the behaviours inherent in those spheres. The tendency toward repetition and 'twice-behaved behaviour' faced a threshold, as the ethereal components of culture and traditions ceased to operate within the realm of these behaviours. The disappearance of both cosmological constructs and the cultural presuppositions associated with them ruptured the coexistence of action and stasis within the same event (Tambiah, 1985; Schechner & Turner, 1985).

An illustrative instance of the vanishing cosmological construct is the disappearance of seasons. The protagonist's anticipation of spring's arrival, laden with cultural significance, contrasts with the reality of a perpetually frozen world. Spring, with its tradition-laden festivities, stands as a vivid embodiment of the intersection between the action of conventions and the stasis of seasons. The recurring cycle of seasons traditionally brought forth a 'restored behaviour' of cultural performativity, underlining the interplay of temporal conventions and embodied memory (Schechner & Turner, 1985; Tomczak, 2017).

Similarly, the *Milkman* novel offers a moving examination of the complexities of power relations and social forgetfulness against the backdrop of a society plagued by enforced disappearance and the political instability it represents. Through the lens of an 18-year-old protagonist, the narrative delves into the nuanced interplay between personal and collective trauma, drawing a haunting parallel between the young individual's experience and the nation's enforced forgetfulness of its own political scars. The persistent stalking endured by the protagonist at the hands of an older man serves as a chilling mirror to the nation's passivity towards its own history of disappearance, reflecting a society's readiness to shift blame onto the victim – a phenomenon often encapsulated as 'blaming the victim'. This narrative technique artfully captures the disquieting convergence of personal and national narratives, weaving a compelling tale that underscores the eerie resemblance between individual and collective silencing.

Accordingly, audio walks have the potential to transform into an avenue for participants to interact with their surroundings through the lens of collective memory by fusing personal narratives with historical and societal threads. It provides a singular platform to navigate the 'twice-behaved behaviour' of societal narratives while instigating dialogues that challenge societal tendencies, such as victim-blaming. The transient and immersive nature of audio walks aligns with the transient nature of disappearance.

Conclusion

The use of an anonymous protagonist helps people feel more connected to one another and creates a new perspective on their environment. Through its immersive nature, this interaction creates a favourable environment for challenging the accessibility of diverse locations, strengthening the participants' sense of solidarity. The strategic shift of agency demonstrated when Andrei disappears, and the player assumes the role of navigating guided by the narrator's voice, reflects the fluidity of the urban environment itself. The city is not merely a physical structure; in its ongoing state of expansion and contraction, it is also an expression of its residents.

This resonates as a *thirdspace* (Soja, 1996) that manifests during the decision-making process for the audio walk sites, the resident interviews performed during the episode preparation, and lastly, when the listener moves through the urban environment while engaging with the audio walk. The participant will experience an epiphany because of the physical exercise and narrative trip coming together in a way that echoes Joseph Campbell's conception of the hero's journey. Participants figuratively embark on a transforming adventure as they progress from point A to point B. The urban experience is best captured by the intersection of disappearance, exploration, and emergence, which transforms abstract ideas into concrete interactions that inspire empowerment and critical thought. These audio walks transform from simple guided tours to participatory platforms that cut through urban, historical, cultural, and discursive settings by relying on themes of disappearance. In this approach, the participants' feeling of agency within these frameworks serves as a catalyst for developing a stronger bond with the urban environment and a sense of place-based agency.

Moreover, in considering the creation of this audio walk, the concept of thirdspace encompasses the interplay between real, imagined, and potential spaces, creating a hybrid realm that transcends the binary of physical and mental spaces. Applied to the urban context, this notion acknowledges the complexity and fluidity of urban experiences, offering an inclusive space where diverse narratives and perspectives converge (Lefebvre, 1991; Harvey, 1996). The audio walk's manipulation of space and time aligns with this concept, as the team navigates the city's historical and evolving dimensions while simultaneously engaging with imagined narratives.

Importantly, the theme of disappearance woven into the audio walk not only embodies urban mutations but also holds therapeutic potential, particularly for trauma survivors. By confronting the idea of disappearance, the participants can engage in a process of reflection and reframing, potentially aiding in their personal healing and resilience (Herman, 1992; Van der Kolk, 1994). The audio walk becomes a conduit for processing loss, acknowledging the ephemeral nature of urban landscapes and personal experiences.

References

Althusser, L. (1970). 'Ideology and Ideological State Apparatuses', *La Pensée* (151).

Bourdieu, P. (1984). *Distinction: A Social Critique of the Judgment of Taste.* Cambridge: Cambridge: Harvard University Press.

Burns, A. (2018). *Milkman.* London: Faber & Faber.

Chu, S-Y. (2010). *'Do Metaphors Dream of Literal Sleep?': A Science-Fictional Theory of Representation.* Cambridge: Harvard University Press.

Foucault, M. (1978). *The History of Sexuality, Volume 1: An Introduction.* New York: Pantheon.

Harvey, D. (1996). *Justice, Nature and the Geography of Difference.* London: Blackwell.

Hassler-Forest, D. (2018). *Science Fiction, Fantasy and Politics: Transmedia World-building Beyond Capitalism. (Radical Cultural Studies).* Lanham: Rowman & Littlefield International.

Hayles, N. K. (2017). *Unthought: The Power of the Cognitive Nonconscious.* Chicago: Chicago University Press.

Herman, J. L. (1992). *Trauma and Recovery: The Aftermath of Violence – from Domestic Abuse to Political Terror.* New York: Basic Books.

Lefebvre, H. (1991). *The Production of Space.* London: Blackwell.

Mayer, I. S. (2009). 'The gaming of policy and the politics of gaming: a review', *Simul. Gaming, 40.*

Mayer, I. S., Carton, L., de Jong, M., Leijten, M. & Dammers, E. (2004). 'Gaming the future of an urban network', *Futures,* 36.

MeetLab. (2021). *How To Disappear Completely,* by Ambrus Ivanyos, Bálint Tóth. Available: https://www.staatsschauspiel-dresden.de/spielplan/archive/h/how-to-disappear-completely/. Accessed: 22 January 2024.

Ogawa, Y. (1994). *The Memory Police.* London: Harvill Secker.

Ogawa, Y. (2019). The Memory Police (S. Snyder, Trans.). Pantheon.

Saunders, M. (2008). 'Life-Writing, Cultural Memory, and Literary Studies', *Cultural memory studies: An international and interdisciplinary handbook,* 8 (2008): 321.

Schechner, R. & Turner, V. (1985). *Between Theater and Anthropology.* Philadelphia: University of Pennsylvania Press.

Soja, E. (1996). *Thirdspace. Journeys to Los Angeles and Other Real-and-Imagined Places.* London: Blackwell Publishers.

Tambiah S. J. (1985). *Culture, Thought, and Social Action. An Anthropological Perspective.* Cambridge: Harvard University Press.

Taylor, D. (1997). *Disappearing Acts: Spectacles of Gender and Nationalism in Argentina's 'Dirty War'.* Durham: Duke University Press.

Tomczak A. M. (2017). On Exile, Memory and Food: Yasmin Alibhai-Brown's The Settler's Cookbook: A Memoir of Love, Migration and Food. In: Onega S., del Río C., Escudero-Alías M.(eds) *Traumatic Memory and the Ethical, Political and Transhistorical Functions of Literature.Palgrave Studies in Cultural Heritage and Conflict.* Cham: Palgrave Macmillan.

Van der Kolk, B. A. (1994). 'The Body Keeps the Score: Memory and the Evolving Psychobiology of Posttraumatic Stress', *Harvard Review of Psychiatry,* 1(5).

18

ABANDONED LANDSCAPES AS PLACES OF POTENTIAL FOR NATURE THERAPY

Glendalough, Ireland

Lyubomira Peycheva

Introduction

Research has been emerging in support of the positive health effects (physical, mental, and social) of interacting with nature (Bratman et al., 2019; Capaldi et al., 2015), with stress restoration, attention restoration, depression and anxiety easing, and mood enhancement, among others, being frequently cited in scientific works accentuating natural environments' positive mental health effects.

The current rupture between nature and society, exacerbated by the concentration of population in urban areas, however, may be impacting people's mental health (Lewis, 2017). The number of people suffering from depression is thought to be approximately 280 million (WHO, 2023). A correlation can be found between growing urban population (270.6 million to 3.82 billion) (Ritchie & Roser, 2018) and increase in all mental disorders by 50% (416 million to 615 million) between 1990 and 2013 (WHO, 2016). This should be a wakeup call that the links between these two phenomena can no longer be ignored. With around 70% of the population expected to live in cities by 2050 (van den Bosch & Meyer-Lindenberg, 2019), what are we to expect in terms of people's mental health? If we are to continue on the path to sustainable development, society's mental health needs to be taken into consideration.

Case study

This chapter presents the case study of an abandoned landscape in Ireland and its potential as an area for nature therapy. While there is a large body of evidence in support of the benefits of natural environments on mental health, not all natural environments have the same effect. Currently most studies are focusing

DOI: 10.4324/9781003371533-22

on evaluating the differences of urban and natural environments (mostly forests) with regard to their health effects (e.g., COST Action E39, 'Forests, Trees, Human Health and Wellbeing'). Whereas this chapter focuses on evaluating the effects on mental health and wellbeing of a particular type of landscape stuck in-between being natural, while still possessing a strong anthropogenic presence – the so-called abandoned landscapes, and in particular, Glendalough, Ireland.

Glendalough (in Gaelic, 'the valley of the two lakes' (Barrow, 1974)) is a glacial valley situated in the hearth of the Wicklow Mountain, eastern Ireland, near the village of Laragh. Two rivers run through the valley, the Gleneala at the western end of the Upper Lake, which joins the Glendasan at the eastern end of the Lower Lake (Barrow, 1974). North from the lakes there is a view towards the Camadery Mountain, which acts as the natural border between Glendalough and neighbouring valleys (Barrow, 1974). The south view from the lake is revealing the Lugduff Mountain, forming the impressive deep valley sidewalls.

The valley has been home for ancient spirituality (the legend of St. Kevin) and a monastic settlement since the 6th century (Wicklow Mountains National Park, 2022). Owing to its unique geology, extensive mining activities from the 13th century up to 1935 altered the landscape, leaving behind the ruins of a once bustling monastic settlement and mining centre that can still be seen today (Mighall et a., 2013). Nowadays, Glendalough is, once again, surrounded by semi-natural oak woodland – an example of the oak woods that once covered most of the area. In 1991, the serene landscape and cultural heritage of the Wicklow mountains gained it the status of National Park (Wicklow Mountains National Park, 2022), attracting around 1 million visitors each year (Wicklow Endless Opportunities, 2017). It has become the embodiment of Irish history, religion, culture, and nature, thus making it a suitable study area – previously abandoned, however, re-discovered by both nature and people.

The study uses four key foundational concepts – nature therapy, biophilia, stress restoration theory (STR) and attention restoration theory (ART), which are briefly outlined below.

1. Nature Therapy

In the scope of the case study presented here, nature therapy refers to any therapeutic activity, from guided professional therapy sessions to self-guided nature-based activities, such as walks, hikes, recreation, and nature observation. It is not limited to the three-dimensional relationship between professional practitioners, patients, and nature. Indeed, it can be a personal experience between an individual and the natural environment by a way of spending time in nature.

2. Biophilia

Biophilia (Wilson, 1984) is the inherent aspect of human evolutionary inheritance manifested in a behavioural and emotional affinity for nature, which contributes to our survival, as well as mental and existential wellbeing (Kellert & Calavrese, 2015; Seymor, 2016). The absence of nature and nature interaction can thus be associated with the onset of mental (e.g., depression) and cognitive issues (e.g., ADHD) (Schmitz, 2012).

3. Stress Restoration Theory (SRT)

SRT is based on the belief that exposure to nature promotes physical and psychological wellbeing (Gullone, 2000). It stimulates a restorative process that affects stress regulation, the level of stimulation and organization, and attention/ fascination. Human capacities to respond in a recuperative manner (e.g., increase in positive emotions, decrease in negative thoughts and stress) in a natural environment are linked to a rooted behaviours arising from the process of adaptation to natural selection and to survival (Ulrich, 1981).

4. Attention Restoration Theory (ART)

A 'restorative environment' is one that promotes rest, healing, and reflection and holds four qualities – Fascination, Being away, Extent, and Compatibility (Kaplan & Kaplan, 1989; Kaplan, 1995).

1. Fascination expresses the fact that our preoccupations vanish in contact with the environment, and we become immersed in the present moment. In such a moment, the very essence of nature subjugates and inspires us. The so-called soft fascination (watching snow fall, trees, listening to birdsong, smelling plants, feeling connected to nature in a natural space) stimulates reflection while promoting the recovery of attention (Herzog et al., 2003).
2. Being away, refers to the physical and/or virtual distance from aspects of everyday life (ibid.; Kaplan, 1995).
3. Extent relates to the balance between the environmental elements which should be extended enough to occupy the mind (Kaplan, 1995), e.g., an environment with rich biodiversity is able to facilitate fascination and provide the opportunity to experience, contemplate or reflect (ibid.).
4. Compatibility is the relationship between the natural environment and the individual, one which allows the individuals to carry out their actions and accomplish their goals without the environment preventing them from doing so, thus creating feelings of freedom (ibid.).

Results

Over a period of three days in July 2021, a group of 30 people, diverse in age, gender, and nationality, were interviewed on their experience of Glendalough, with eight thematic topics identified: (1) nature escape; (2) freedom; (3) sensory heaven;

(4) calming and relaxing nature; (5) appreciation; (6) change of perspective; (7) memories; (8) life satisfaction (Table 18.1).

Theme 1: Nature Escape

When talking about how their visit to the area made them feel, participants perceived themselves as being away from something they considered somewhat negative, such as 'get out of the city', 'disconnect from work', 'disconnect from the screen', 'a break from work' and 'distance from work', or even 'away from the wife'. The notion of being away can be perceived as escape from one's usual setting, situation, or routine: 'I wanted to see around Ireland and also to get out of the city'.

Participants also discussed 'getting away' from situations such as stressful work, or the digital reality of their work life, referring to the fact that they have been working from home, and all their work was done virtually. 'Disconnect from work' can be seen as escape from a stressful situation, or an escape towards a much-needed mental restoration: 'It was good to get away from the screens. Because of

TABLE 18.1 Description of the identified themes

Theme	Description
1. Nature escape	Reports of nature being a means of escape and distance from one's worries, problems and stress associated with everyday life, as well as physical locations such as the city.
2. Freedom	Reports of nature being able to facilitate feelings of freedom.
3. Sensory heaven	Sensory (visual, auditory, tactile, and olfactory) input from nature as a driver of positive reactions such as relaxation and happiness.
4. Calming and relaxing nature	Interactions with nature and the reported calming and relaxing effect they had on visitors.
5. Appreciation	Reports of heightened appreciation of nature and life after visiting natural environments.
6. Change of perspective	Reports of viewing negative situations in a more positive light, thus having a better (or changed) perspective.
7. Memories	Experiences of the area were often linked with memories and past experiences, as a testimony to the influence memories have on shaping perceptions and representations.
8. Life satisfaction	Reported strong correlation between physical activity and resulting feelings of happiness, acknowledging self-worth and satisfaction.

COVID we were confined at home, working from home, and started feeling a bit disconnected'.

On the other hand, 'forget all my troubles' can be regarded as escaping our internal environment or mental state. In this case natural environments were seen to have positive effect on people's ability to recuperate after stressful situations. This was reflected in Stress Restoration Theory and Attention Restoration Theory. Thus, restoration, being attention or stress, is deemed to be essential for a healthy mind.

While the process of escaping reality by spending time in nature was a common topic among participants, it raised the question: what were they escaping to?

Theme 2: Freedom

Interviewees reported feeling free for the first time since the COVID-19 pandemic started. They reported previously feeling stuck at home, not being able to do what they wanted and feeling the need to get out. Feeling free to get out of home and venture into the nature was a way to avoid government-imposed travel and domestic limitations: people were feeling in charge of their own actions again, and they found this freedom through exploring open natural spaces. For example, when asked to describe what they felt is their state of mind at the time of the interview: 'Free, if that makes any sense, especially after COVID, not being able to get out for so long...' and 'Some people had cabin fever before the Pandemic but will go further out now just to get away from home, to break free'.

Being free is an essential aspect of psychological wellbeing (Gao & McLellan, 2018). The importance of freedom hinted through the compatibility factor of ART suggests that people should be able to do what they desire without being prevented from the environment, thus being free to carry out actions and accomplishing their goals. Visiting places such as Glendalough allowed people to do as they please, thereby creating a sense of freedom: 'You enjoy it [nature] more. Especially last year with the 5 km limitation it was very rough, not being able to get out anywhere. You are not just stuck at home'.

Theme 3: Sensory Heaven

Interviewees reported pleasant visual, auditory, tactile, and olfactory sensory input from nature. Some highlighted how the colours of the valley were magnified by the sun that day; others talked about the almost melodic sound of the river behind them; whereas others appreciated the absence of any sounds in the woods beside the lake and the relaxation this facilitated. The 'beautiful lakes' and 'spectacular mountain views' were also mentioned.

Some people appeared to enjoy the simple things nature could give them, such as the aesthetic pleasure, or the quietness of the area that facilitated a sense of calmness and relaxation. This can be related to the minimum cognitive effort required for mental processing during simple nature activities, such as 'walking around'. It can be said that in a way the environment facilitated fascination which in turn allowed the participants to focus their attention on simply enjoying the view, and completely pushing away feelings of stress and anxiety: 'It eases anxiety, gives me a break from work'.

These findings are in line with a study by Brymer, Crabtree and King (2020) which talks about immersion in the present moment and the ability of nature to help people disconnect from stress and worries through perceptual experiences: 'Sun, air, natural beauty – they all have positive effects on my mental health. I forget about work and just focus on the present moment'.

When asked to describe their state of mind, some of the responses were: 'visually happy', 'happier than before' and 'in a better mood'. These responses are a testimony to the ability of natural surroundings to alleviate stress and act as mood enhancers – natural therapists. This could be attributed to colour psychology and assumptions that the colour green evokes positive mood enhancement due to its direct association with nature, life, and regeneration. Or, it can be directly tied to the biophilia hypothesis, which states that we are and always have been innately drawn to nature and all things natural. By fulfilling this inherent need to be in contact with nature, we nourish our survival, as well as mental and existential wellbeing.

Theme 4: Calming Nature

The words 'calm' and 'relaxed' came up frequently when respondents described what they felt was their state of mind at the time of the interview. The landscape and even the effects on their mental health were described as calm and relaxing. In fact, two of the most common states of mind identified during the interviews were 'relaxed' (19 times) and 'calm' (9 times). Feeling 'de-stressed' was reported as both a mental state and a positive effect of people's visit. People experienced a new-found connection with nature as they saw its tranquillity and stillness being reflected in their own mental state: 'Well, I feel relaxed here. It's so quiet here by the lake that it makes me quiet down and release all the stress' and excitement when the surrounding environment was filled with their loved ones' laughter and happiness: 'I'm out with the kids. We just wanted to have fun and have a good time out together. I'm happy when they are'.

Theme 5: Appreciation

The term appreciation often emerged to describe the positive effects of visiting Glendalough. There was reflection over the beauty of the area, appreciation of the effect it had on wellbeing, and a newfound gratefulness to and for nature: 'I feel grateful for what we have on our doorstep, grateful that we are so lucky to live close by, 30–40 minutes away. I feel appreciation. You don't have to go far to see something so beautiful'; 'You forget how beautiful it is here and every time you come back you realize it all over again. You appreciate – no, re-appreciate – the beauty of the place'; and 'Not being able to enjoy what's on your doorstep [due to COVID-19] resulted in heightened appreciation of natural beauty and resources'.

Being grateful is essentially appreciating the benefits that people received from their time in Glendalough, thus resulting in an increase in 'life satisfaction, vitality, happiness, optimism and positive affect', and less 'anxiety and depression' (Alkozei et al., 2018).

Theme 6: Change of Perspective

With gratitude and appreciation came a changed perspective. Participants often expressed having their perspective changed after their visit to Glendalough. What they meant by that was a complex shift of perceiving a certain situation as very troubling and stressful to viewing it from a different perspective or being less problematic. People generally felt better after engaging in nature-based activities which allowed them to have a more positive view of their life and situation. They were comparing the extent of the valley, or the size of the boulders scattered around it, against the size and severity of their problems, and the result was a more optimistic view of their situation and hope that things would get better: 'You feel so small here. Everything around you is so big. Look at these boulders... You know, it puts things into perspective. Your problems seem so small, insignificant...' and 'The place changes my perspective, and my problems fade away here, they seem insignificant'.

The ability of nature to facilitate a state of reflection was reported by many people as feelings of being 'in my thoughts' or being 'reflective'. People were able to be nothing but mere observers of their own situation through a process of personal reflection which allowed them to see their problems from a different, more positive angle. After that they reported being able to solve complicated tasks or situations and being more productive in their endeavours.

Theme 7: Memories

A common theme among the interviewees was the one of memories connected to the place. People talked about how they came here as children and have very fond memories of the area, which always brings them back. Some spoke about 'nostalgia' for their childhood awoken by the place, others mentioned 'sentimental' when describing the reason for their visit, and some said they were familiar with the area and felt comfortable coming here again. All this is related to our memories of the place. We no longer see the landscape simply as a natural territory; we see it as a lived space refracted through the prism of our memory, thus evoking different emotions in us: 'Well, it's funny, but our first date was here. Also, it's her birthday so we wanted to do something special' and 'There are so many memories connected to this place. We've been coming here since we were kids. And always happy memories. I feel connected to the place as it's been present while I was growing up'. Some people even reported feeling a special attachment to the area due to fond memories of past experiences. Thus, finding themselves there again gave them a surplus of wellbeing, which could be similar to the feeling of being 'at home'.

Theme 8: Life Satisfaction

Much of the findings were related to people feeling a sense of self-satisfaction after spending time in nature, often attributed to the activities performed in nature as well as the simple presence of nature and the act of reconnecting to it. People talked about feeling 'energised', a 'good type of tiredness', 'focusing on the exercise and forgetting all problems', 'achieved something with your day', being 'fulfilled'. For interviewees, life satisfaction was brought about by a combination of physical activities, aesthetics, and sensory inputs from nature with their effect on cognitive functions, as well as biological processes occurring in the human body after physical activity. Which is why majority of the people interviewed reported feeling happy, feeling well, feeling satisfied, even changing their perspective after visiting Glendalough. Some of the interviewees mentioned 'clear my mind' as a positive effect of their visit to Glendalough, referring to the fact that the area was facilitating relaxation and allowing the visitor to be away from places associated with stress. One interviewee mentioned specifically 'the dopamine effect' and explained that they felt stress relief and satisfied with their effort. People felt like they were doing something productive with their day and that ultimately had positive effect on their wellbeing: 'It's good to get out in nature. You feel like you've achieved something with your day. My day feels fulfilled and I feel better' and '...physically energised after walking the three-hour loop. Tired because of the heat, but happy with myself'.

Overall, the practice of physical activity had the power of distraction in the face of everyday-life difficulties; of developing the feeling of efficiency and self-esteem; and an energising power which promoted mood enhancement, improved the attentional and cognitive capacity (Lawton et al., 2017) and resulted in participants reporting feelings of personal goal fulfilment and life satisfaction.

Glendalough – an Area of Potential for Nature Therapy?

The responses to the final survey question, Q9 – 'If there was Nature therapy here, would you come?' – outlined the opportunities and challenges when considering abandoned landscapes as places of potential for nature therapy. As mentioned earlier, Nature therapy is a broad concept, encompassing various activities in nature, having no uniform application and no expected outcomes, thus it can be guided or self-guided nature experience.

However, during the interviews, it became evident that people were not completely familiar with the concept of 'nature therapy'. Confusion and negative associations with the word 'therapy' emerged. Majority of the people expressed concerns about the concept of guided therapy and reflected on the fact that nature is free, and they shouldn't pay for receiving its benefits. The overall feeling around the topic was that people were happy to continue doing nature-based activities on their own, however, not so much in groups and not having to pay for it: 'It would depend on what it is. I wouldn't do something guided, but more something myself'.

The social aspect of guided nature therapy was identified as a setback rather than encouragement for people to participate. Also, it seemed as though guided therapy was perceived as something that would limit people's freedom in the area thus impacting on their subjective wellbeing:

I am not a big fan of somebody telling me what to do and how to spend my day outdoors. I wouldn't be able to properly experience the nature, also I enjoy being in nature with the least amount of people possible. But I think for a lot of people it would work. Maybe I would only do it if it would teach me something I can apply on my own.

Nevertheless, the word 'therapeutic' was used to describe the time spend in Glendalough by several participants. And the therapeutic effect of the area was always attributed to the presence of nature. Glendalough's past as a mining centre did not impact people's perceptions of the area as natural, wild, and beautiful. Several people didn't even think about it until it came up during the interviews, saying: 'Nature has reclaimed the place' and 'It [the mining site and remnants] has become a part of the landscape, just as natural as the trees around. They do not impact on my idea of the place as natural. It is wild and rugged'.

Nature allowed people to overlook the history of the place as a centre of industrial activities, the perceptions of which are almost always negative, by masterfully incorporating the remainders of these activities into the landscape, thus making it almost difficult for the visitor to imagine the place without them.

These discoveries can be used as an example for answering the bigger question: are abandoned landscapes places of potential for nature therapy? They show that through good restoration practices, environmental responsibility, and awareness, once abandoned landscapes can be rejuvenated and re-invented into the future centres of human-nature connectedness, places of solace, refuge, wilderness, and areas for nature therapy, nurturing the future healthy, happy, and sustainable generations.

Conclusion

In summary, the interviewees described Glendalough as a beautiful, peaceful, natural environment providing nature experiences as an alternative to everyday life, which allowed temporary rest from the attention efforts that certain activities require (work, problems, and worries, etc.) associated with stress. By immersing themselves in a fascinating natural environment, consistent and compatible with their innate needs, the participants mentally moved away from their preoccupations. Restoration of attention and healing of the mind was based on: fulfilling our inherent need to be in contact with nature (i.e., biophilia), thus contributing to our survival, psychic development and adaptation to an ever-changing world; cleaning the mind of residual cognitive noise produced during daily tasks by focusing it on the sensory inputs Glendalough provided; recovering attention fatigue by focusing on physical activities, such as hikes and walks; feeling life satisfaction by focusing on fulfilling complicated tasks ('hiking the Spink'); reflecting on existential questions such as life perspective, priorities, purpose, and one's place in the universe, thus changing their perspective to a more positive one, which resulted in general happiness: 'I definitely feel happier than before!'

People reached a state of relaxation which allowed them to get out of their heads and gain a new perspective on their life and their problems. Thanks to this new-found (re)connection to nature people developed a greater appreciation for it. And during times, such as the COVID-19 pandemic, people found their way back to nature, immersed into it, became part of it and found solace in it, which resulted in an enhanced sense of eco-responsibility.

Given the results of this study, Glendalough, indeed, promoted mental wellbeing. This makes it an area of great potential for nature therapy, aimed in restoring the human psyche as well as the natural environment, as they are both inherently connected and co-dependant. These findings are a testimony to the hidden potential abandoned landscapes have in providing an environment that nurtures our mental health. This is in contrast with the common misconception of these areas as wastelands, hostile to nature and having little or no value as ecosystems. For those who foresee their restoration and potential reuse, they have the potential to boost biodiversity, people-nature connectedness, and overall wellbeing.

Recommendations

Some forms of nature therapy, such as the Japanese practice of *Shinrin-yoku* ('forest bathing') can be of interest (Kotera et al., 2020). This practice consists of mindful walks in any natural landscape, where one is focusing on their senses and what surrounds them. It can be organised by professionals or self-guided. Most importantly, the practice can be freely applied by anyone looking for a restoring and therapeutic experience. It is individualistic, non-invasive, self-paced, trauma-informed experience, where we and nature are the guides towards finding inner peace.

Gardening, or horticultural therapy, is considered to have similar effects. The impacts of gardening on wellbeing have been related to attention restoration and stress restoration (Clatworthy et al., 2013). Gardening can be practiced at home or in community gardens, alone or in groups. It can consist of simply walking around and observing gardens, or in performing the act of gardening itself. It provides a safe space for exploring nature and has been found to result in increased overall wellbeing and social inclusion for people suffering from mental health illnesses and traumas (Söderback et al., 2004).

If cities want to provide their residents with safe spaces to explore and enhance their wellbeing, greater focus should be put towards the planning and inclusion of nature pockets – forested urban parks and community gardens – in the urban landscape. By recognising and integrating our inherent affinity for nature, informed placemaking can create more harmonious and sustainable spaces that promote both physical and psychological wellbeing. This will be a step towards the move to more sustainable cities and communities, as well as a steppingstone towards overcoming traumas and healing of the collective mind.

Based on these conclusions it is highly advised that you spend more time outdoors, in natural spaces. Go and explore that mountain/forest/park/garden you so often pass by on your way home. Take in the view. Smell the flowers. Get your hands in the moist earth. Find shelter under the trees. Swim in the cold lake's waters. Rest your head on the grass. Listen to the birds' songs. Breath in… and enjoy the process that begins inside of you!

References

Alkozei, A., Smith, R. & Killgore, W. D. (2018). 'Gratitude and subjective wellbeing: A proposal of two causal frameworks', *Journal of Happiness Studies*, 19(5).

Barrow, L. (1974). Glendalough and St. Kevin. *Dublin Historical Record*, 27(2).

Bratman, G. N., Anderson, C. B., Berman, M. G., Cochran, B., De Vries, S., Flanders, J., Folke, C., Frumkin, H., Gross, J. J., Hartig, T. & Kahn, P. H. (2019). 'Nature and mental health: An ecosystem service perspective', *Science advances*, 5(7), p.eaax0903.

Brymer, E., Crabtree, J. & King, R. (2020). 'Exploring perceptions of how nature recreation benefits mental wellbeing: a qualitative enquiry', *Annals of Leisure Research*, pp. 1–20.

Capaldi, C. A., Passmore, H. A., Nisbet, E. K., Zelenski, J. M. & Dopko, R. L. (2015). 'Flourishing in nature: A review of the benefits of connecting with nature and its application as a wellbeing intervention', *International Journal of Wellbeing*, 5(4).

Clatworthy, J., Hinds, J. & Camic, P. M. (2013). 'Gardening as a mental health intervention: A review', *Mental Health Review Journal*, 18(4).

Gao, J. & McLellan, R. (2018). 'Using Ryff's scales of psychological wellbeing in adolescents in mainland China', *BMC psychology*, 6(1).

Gullone, E. (2000). 'The biophilia hypothesis and life in the 21st century: increasing mental health or increasing pathology?', *Journal of happiness studies*, 1(3).

Herzog, T. R., Maguire, P. & Nebel, M. B. (2003). 'Assessing the restorative components of environments', *Journal of environmental psychology*, 23(2).

Kaplan, R. & Kaplan, S. (1989). *The experience of nature: A psychological perspective*. Cambridge: Cambridge University Press.

Kaplan, S. (1995). 'The restorative benefits of nature: Toward an integrative framework', *Journal of environmental psychology*, 15(3).

Kellert, S. & Calabrese, E. (2015). *The practice of biophilic design*. London: Terrapin Bright LLC.

Kotera, Y., Richardson, M. & Sheffield, D. (2020). 'Effects of shinrin-yoku (forest bathing) and nature therapy on mental health: A systematic review and meta-analysis', *International journal of mental health and addiction*, pp. 1–25.

Lawton, E., Brymer, E., Clough, P. & Denovan, A. (2017). 'The relationship between the physical activity environment, nature relatedness, anxiety, and the psychological wellbeing benefits of regular exercisers', *Frontiers in psychology*, 8.

Lewis, R. (2017). 'A qualitative study of psychotherapists' experience of practising psychotherapy outdoors' (Doctoral dissertation, Dublin Business School) (dbs.ie). [Accessed: 3 May 2021].

Mighall, T. M., Timpany, S., Critchley, M. F., Martinez Cortizas, A. & Silva Sanchez, N. (2013). 'A palaeoecological assessment of the impact of former metal mining at Glendalough, County Wicklow Ireland', *Mining Heritage Trust of Ireland*, 13.

Ritchie, H. & Roser, M. (2018). *Urbanization. Our World in Data.* Available from: Urbanization - Our World in Data Accessed 29 April 2021.

Schmitz, B. R. (2012). *Nature-deficit disorder and the effects on ADHD.*

Seymour, V. (2016). 'The human–nature relationship and its impact on health: a critical review', *Frontiers in public health*, 4.

Söderback, I., Söderström, M. & Schälander, E. (2004). 'Horticultural therapy: the 'healing garden' and gardening in rehabilitation measures at Danderyd Hospital Rehabilitation Clinic, Sweden', *Paediatric rehabilitation*, 7(4).

Ulrich, R. S. (1981). 'Natural versus urban scenes: Some psychophysiological effects', *Environment and behavior*, 13(5).

van den Bosch, M. & Meyer-Lindenberg, A. (2019). 'Environmental exposures and depression: biological mechanisms and epidemiological evidence', *Annual review of public health*, 40.

Wicklow Endless Opportunities (2017). 'Wicklow Tourism and Strategy Plan'. Available: www.wicklow.ie/Portals/0/Documents/Strategy%20Documents/County%20Wicklow%20Tourism%20Strategy%20and%20Marketing%20Plan%202018%20-%202023.pdf. [Accessed: 1 August 2021].

Wicklow Mountains National Park (2022). *The Geography of Glendalough: Notes for Teachers.* Available: www.nationalparks.ie/app/uploads/2022/09/JC-Geog-Teacher-Notes.pdf. [Accessed: 14 July 2023].

World Health Organisation (2016). 'Investing in treatment for depression and anxiety leads to fourfold return. Available from: Investing in treatment for depression and anxiety leads to fourfold return' [online], 13 April 2016. Available: www.who.int/news/item/13-04-2016-investing-in-treatment-for-depression-and-anxiety-leads-to-fourfold-return. [Accessed: 20 April 2021].

World Health Organisation (2023). 'Depressive disorder (depression)' [online], 31 March 2023. Available: www.who.int/news-room/fact-sheets/detail/depression. [Accessed: 13 July 2023].

19

THE PROMISE OF TRAUMA-INFORMED MIGRANT PLACEMAKING

Arts-based Strategies for Compassion and Resilience

John C. Arroyo and Iliana Lang Lundgren

Trauma-informed migrant placemaking is a distinct type of placemaking occurring around the globe in local communities undergoing population growth and demographic transition. As borders continue to blur, push factors, such as political instability and religious persecution, and pull factors, such as improved economic opportunities, are increasing migration rates. The reception of migrants, however, is a wide and varied spectrum. All too often newcomers of all kinds face discrimination, legal threats, and harsh living conditions in the tense places they intended to be their refuge.

Trauma-informed migrant placemaking considers the type of trauma migrants experience in particular contexts after migrating from their home country to their host country. It responds, often through participatory and community-based art, to build or create a sense of place that addresses trauma and supports newcomers through their unique experiences as forced or voluntary immigrants, refugees, or asylees. While immigrants typically move to a new country for safety, economic opportunity, family reunification, or other similar reasons, refugees and asylees have been forced to leave their country due to hostile circumstances including war, violence, or climate-related emergencies. The value of trauma-informed migrant placemaking lies in how it combats isolation, anxiety, and pain by building compassionate and resilient alliances across variegated migrant communities and longer-term (extant) residents in the same region.

Our analysis is based on inductive reviews of trauma-informed migrant placemaking examples in creative placemaking literature, local media outlets, creative placemaking and social practice grant awardee databases, the American Planning Association's award winners, and online community forums. This review drew from an initial database of 250 creative placemaking cases (200 in the US and 50 international cases) over the last 13 years; narrowing down to four representative

DOI: 10.4324/9781003371533-23

cases of trauma-informed migrant placemaking within the larger pool. This chapter presents the findings of this study across four core themes, with short illustrative examples of practice given. It concludes by offering brief recommendations for planners, designers, policymakers, and cultural workers aiming to include newcomers in arts-based interventions across rural and urban areas (Arroyo, 2020). Four key lessons from trauma-informed migrant placemaking projects highlight the need for art projects to serve as forums to discuss difficult topics (such as family separation and deportation), to provide an entry into democratic community-based processes that may be unfamiliar to newcomer populations, to uncover the power of food and language stewardship as a bridge for cultural adaptation, and to identify when place-based projects must transition to broader, comparative scales and activism.

Trauma-Informed Placemaking Theme One: Forced Family Separation

Forced family separation occurs when a family member must live in a separate geographic (typically country) location from other family members. It is involuntary, often the result of legal enforcement due to deportation, and when without legal permission to migrate to and live in a new country, migrants often live in fear of detainment or deportation. In the US, children born to immigrant parents have legal status to remain in the country, while their parents may not have this permission. In these cases, parents and sometimes children are separated from each other. In other instances, young children migrated with their parents and grew up in the US as undocumented individuals, leading to nearly 12 million people, many of whom have registered under the Deferred Action for Childhood Arrivals (DACA) or their parents as Deferred Action for Parents of Americans (DAPA) (Kamarck & Stenglein, 2019).

Separation can also occur during the migration process. For example, the 'zero tolerance' policy in the United States (implemented in May 2018, later revoked by the Justice Department in January 2021) separated many parents from their children while families were attempting to cross the US-Mexico border (Diaz, 2021). Policies such as 287(g) in the US continue to delegate federal authority to state and local law enforcement officers to perform specified immigration officer functions (such as the largest number of family separations) under the direction and oversight of the United States Immigration and Customs Enforcement (US ICE, 2023). More recent US policies such as now concluded Title 42 of the Public Health Services Act authorized the Director of the Centers for Disease Control and Prevention (CDC) to suspend entry of individuals into the US to protect public health due to high rates of COVID-19 transmission (Pillai & Artiga, 2022). Trauma can result from the actual separation of family members as well as fear of potential future temporary or permanent separation when family members do not have the legal status to remain in the country.

The Papalote Project is a trauma-informed art initiative supporting children impacted by the deportation of their parent(s) from the US by increasing awareness and understanding of the realities of immigration challenges for many families. In 2013, Artist Rosalia Torres Weiner worked with young people to design kites as an outlet to process emotional trauma caused by separation from their parents. During one workshop hosted by a church in Charlotte, North Carolina, children included pieces of their family members' clothing in the kite designs. Rosalia wrote, 'The concept of the art installation is to simulate the kites in flight conveying that by attaching their emotions to the kites, they are released, and the children can begin to heal within a caring community' (Garza, 2013). The kites supported immigrant youth by creating a place for them to share and release their emotions in a playful way, while the exhibit at the Levine Museum displaying the kites, *El Papalote Mágico*, brought attention to family separation and its devastating impacts in a series of community conversations and engagement around this topic through public programs hosted at the museum and in the community. This exhibition brought attention to the individualized trauma of family separation that some immigrant families face and highlighted the voices of immigrant youth, voices that are often silenced in mainstream immigration debates.

While *The Papalote Project* had substantial organizational support and received attention from the public, other placemaking initiatives for separated immigrant youth are less obvious or not commonly categorized as placemaking. Victoria 'Vicko' Alvarez's comic book for youth, *Rosita Gets Scared,* is an act of trauma-informed placemaking as it tells the story of a young immigrant girl in the US, articulating the fear of deportation and family separation while sharing how Rosita processes her feelings (Moreno, n.d.). The book was published in 2017, partly in response to the rising fear and tension in immigrant communities due to the Trump Administration's anti-immigrant agenda, and is available in multiple languages, including Spanish. Alvarez collaborated with Organized Communities Against Deportations to incorporate *Know Your Rights* information in the book at a child-appropriate level. The book includes a glossary of key terms and concepts for children as part of the fear stems from a lack of understanding of how and why people are deported and separated. This book allows immigrant and non-immigrant children at home and in school to better understand the complexities around deportation, encourages feelings of empathy and understanding, and provides children and their families with useful strategies to process emotions (Berthet Garcia, 2017; Moreno, n.d.). Immigrant youth can see themselves in Rosita, understand that they are not alone, and feel that they have a rightful place in the world despite the narratives of hate and fear that are often connected to immigrant families.

Trauma-Informed Placemaking Theme Two: Persecutive Displacement

Persecutive displacement is specific to experiences of hostility that result in the movement of migrants from one place to another. This displacement, like family separation, is forced because a community of people is not accepting a person or group of people. In persecutive displacement, the lack of acceptance and ill-treatment can be connected to a person or family's religious beliefs, ethnicity, cultural background, political views, gender, sexual orientation, or any other part of a person or family's identity. Trauma from persecutive displacement can be associated with past experiences of violence during a migration journey or in the host society.

The *Victoria Square Project* (VSP), located on Elpidos Street in Athens, Greece, focuses on cultural exchange in its placemaking initiative. Artists Rick Lowe (US) and Maria Papadimitriou (Greece) designed the project in 2017 in response to the influx of Syrian refugees in Greece fleeing violence and persecution associated with the Syrian Civil War (Victoria Square Project, n.d.). It is a physical space, described as an 'evolving social sculpture' and the 'Living Room space', that facilitates cross-cultural exchange, social gathering, and empowerment of residents including the immigrant population (Rava Films, 2018). The space is operated by and for the community. It hosts young artists, workshops centered around the concept 'Who is a contemporary Athenian?', book clubs, workshops about textiles from different cultures, and youth and adult programs to educate each other about different countries.

In 2022, VSP ran a program called *Victoria travels to…* and invited youth to free workshops to learn about Nigeria and Albania through art and through children who had immigrated from these countries. While VSP's founding was funded by the Stavros Niarchos Foundation (a member of ArtPlace America's funding consortium) in collaboration with Allianz Kulturstiftung, specific projects are often supported by the Greek Ministry of Culture or the US Embassy in Athens. A key element of VSP is community leadership, including recent immigrants, who design the programming and space with a participatory model.

Experiences of trauma for migrants often stem from the past, however, we found that many migrants also experience new trauma from persecutive displacement. *NakivArt* and *Colors of Connection* are trauma-informed placemaking projects working with refugees and conflict-affected populations. *NakivArt* hosts art therapy workshops for refugees at the Nakivale Refugee Settlement in Insingiro, Uganda. The workshops support youth to talk about past traumas and use art to share feelings and process stress. Youth face challenges with current feelings of isolation, loss of their homes, and trauma from their migration experiences. Workshops utilize drawing, filmmaking, and theater/skits to create a temporary space for youth to create new bonds and a sense of home (Lindrio, 2018).

Colors of Connection runs arts-based programs for girls and young women, such as the Courage in Congo (Figure 19.1) program in Goma, a region heavily impacted by the conflict in Eastern Congo. The organization has run similar programs within refugee camps in Burkina Faso and a post-war town in Liberia to support young people who have experienced gender-based violence and abuse and are managing depression, anxiety, stigma, and social isolation (One Global

FIGURE 19.1 Community and participants viewing completed murals, April 2016. Image credit: Colors of Connection.

Voice, 2021). *Colors of Connection* relies on art-based activities for individuals and groups such as community mural-making to support participants to reduce anxiety, boost self-esteem, and build healthy relationships.

Outside of the refugee context, voluntary immigrants also face past and current trauma from persecutive displacement. The *'Tied Together by a Thousand Threads'* mural in Phillips Square, Boston, reflects themes of displacement and gentrification – both past and current – for the Chinese immigrant community (Sun, 2020). Multiple generations of Chinese residents came together to paint the mural, led by artist Shaina Lu, to highlight Chinatown's history as a garment-making hub and capture the challenges the community is facing today (Deng, 2019). The act of painting this mural and the existence of the mural itself is an example of trauma-informed placemaking by multiple generations of immigrant families.

Trauma-Informed Placemaking Theme Three: Cultural Displacement

A third type of trauma for migrants is associated with cultural displacement. This type of displacement can include loss of culture, home, or traditions during or after a migration experience. It conjures the same effects of 'assimilation', a highly criticized process newcomers undergo when they replace their cultural customs and traits to adapt to traditional standards of a host country's society. In an immigration context, the term implies newcomers are unable to maintain aspects of their native culture while simultaneously adopting new ones where they've settled. Oftentimes, calls for assimilation are synonymous with racial and ethnic othering – a concern about people who are different from us and the foundation of xenophobia. In our exploration of trauma-informed migrant placemaking cases identity loss, isolation, feelings of 'homelessness', lack of agency, and the devaluation of cultural knowledge and traditions emerged as prevalent themes.

The *New Roots Farm Incubator* (2010) program in Salt Lake County, Utah works to address the devaluation of cultural knowledge and loss of connection to cultural traditions by actively celebrating the traditional cultural assets and food/cooking of Bhutanese refugees and other migrants. Supported by Salt Lake County's Office of New Americans and in collaboration with the International Rescue Committee, one of the key refugee relocation organizations approved by the US government, this program works with 150 farming families and encourages longer-term residents to learn new skills from the refugee community and try traditional food from their cultures while refugees retain a connection to their culture (Cortez, 2011; New Roots, n.d.). Many refugees have experience with agriculture in their homeland, and the New Roots Farm provides an opportunity to tend to traditional crops from their regions such as African eggplant, Thai chili peppers, *molokhia* (Egyptian Spinach), and long beans (Utah Department of Agriculture, 2022). The opportunity to build community (with other migrants as well as longer-term residents), grow familiar food, and earn income supports migrants in remaining connected to their homeland while building healthy, sustainable lives for themselves and their families in a new country (Nalewicki, 2021). Most of all, *New Roots* uses food as a common convener to bridge commonalities between newcomers and extant populations.

Just as food and agriculture traditions are foundational to culture, language can help create a sense of place for migrants. To prevent and reduce cultural displacement, Pablo Helguera created *Librería Donceles*, a traveling bookstore and art project with Spanish-language literature. The bookstore served as a community meeting place, hosted events to foster cultural understanding and tolerance, and addressed the gap in literary outlets for the Spanish-speaking community in the US (Freitas, 2015). It was first installed in Brooklyn, New York, in 2013, then traveled to Phoenix, San Francisco, Seattle, Chicago, Indianapolis, and Boston over the course of four years. This form of mobile placemaking is temporary, and while it is not a permanent solution, its purpose is to engage local communities across the country with new ideas for placemaking and opportunities to keep language and culture alive for immigrant families and future generations.

Trauma-Informed Placemaking Theme Four: Lack of Visibility and Misrepresentation

As seen in some of the previous examples, trauma-informed placemaking in migrant communities also addresses the lack of visibility migrants experience when their stories and voices are suppressed. Our analysis discovered placemaking

FIGURE 19.2   *Little Amal* in Lublin, part of her journey in Ukraine and Poland to visit Ukrainian refugee children and families. Image credit: The Walk Productions and Handspring Puppet Company, photograph by Ignac Tokarczyk.

initiatives that address the misrepresentation of immigrant stories and create places and platforms for immigrants to tell their own stories.

In 2021, Good Chance Theater organized *The Walk with Little Amal* (Figure 19.2), a mobile placemaking initiative, and created a platform for the voices of young refugees. The moving arts festival traveled through Turkey, Greece, Italy, Switzerland, Germany, Belgium, France, and the UK, following a common path for migrants in this part of the world, and offered the opportunity for refugees to share their stories and connect and build relationships with new communities (Futurecity, 2021). *Little Amal*, a sculpture/giant puppet of a young Syrian refugee child, was the symbol of the walk, and music, dance, and theater were incorporated into this mobile placemaking initiative. She stands for 'children fleeing war, violence and persecution and each with their own story' and '*Little Amal*'s urgent message to the world is ' "Don't forget about us" ' (Walk with Amal, n.d., n.p.). While the journey across Europe is complete, this project and the symbol of *Little Amal* continue as she visited New York City from 14 September to 2 October 2022.

Little Amal is 'an international symbol of compassion and of human rights' and 'carries a message of hope for displaced people everywhere, especially children who have been separated from their families' (Walk with Amal, n.d., n.p.). This placemaking initiative is unique in its global scale, yet hyper-local focus in communities around the world. This is possible through partnerships created with local communities that host 'Events of Welcome' for *Little Amal* (ibid.). In New York, *Little Amal* met with artists, community leaders and groups, and residents of New York from diverse backgrounds. Her journey in New York tells the story of a potential refugee's experience, from arriving at JFK airport to meeting a new friend at Central Park to feelings of loneliness and a dance experience in Washington Heights to looking for shelter with others in the East Village. In the UK, the story of *Little Amal* is told similarly but in the context of the local communities in this country (Futurecity, 2021). The stories of *Little Amal*'s journey are made possible by partnerships with theaters, art organizations, and public institutions. This placemaking initiative increases awareness around the experiences of young refugees and communicates the challenges they face, feelings of fear and loneliness, and the value of new friendships while promoting a movement to welcome all newcomers.

Lessons from Trauma-informed Migrant Placemaking

Our review of arts-based trauma-informed migrant placemaking projects revealed a series of lessons shared across varying migrant reception conditions. First, several migrant examples focused entirely on the family unit, while others used craft and writing to focus on the delicate nature of communicating forced family separation to youth. In The Papalote Project, the use of a playful artistic medium such as kites provided undocumented youth in North Carolina an opportunity for their entire family to discuss their concerns about deportation and safety. In *Rosita Gets Scared*, the story uses illustrations as a tool to explain the complex impacts of immigration and deportation on the family unit, including extended families. Rosita humanizes undocumented youth to be more than just a statistic or scare tactic for anti-immigrant rhetoric. Both projects signal the need for a broader, non-traditional form to get to the heart of a difficult aspect of contemporary migration challenges in the US.

Second, we found that persecutive displacement is often associated with violence and this can occur before, during, and/or after an initial experience of displacement. This violence includes physical and emotional abuse and is enacted by different actors – from detention center employees, family members, community members, police officials, and other 'so-called' trusted officials often at the community level, making streets or cities unsafe. In the *Victoria Square Project*, the living room and sculpture spaces were necessary to make Nigerian and Albanian youth feel welcome. In many countries, elected officials and law enforcement are not seen as trusted members and are often the reason refugee communities have fled their home communities. The *Colors of Connection* project offered a safe forum unavailable in other elements of society to support women and young girls experiencing gender-based violence. *Tied Together by a Thousand Threads* highlighted the importance of history as a marker to show how some of the challenges facing earlier waves of immigrants endure. In all three instances, the visual arts and sculpture served as introductions to the democratic process previously unknown to many newcomers in their host communities. Each project hoped to inspire a new generation of community leadership, whether in the specific art project or in broader governing bodies.

Third, our analysis found that ephemeral and mobile projects on food and language preservation serve as an essential bridge between newcomers and their experiences of cultural adaptation. Outdated expectations of assimilation have often prevailed in migration trajectories. By transitioning to cultural adaptation, newcomers are invited to maintain all their preferred cultural traditions while adapting new ones from their host society, especially in second and consecutive generations. In *New Roots*, the Salt Lake County Office of New Americans relied on a seasonal local farm as a common denominator to bring together Salt Lake County's growing refugee resettlement community. The value of connecting through crops sowed the seeds for supporting both immigrant and non-immigrant-owned restaurants in the area. In *Librería Donceles*, a temporary, traveling bookstore filled a gap in Spanish-language literature as a supplement to any English as a Second Language acquisition or as a strategy for maintaining Spanish language proficiency through native speakers, heritage speakers, and novice learners. By having a footprint – even when mobile or seasonal – newcomers participating in *New Roots* and *Librería Donceles* were invited to lay unique claim to physical space.

Finally, we found that theater was a valuable medium to transition the lack of visibility and misrepresentation from place-based work to a regional scale. For example, through the *Walk with Little Amal*, the Good Chance Theater became an international symbol of human rights, in Turkey, Italy, and Germany, countries undergoing a refugee crisis only exacerbated by political division standing against their welfare. In the US, Little Amal's visit touched upon the loneliness and accommodation precarity refugees often experience. Trauma-informed placemaking projects such as this one illustrate that some of the most vulnerable communities in society yearn for creative opportunities to share their voice and story. From the *Undocumented and Unafraid* movement in the US to global *Refugees Welcome Here* initiatives, trauma-informed placemaking projects provide a safe and necessary venue for expression and policy change for newcomers unable to find similar outlets in other facets of their life.

Conclusion: A Creative Space for Healing

Trauma-informed migrant placemaking exists in many different forms as it responds to trauma that is unique to the migrant experience: (1) trauma caused by forced family separation; (2) trauma resulting from persecutive displacement; (3) trauma associated with cultural displacement; and (4) trauma stemming from lack of visibility or misrepresentation. The cases shared here strive to center migrants of all types in art initiatives. Each addresses relationships with longer-term residents to create more welcoming, inclusive spaces that allow people to process past trauma and prevent future experiences of pain.

While every migrant's story is unique, it is important for planners, designers, policymakers, and cultural workers to understand that some experiences of trauma come from collective circumstances. Trauma-informed migrant placemaking commonly occurs in group settings while also allowing space for people to process individual emotions. It creates a more casual and creative place, physical space, or a moment in time – and requires less threatening or re-traumatizing energy – where migrants express themselves through various art forms and media, release pain, and build community. Hyper-local projects like The Papalote Project help diffuse the othering that often brews tension in traditionally conservative communities like the US southeast. Broader initiatives, like *Walk with Little Amal* move outside of the migrant community to educate others about the experiences of immigration in efforts to foster environments that welcome, accept, and honor newcomers of all strides.

By choice or by force, as migrants continue to arrive in new parts of the world, it is imperative that supportive ecosystems in communities, cities, and countries actively support trauma-informed migrant placemaking. Planners and designers must work with municipal local arts agencies and economic development bureaus to create policies, programs, and formal guidelines for trauma-informed migrant placemaking in general plans beyond typical arts district overlay zones or specific cultural plans not linked to the comprehensive plan. Non-profit leaders can use lessons from trauma-informed projects to provide new resources, advocacy, and initiatives based on the direct experiences and requests of the migrants they serve. Cultural workers can play a key role in creating arts-based projects based on positive narrations about migration. The right to migrate is as much of a human right as the right not to migrate. As a result, there is tremendous value in supporting participatory art initiatives that create a space for migrants to process trauma and rebuild their lives on their own terms. The arts can educate, heal, inform, and support communities. Artists and cultural workers invite the public to visualize trauma and pain that cannot be articulated in other forms and break down barriers in communication between diverse groups. Above all, trauma-informed migrant placemaking creates a symbolic and physical space, in and of itself, that frees migrants from generations of trauma.

References

Arroyo, J. (2020). *Bridging Divides, Creating Community: Arts, Culture, And Immigration.* A Creative Placemaking Field Scan, ArtPlace America, LLC and Welcoming America. Available: https://welcomingamerica.org/resource/bridging-divides-creating-community-arts-culture-and-immigration/. [Accessed: 12 August 2023].

Berthet Garcia, N. (2017). 'The Artist Behind Scholar Comics has a New Project to Talk to Kids About Deportation Fears', *Mijente*, 4 August. Available: https://mijente.net/2017/08/artist-behind-scholar-comics-new-project-talk-kids-deportation-fears/. [Accessed: 12 August 2023].

Cortez, M. (2011). 'New Roots Farm to Nourish Bodies and Souls of Refugee Families' [online], *KSL.com*, 21 April 2011. Available: www.ksl.com/article/15232004/new-roots-farm-to-nourish-bodies-and-souls-of-refugee-families. [Accessed: 12 August 2023].

Deng, O. (2019). 'Murals, Massages and Mandarin: Chinatown Uncovered', *Boston Magazine* [online], 5 February 2019. Available: www.bostonmagazine.com/news/2019/02/05/murals-massages-and-mandarin-chinatown-uncovered/. [Accessed: 12 August 2023].

Diaz, J. (2021). 'Justice Department rescinds Trump's 'Zero tolerance' immigration policy', *NPR*, 27 January 2021. Available: www.npr.org/2021/01/27/961048895/justice-department-rescinds-trumps-zero-tolerance-immigration-policy. [Accessed: 12 August 2023].

Freitas, A. (2015). 'A glimpse inside Librería Donceles, NYC's only Spanish-language used book store', *Remezcla* [online], 2 April 2015. Available: https://remezcla.com/features/culture/a-glimpse-inside-libreria-donceles-nycs-only-spanish-language-book-store/. [Accessed: 12 August 2023].

Futurecity (2021). 'Little Amal inspires support for refugees along #thewalk–an epic 8,000 kilometres theatrical journey across Europe' [online]. Available: https://futurecity.co.uk/little-amal-inspires-support-for-refugees-along-thewalk-an-epic-8000-kilometres-theatrical-journey-across-europe/. [Accessed: 12 August 2023].

Garza, R. (2013). 'Hope after deportation in Charlotte, North Carolina: Papalote Project and Latin American coalition support immigrant communities', *UnidosUS* [online], 15 February 2013. Available: www.unidosus.org/blog/2013/02/15/hope-after-deportation-in-charlotte-north-carolina-papalote-project-and-latin-american-coalition-support-immigrant-communities/. [Accessed: 12 August 2023].

Kamarck, E. & Stenglein, C. (2019). 'How many undocumented immigrants are in the United States and who are they?', *Brookings Institute* [online], 12 November 2019. Available: www.brookings.edu/policy2020/votervital/how-many-undocumented-immigrants-are-in-the-united-states-and-who-are-they/. [Accessed: 12 August 2023].

Lindrio, P. (2018). 'Art Therapy: Refugees in Ugandan Camp Using Drawing and Drama to Heal Trauma', *Global Press Journal* [online], 16 November 2018. Available: https://globalpressjournal.com/africa/uganda/art-therapy-refugees-ugandan-camp-using-drawing-drama-heal-trauma/. [Accessed: 12 August 2023].

Little Amal (n.d.) [online]. Available: https://walkwithamal.org/about-us/little-amal-the-walk/. [Accessed: 12 August 2023].

Moreno, N. (n.d.). 'Book addresses kids' deportation fear', *Chicago Tribune* [online]. Available: https://digitaledition.chicagotribune.com/tribune/article_popover.aspx?guid=ea3fbded-d6c7-4f56-a44e-d2969ca25f9e. [Accessed: 12 August 2023].

Nalewicki, J. (2021). 'These Refugees Are Planting New Roots in Salt Lake City', *Modern Farmer* [online], 5 June 2021. Available: https://modernfarmer.com/2021/06/these-refugees-are-planting-new-roots-in-salt-lake-city/. [Accessed: 12 August 2023].

New Roots SLC (n.d., 2010). Available: https://newrootsslc.org/. [Accessed: 12 august 2023].

One Global Voice (2021). 'Overcoming post-conflict trauma with art: Three African case studies' [online], 17 September 2021. Available: https://oneglobalvoice.it/arts/overcoming-post-conflict-trauma-with-art-case-studies-from-drc-rwanda-and-sierra-leone/. [Accessed: 12 August 2023].

Pillai, D. & Artiga, S. (2022). 'Title 42 and its Impact on Migrant Families', *Kaiser Family Foundation* [online], 26 May 2022. Available: www.kff.org/racial-equity-and-health-policy/issue-brief/title-42-and-its-impact-on-migrant-families/. [Accessed: 12 August 2023].

Rave Films (2018). *FIELDWORKS: Rick Lowe, the Victoria Square Project* 2018 [online]. Available: www.ravafilms.com/abog-rick-lowe. [Accessed: 12 August 20323].

Red Calaca Studio (2022). 'The Papalote Project' [online]. Available: https://redcalacastudio.com/the-papalote-project. [Accessed: 12 August 2023].

Sun, M. (2020). 'Lifting up Chinatown through Creative Placemaking' in *Sampan* [online] 18 September 2020. Available: https://sampan.org/2020/boston/lifting-up-chinatown-through-creative-placemaking/. [Accessed: 12 August 2023].

US ICE (2023). 'Delegation of Immigration Authority Section 287(g) Immigration and Nationality Act'. ICE, U.S. Department of Homeland Security, www.ice.gov/identify-and-arrest/287g. Accessed 18 Dec. 2023.

Utah Department of Agriculture and Food. (2022). 'New Roots Farm', Utah's Own. 25 August, viewed 12 September 2022, www.utahsown.org/profile/new-roots-farm

Victoria Square Project (n.d). [online]. Available: www.victoriasquareproject.gr. [Accessed: 12 August 2023].

20

PAINTING BACK

Creative Placemaking in Vancouver's Hogan's Alley

Friederike Landau-Donnelly

Introduction: Black Placemaking in Vancouver

The quest of this chapter is to unpack some of the historical and ongoing intersectional socio-spatial violences in the City of Vancouver, British Columbia, Canada, located on the traditional unceded territories of the xʷməθkʷəy̓əm (Musqueam), Sḵwx̱wú7mesh (Squamish), and səlilwətaɬ (Tsleil-Waututh) Nations. With over 95 percent of the province of British Columbia being located on unceded Indigenous lands, this geographic area holds various inter-generational conflicts or 'traumascapes' (Tumarkin, 2005), that are inscribed into negotiations about cultural identity, belonging, and ownership. My specific lens lies through creative practices of placemaking in Vancouver's downtown neighborhood of Hogan's Alley, a site of significant displacement of a large proportion of local Black residents.

I am neither a Canadian citizen nor a current resident anymore. Hence, importantly, my indirect connection to this conflict-laden place differs from those who have grown up close(r) to the local histories and places striated by both voluntary and forced migration (or have failed to shed light on this weight, which is co-implicated with white settler colonialism). I approach the places discussed here as sites of memory that carry both trauma and joy, bearing traces of multiple pasts and presents – considering the work of memory, or memory work, as a crucial medium 'to speak to past and present realities of intersectional violences' (Courtheyn, 2016, p. 936; Rose-Redwood et al., 2022).

I first investigate how official rationales and aspirations toward reconciliation and redress in the urban cultural context translate into concrete policies, measures, programs and support (Henderson & Wakeham, 2013; Gurstein & Hutton, 2019) and clarify terminological differences between reconciliation, redress, resurgence and reclamation of visibility. Secondly, I examine the commissioned mural *Hope*

DOI: 10.4324/9781003371533-24

FIGURE 20.1   Anthony Joseph (2020), *Hope Through Ashes: A Requiem for Hogan's Alley* [mural], Vancouver Mural Festival, Vancouver, March to September 2020. Image: 13 April 2023.

Through Ashes: A Requiem for Hogan's Alley (2020) by local artist Anthony Joseph (Figure 20.1) and go on to read the mural as an act of creative placemaking that takes, makes and keeps place. Lastly, the chapter concludes with a reflection on how trauma-informed public art can stimulate processes of healing in colonially wounded places. The take-away from this chapter is, ultimately, that public art can help places to heal, if artworks are paired with long-term institutional commitments to fight racism and discrimination.

Reconciliation, Redress, Resurgence, and Reclaiming Visibility

While 'reconciliation' and 'cultural redress' are commonly used terms within Vancouver's local discourses on socio-spatial justice, I contend with Henderson and Wakeham (2013, p. 9) to carefully consider the 'slippery relation' between the two. Problematically, 'reconciliation' might overshadow cosmetic initiatives that maintain uneven power relations between white settlers, settler institutions, and marginalized communities. Similar to critical museum scholar Marstine (2017, p. 43), I view reconciliation (and redress) as 'a pluralistic and conflicted stance without closure', thus remaining radically open, and partially insoluble in the face of ineradicable conflicts. Another 'R word' that surfaces in this debate is 'resurgence', 'a new increase of activity or interest in a particular subject or idea that had been forgotten for some time' (Cambridge Dictionary, 2023). The term was used in the programming of the local, public and privately funded 2020 Vancouver Mural Festival (VMF) promoting murals produced by Black and Indigenous artists under the header of the *Black Strathcona Resurgence Project* (BSRP). It was in this context that the *Hope Through Ashes* mural emerged and reshuffled socio-spatial relations between present and past communities and memories throughout the Hogan's Alley neighborhood.

BSRP endeavored to be a part of the ongoing process of reclaiming Black visibility in Hogan's Alley, with an aim of 'reclaim[ing] visibility and reconcil[ing] the erasure and systemic racism endured over time by Black people in Vancouver' (VMF, 2021), at a time that paralleled by the global upsurge of Black Lives Matter (BLM) protests. VMF had commissioned the *Hope Through Ashes* mural, which stretches almost 45 meters long to commemorate local personalities from Hogan's Alley and beyond, and is located on the very grounds of the demolition of Hogan's Alley's cultural richness – the Georgia Viaduct, built in 1972.

Placing Trauma in Wounded Cities

Within the growing body of literature on placemaking (Courage, 2017; Courage et al., 2021), creative placemaking (Loh et al., 2022) and Black placemaking (Hunter et al., 2016; McKittrick, 2016) in particular, concerns about dispossession and engagement have percolated into a critique of placemaking as potentially counter-productive to emancipatory socio-spatial struggles. Hence, Loh et al. (2022, p. 9) suggest placekeeping as a counter-tool to placemaking, to 'avoid displacing an immigrant community while branding that area for the immigrants' culture'. Placekeeping, in contrast to placemaking, aims to consider and attend to a place's multi-layered histories, including tensions and trauma (Loh et al., 2022).

To strengthen the relation between trauma and place, Drozdzewski et al. (2016, p. 3) point out that 'the mobilisation of memory has the capacity to transform places and keep our articulations of places of, and in memory, fluid'. Let us turn to this place with fluid memories that are being transformed through painting (back) stories from the past. Within a trauma-sensitive approach, scholars such as Sharpe (2016) interconnect historical traumas of colonization, slavery and exploitation with the ghostly absence-presence of past bodies, voices and struggles. Anderson and Daya's (2022, p. 1691) account of 'memory justice' places emphasis on 'small scale, ephemeral interventions [that] honour the mobility and fragmentation, the softness and the ordinariness of much historico-spatial pain'. Interconnecting trauma and healing, they establish trauma as a processual category to be localized in concrete places that suffer from spatial injustice, encouraging trauma-informed urban practices of placemaking to take up both concrete physical as well as discursive or imaginative space to tell stories of spatial belonging and displacement. Courtheyn (2016, p. 941) draws attention to the (soft) potential of memory work, claiming that 'memory is the strength of our resistance'. Similarly, considering memory as a political practice allows for an understanding of the mural not merely as a commemorative canvas, but as a political medium and materiality, a matter of concern and care.

In a place-historical context, trauma 'refers to the collective and cumulative emotional wounding across generations that results from cataclysmic events targeting a community'. Similarly, Falkenburger et al. (2018, p. 1) describe community trauma as affecting 'social groups or neighborhoods long subjected to interpersonal violence, structural violence, and historical harms'. Traumatic experiences and memories remain crucially context-dependent. Hence, the following empirical vignette of the *Hope Through Ashes* mural considers trauma as a constitutive dimension of marginalized communities' struggles and identities to heal and repair (Till, 2012, p. 6; Thompson, 2018).

Hogan's Alley on a Porous Map

Located on four blocks between Prior and Union Streets and adjacent to Vancouver's Chinatown and Strathcona, Hogan's Alley was once home to more than 800 members of Vancouver's Black community, as well as Italian, Japanese and Chinese immigrants (*Places That Matter*, 2023). Often referred to as 'dirt lane' (CBC News, 2020), Hogan's Alley was formally known as a cultural hub for music, food, entertainment and informal economies. Devalued by urban planners since the early 20th century, who problematized poverty and poor living conditions, Hogan's Alley and neighboring Strathcona were described as an 'area of poor moral and physical health; a decrepit neighbourhood that 'could spread like a disease throughout the city if it is not destroyed and redeveloped' (Scott, 2013). In 1972, the construction of the Georgia Viaduct finalized the displacement of not only Black residents, but also Italians, Asians, and First Nations people, who had gradually been living there in greater numbers since the 1950s (Compton, 2011).

Considering Hogan's Alley's diverse former residents, any assumption of *one* unified Black community becomes complicated by other types of racial discrimination and spatial dispossession. The Black community, notably, 'has never held one name or shade. It exists and it doesn't. It is, and has been, a tool for survival. It can turn an apocalypse into a future' (de Barros, 2021, n. p.). Furthermore, while narratives of Hogan's Alley often tilt to center loss and absence, it has also been described as 'vital to place-based community nurturance, preventing the impacts of social isolation' (Madden, 2021, p. 7). In line with accounts of Black planning and redress in Vancouver (Allen, 2019; Rudder, 2004), the contestations around who was living and who was being displaced from Hogan's Alley become entangled in more complex struggles amongst variously marginalized communities. To bring the memory work of Hogan's Alley's cultural life (back), which is embedded in the 'defence of land and livelihood' (Courtheyn, 2016, p. 948), it is important to mention urban development requests such as the Hogan's Alley Land Trust or a Black Cultural Centre, advocated by Black-run community organizations such as Hogan's Alley Society. These places are imagined to be reinforcing Black community, housing safety and cultural exchange in the place where Hogan's Alley once flourished.

Painting Memory Back: *Hope Through Ashes*: A Requiem for Hogan's Alley

In 2019, Toronto-born curator Krystal Paraboo was hired as guest curator of the first-ever VMF artist call for Black Vancouver-based artists, which later transformed into the *Black Strathcona Resurgence Project* (Vancouver Mural Festival, 2021). Paraboo's curatorial portfolio was to 'speak to VMF's lack of engaging with black history, and black artists and blackness in general' (November 11, 2021) and to encourage visibility and educate about Hogan's Alley as an area full of Black art, culture, entertainment, food and the many un-commemorated everyday stories. Paraboo commissioned US-born artist Anthony Joseph to create a mural in an alternative site to that originally assigned by VMF, suggesting placing the mural directly on the walls of the Georgia viaduct, which is, as Paraboo argues, 'the very instrument that is responsible for destroying the community' (of Hogan's Alley; November 11, 2021). Even before the mural went up, these spatial politics of the mural's location set the stage for a complicated process of making, taking or keeping place. Joseph explains the quest for this location as follows:

> *Because we need to claim a part, if not all, of this area [of Hogan's Alley] back, not in terms of only Black people can live here (…) this history being so easy to miss. People need to know what used to be here. People need to know that this was once a thriving, blossoming Black community that everybody benefitted from and enjoyed.*
>
> *(Joseph, March 4, 2021)*

Hence, the choice of the mural's location mattered as part of BSRP's educational trajectory to illuminate, educate and bring back some of Hogan's Alley's multiple histories. As Crang and Travlou (2001, p. 161) pointedly put it, the 'spatiality of memory links the social and the personal' – hence, the place of a commemorative mural giving voice to forgotten, displaced and excluded stories could link social and personal trauma. As Joseph says, 'every mural does bring a resurgence' (July 26, 2021). As there are so many people, places, things, experiences to remember, the kaleidoscopic nature of Joseph's mural sheds light on a lot of cultural, political, religious and sports-related characters. In reference to Black porters who worked at the Canadian Railway station nearby, Joseph explains the choice of motif as follows:

> *I decided to use the train as the narrative throughout the piece, so that it's like you're on a train ride and as you're going through the mural (...) the sky train [Vancouver's local transit] has different stops, every section of the viaduct would have a different slice of Hogan's Alley history. And from there I just wanted to mix it up. I won't be able to cover everybody, there's not enough space and there's not enough time (...) especially because of Black Lives Matter, the boiling point that the Black Lives Matter movement had reached and the killing of George Floyd.*
>
> *(Joseph, March 4, 2021)*

Spatial planning decisions like the erection of the Georgia viaduct doubtlessly cut apart families, friendships, businesses and culture, and thus fragmented space. And memory fragments space, too. Such a notion of spatially transformative memory points to the openness and contestability of both memory and space. In this vein, Joseph's mural variously splinters space – on the one hand, it de-fragments an existing fragmented, or wounded, space by bringing together mosaic memories, stories and personalities. On the other hand, it *further* fragments the already dispersed culture of memory in Hogan's Alley to make more space for hitherto untold, other movements of memories – meandering along the Georgia viaduct like a fully-packed train with people, stories and goods of the past and present.

In light of the multiple communities that have inhabited Hogan's Alley in the here and now, but also back in the day, artists, curators and Hogan's Alley Society community activists speak of the need to interconnect marginalized communities in their specific struggles for cultural redress. Specifically, bonds of solidarity are forged amongst Black people and people of color as well as other historically marginalized communities such as urban Indigenous peoples and Chinese-Canadians with Chinatown bordering Hogan's Alley and the Downtown Eastside, which is (in)famous for homelessness, poverty, and the use of harmful drugs (Robertson, 2007). Vis-à-vis the intersectionally interwoven, traumatic experiences of socio-spatial displacement and racism, Joseph notes that BSRP is also a means to 'show the neighboring communities and cultures that we are hand and hand with including the community, Chinatown, which is literally right next door' (July 26, 2021). Part of reclaiming Hogan's Alley's histories, then, is also to acknowledge the variety of experiences of displacement and trauma in relation to white settler colonialism. By working together in solidarity, organizers from BSRP set out to fight against structural discrimination. In this vein, the concluding BSRP report claims that 'there is a chance to make history by bringing the message of solidarity together through mural art' (BSRP, 2021 p. 9). Yet, in this context, it is particularly hard to bear that Paraboo herself felt treated unfairly and traumatized due to organizational controversies with the new VMF managing director (November 11, 2021).

Besides Joseph's mural, in 2021 a sibling mural was commissioned, called *Solidarity Storytelling*, by Chinese artist Emma Xie, Indigenous artist Chase Gray, and Black artist John Sebastian (Klovance, 2021). BSRP's vision for the three-panel mural was to

> interweave Chinese, Black and Indigenous culture and presence in a vibrant and dynamic manner that honors the past, present and future generations of these communities building towards a decolonized future collectively. In solidarity with one another, each panel depicts layers of storytelling from the aforementioned communities that reflect 'elders, youth, queer & 2-Spirit identities [sic]'.
>
> *(BSRP, 2021 p. 17)*

The attempt at visibilization relates to curatorial difficulties that oscillate between making invisibilities and erasures visible in self-directed ways, instead of lip service 'reparative' or tokenistic commissions. For example, Paraboo speaks of the challenge to create 'a type of visibility for this past that is absolutely there' (November 11, 2021). Joseph argues against invisibilization that might creep up (again) in case murals were decommissioned after the two years of maintaining the mural that VMF and wall owners contractually agree to. He states: 'You're not going to erase these pieces, and if you do, there will be hell to pay and you will hear from us' (July 26, 2021). Considering oft-temporary artistic media like murals as forms of heritage, Joseph is keen to establish a longer-term consciousness of 'what used to be here, and let it be known for what happens afterwards […] the more we forget history, the more it repeats itself' (March 4, 2021). In reaction to this, Joseph called the City of Vancouver 'very vocal about wanting to maintain the mural especially with the national attention it received' (July 26, 2021). Hence, the mural fulfils a double function of (re)activating knowledge about past usages of the space and keeping place to let this historical acknowledgement linger. Paraboo considers murals as places of memory work in terms of their 'educational component' but, more generally, appreciates the impermanence of artwork in public space (November 11, 2021). Paraboo underscores:

> *The mural's primary functionality is not necessarily to be designated as a heritage site or a heritage object (…) there's no one way or blueprint for how to interact with this work or what this work means, and that's the beauty of what art is. Everyone has their own relationship with it and while I appreciate that it was acknowledged as a salute to heritage, I feel like its functionality is beyond that.*
>
> *(Paraboo, November 11, 2021)*

Another facet of inter-community solidarity is artists' relations to the graffiti scene and people without a home dwelling in public space. Joseph describes everyday contact with graffiti writers as predominantly respectful. He remembers that passers-by and social media commentaries thanked him for adding color to the city (March 4, 2021). Seeing graffiti sprayers one morning during the process of painting the mural, Joseph realized they were not spray-painting on the mural itself, but merely the protective scaffolding, hence not violating the unwritten code of conduct amongst street artists (i.e., not to go after each other's work). Joseph re-narrates the writers' response: '"Hey, we're just doing it on the scaffolding, hope you don't mind?" I said, "Go nuts on the scaffolding. That's going to be taken down anyway." "Yeah, we'd never mess with the piece." So, that's fine. Never had any trouble' (March 4, 2021). Hence, while tensions can arise between graffiti and mural artists (Landau-Donnelly, 2022), in this case, the different users of the space (or may we call them placemakers alike?) co-exist in a non-violent way.

Relations with homeless people were more complicated. Joseph himself found interactions with homeless people dwelling and sleeping in the nearby area of the mural to be a 'touchy subject' (July 26, 2021). Joseph (March 4, 2021) was mindful not to (further) displace people already struggling in Vancouver:

I'm not one to tell people who live on the streets or where they can lay their head. There have been a few people who when I came in the morning, they were shooting up (...) one person was kind of annoyed: 'What? I can't be anywhere'. 'Listen, I just want to paint, if you could just move over a bit?' I caught him in a bad mood.

(Joseph, March 4, 2021)

The discourse to keep the area clean, for Joseph, has no ring of social cleansing. Instead, he argues for the removal of needles used for substance consumption for safety reasons (ibid.). In these everyday interactions, workings of solidarity subtly shine through. In a place like Hogan's Alley, which has forcefully kept people from placemaking or placekeeping, Joseph's permissive approach of co-existence succinctly rubs against the City of Vancouver's initiative to place rock installations in front of the mural to keep people from camping there. This defensive or hostile architecture (Chellew, 2019) possibly reinforces existing conflicts between past and present communities, and does not assist in efforts at reconciliation or restoration, let alone redress or resurgence in the area. In light of encampments physically damaging the mural, Joseph considers it 'kind of ironic that smoke and ash is defacing a mural that is entitled *Hope Through Ashes*' (March 4, 2021). In sum, bonds of solidarity between very different people moving through Hogan's Alley are complex and consistently in the making. Some of these bodies have experienced displacement in the past, and now resurge by (re)appearing in public space and public art; others experience, or continue to experience, displacement-related trauma in the contemporary housing and opioid crises of wounded

Vancouver. Between these diverse experiences of placemaking and placekeeping, the commemorative mural of *Hope through Ashes* hovers between hope, hurt and possibilities for healing.

Outlook: Ghostly Memories in (Post-)traumatic Cities

By reading official narratives of reconciliation against community-led initiatives and enactments of resurgence, this chapter has touched on the political underpinnings of public art in reconciliatory contexts. It has described how public art pieces can assist in placemaking or placekeeping exercises that (re)paint inter-generational stories and memories of systemic racism and intergenerational trauma into wounded cities. Lahoud et al.'s (2010, p. 19) poetic reflections on post-traumatic urbanism help to outline the potential of a trauma-informed approach to art in public space: 'trauma is the drama in which both history and the future are at stake, held in a suspended crisis; the cards have been thrown up in the air but they have not yet landed'.

In summary, a trauma-informed approach to public space, public art and the encounters of heterogeneous, marginalized communities therein, benefits from being attuned to the always-ruptured politics of absence and presence in memory and urban space. No commemorative piece of public art can ever, nor should it ever, be expected to replace or fully recover long-standing and inter-generational experiences of institutionalized racism, and concomitant trauma. Public art, and creative placekeeping, can create space for the simultaneous joy and pain of complicated histories. Yet, for the enactment of longer-term healing towards socio-spatial equity, a shift in institutional mindsets to combat racism and discrimination is needed, too. At last, because ghosts of difficult pasts will linger (whether we like it or not), they might need to be greeted with open arms to further mend their wounds.

> *give the ghosts a voice*
> *give the ghosts a place*
> *but this place is not enough*
> *not all ghosts are from the past*
> *'post' is an undeliverable parcel*
> *the present is further away than the present, and it always will be*
> *breathe, breathe, scream* (Friederike Landau-Donnelly, #poeticacademic).

References

Allen, S. (2019). *Fight the Power: Redressing Displacement and Building a Just City for Black Lives in Vancouver.* Master's Thesis, Simon Fraser University. Available: https://summit.sfu.ca/item/19420. [Accessed: 12 August 2023].

Anderson, M. & Daya, S. (2022). 'Memory Justice in Ordinary Urban Spaces: The Politics of Remembering and Forgetting in a Post-Apartheid Neighbourhood', *Antipode*, 54(6).

Black Strathcona Resurgence Project (BSRP) (2021). Available: https://vanmuralfest.ca/community-projects/bsrp. [Accessed: 12 August 2023].

Cambridge Dictionary: *Resurgence* (2023). Available: https://dictionary.cambridge.org/de/worterbuch/englisch/resurgence. [Accessed: 24 March 2023].

CBC News (2020). 'Why the Vancouver viaducts are a symbolically important place for an anti-Black racism protest' [online], 15 June 2020. Available: www.cbc.ca/news/canada/british-columbia/viaduct-hogan-s-alley-significance-1.5612399. [Accessed: 24 March 2023].

Chellew, C. (2019). 'Defending Suburbia: Exploring the Use of Defensive Urban Design Outside of the City Centre', *Canadian Journal of Urban Research*, 28(1).

Compton, W. (2011). *After Canaan: Essays on Race, Writing, and Region*.

Courage, C. (2017). *Arts in place: The arts, the urban and social practice*. Abingdon: Routledge.

Courage, C., Borrup, T., Jackson, M.R., Legge, K., McKeown, A., Platt, L. and Schupbach, J. eds.,2020. *The Routledge Handbook of Placemaking*. Routledge.

Courtheyn, C. (2016). '"Memory is the strength of our resistance": an "other politics" through embodied and material commemoration in the San José Peace Community, Colombia', *Social & Cultural Geography*, 17(7).

Crang, M. & Travlou, P. (2001). 'The City and Topologies of Memory', *Environment and Planning D: Society and Space*, 19(2).

De Barros, K. C. B. (2021). 'Black History Month: What is Hogan's Alley?' in *ROOM Magazine*, 17 February 2021 [online]. Available: https://roommagazine.com/what-is-hogans-alley/#:~:text=Hogan's%20Alley%2C%20located%20in%20what,%2C%20and%20Tsleil%2DWaututh%20peoples. [Accessed: 24 March 2023].

Drozdzewski, D., De Nardi, S., & Waterton, E. (2016). *Memory, place and identity: Commemoration and remembrance of war and conflict*. Abingdon: Routledge.

Falkenburger, E., & Arena, O. (2018). *Trauma-Informed Community Building and Engagement*. Urban Institute, The Kresge Foundation.

Gurstein, P. & Hutton, T. (2019). *Planning on the edge: Vancouver and the challenges of reconciliation, social justice, and sustainable development*. Vancouver: UBC Press.

Henderson, J. & Wakeham, P. (2013). *Reconciling Canada: Critical Perspectives on the Culture of Redress*. Toronto: University of Toronto Press.

Hunter, M. A. et al. (2016). 'Black Placemaking: Celebration, Play, and Poetry', *Theory, Culture & Society*, 33(7–8).

Klovance, R. (2021). 'Murrin substation murals help unveil Hogan's Alley history', *BC Hydro*, 16 August 2021. Available: www.bchydro.com/news/conservation/2021/murrin-substation-murals-history.html. [Accessed: 24 March 2023].

Lahoud, A., Rice, C. & Burke, A. (2010). *Post-Traumatic Urbanism*. Lagos: Academy Press.

Landau-Donnelly, F. (2022). 'Contentious Walls', *CAP–Public Art Journal*, 3(2).

Loh, C. G. L., Ashley, A. J., Kim, R., Durham, L., Bubb, K. (2022). 'Placemaking in practice: Municipal arts and cultural plans' approaches to placemaking and creative placemaking', *Journal of Planning Education and Research*. 1–12.

Madden, C. (2021). 'Race and Contemporary Architecture: Making a Case for a Hogan's Alley Cultural Centre at 898 Main Street, Vancouver, BC', *Canadian Sociological Association Conference 2019*. Available: www.csa-scs.ca/conference/paper/race-and-contemporary-architecture-making-a-case-for-a-hogans-alley-cultural-centre-at-898-main-street-vancouver-bc-2/. [Accessed: 24 March 2023].

Marstine, J. (2017). *Critical Practice*. Routledge eBooks [Preprint]. Available: https://doi.org/10.4324/9781315272016. [Accessed on: 24 March 2023].

McKittrick, K. (2016). 'Rebellion/Invention/Groove', *Small Axe: A Caribbean Journal of Criticism*, 20(1).

Places that matter. (2023). *Hogan's Alley*. Available https://placesthatmatter.ca/location/hogans-alley/ [Accessed on: 24 March 2023].

Robertson, L. (2007). 'Taming Space: Drug use, HIV, and homemaking in Downtown Eastside Vancouver', *Gender Place and Culture*, 14(5).

Rose-Redwood, R. et al. (2022). 'Monumentality, Memoryscapes, and the Politics of Place', *ACME: An International Journal for Critical Geographies*, 21(5).

Rudder, A. J. (2004). *A Black Community in Vancouver?: A History of Invisibility*. MA Thesis. University of Victoria. Available: https://dspace.library.uvic.ca/bitstream/handle/1828/733/rudder_2004.pdf?sequence=1. [Accessed: 24 March 2023].

Scott, C. (2013). *History The End of Hogan's Alley – Part 1*. Toronto, ON: Spacing Magazine. Available at: http://spacing.ca/vancouver/2013/08/12/the-end-of-hogans-alley-part-1/

Sharpe, C. (2016). *In the Wake: On Blackness and Being*. Durham: Duke University Press.

Thompson, V. M. (2018). 'Repairing Worlds: On Radical Openness beyond Fugitivity and the Politics of Care: Comments on David Goldberg's Conversation with Achille Mbembe', *Theory, Culture & Society* [Preprint].

Till, K. E. (2012). 'Wounded cities: Memory-work and a place-based ethics of care', *Political Geography*, 31(1).

Tumarkin, M. M. (2005). *Traumascapes: The Power and Fate of Places Transformed by Tragedy*. Melbourne: Melbourne University Publishing.

Vancouver Heritage Foundation (2021). *Hogan's Alley*. Available: https://placesthatmatter.ca/location/hogans-alley/. [Accessed: 24 March 2023].

Vancouver Mural Festival (2021). *Black Strathcona Resurgence Project*. Available: https://vanmuralfest.ca/community-projects/bsrp. [Accessed on: 24 March 2023].

21

WANNA DANCE?

Using Creative Placemaking Value Indicators to Identify COVID-lockdown-related Solastalgia in Sydney, Australia

Cathy Smith, Josephine Vaughan, Justine Lloyd and Michael Cohen

Introduction: Lockdown as Urban Trauma

This chapter explores creative placemaking and its valuation as one way to address place-related trauma. It proposes that some urban public domain within COVID-19 lockdowns became sites of place-trauma (Donovan, 2013), or 'traumascapes' (Tumarkin, 2019, p. 5), resulting in feelings of 'solastalgia' by local communities (Albrecht, 2006). Understanding and evaluating the promise and potential of trauma-informed creative placemaking through the lens of solastalgia requires thinking beyond place as specific site. To address this, the chapter recognises place as people's layered experience of time and space that includes events, memory, affect, social ties and representations. It is organised around a case study of the 2021 temporary art installation and dance performance series, *Wanna Dance* (Figure 21.1). Using the value indicators contained in our *Valuing Creative Placemaking (VCPM) Toolkit* (Cohen et al., 2023), we use this case study to foreground the less tangible aspects of creative placemaking as they relate to forms of trauma – in this case the experiences of loss and isolation resulting from COVID-19 restrictions, especially the loss of access to public space.

Located in a Sydney city laneway in a declining part of the southern end of Sydney's central business district, *Wanna Dance* was a site-specific collaboration between designers, prop-makers, sound artists, choreographers, and dancers. As a project case study, the temporary placemaking project highlights the highly charged nature of public encounters and everyday experiences of the arts in post-COVID public space: not only through its co-option of a 'leftover', unused public space, but because it demonstrates how a collaborative installation celebrating simple acts of moving and breathing in outdoor spaces can recuperate a city where breathing itself is under threat, due to the COVID-19 pandemic. When

DOI: 10.4324/9781003371533-25

FIGURE 21.1   Professional dancers and the general public interact with *Wanna Dance*, a temporary installation in a laneway in Sydney, Australia. Image credit: City People/Anna Kucera (2021).

evaluating trauma-informed creative placemaking, it is important to contextualise benefits in the context of the crisis itself, especially when activities are trauma- and community-specific. The experience of the community in the trauma is significant because negotiating place-based trauma such as solastalgia relies upon 'the active participation of the people who will live, own and otherwise care for that place' (Donovan, 2013, p. 240). Thus, in the context of Sydney's ongoing pandemic lockdowns, *Wanna Dance* provided an immensely valuable, unique opportunity for artists and strangers alike to physically express themselves and socialise in an otherwise neglected urban space.

After introducing the *Wanna Dance* project as an example of creative placemaking responding to place-related trauma, the chapter explores the discourses of trauma-informed placemaking, and their intersections with broader theories and practices of creative placemaking, focusing particularly on the theory of solastalgia and COVID-19 lockdown sites as traumascapes. Approaches to identifying and evaluating impacts of creative placemaking, in the context of demonstrating benefits to funding bodies and communities, are also discussed and then applied to the *Wanna Dance* case study. We conclude the chapter with a discussion about the benefits and challenges of creative placemaking projects as a means of encouraging healing and improving people–place relationships following environmental trauma.

In the context of trauma-informed design, the ability to value and evaluate placemaking projects can be challenging to administer within active crisis scenarios, where urgent issues of health and security may raise questions about the relevance of creative enterprises. Although the emotional, mental, spiritual, and health implications of creative placemaking projects are understudied (Galway et al., 2019), advocates argue they have an important role to play in helping affected victims negotiate the mental and physical dimensions of place-linked trauma during recovery phases (Donovan, 2013). This research uses a practice-led methodology which has dynamism, moving between the different positions of theory and practice (Smith & Dean, 2009). A practice-led methodology aligns with the view of placemaking as 'a project of mutual learning and sustained yet productive disagreement' (Zitcer, 2018). We also use our co-authored *Valuing Creative Placemaking Toolkit* (Cohen et al., 2023) to frame our exploration of the case study in the context of place-trauma. The *VCPM Toolkit* incorporates social and environmental value indicators alongside more conventional economic indicators related to improved business activity and perceptions of property to evaluate creative placemaking initiatives. A selection of questions from the *VCPM Toolkit* were used to develop discussion points on the *Wanna Dance* project as a case study of trauma-informed design. Lloyd, Smith, and Vaughan used an open-ended question approach to collect 'insights revealed by the artist-practitioner' Cohen on the *Wanna Dance* project (Sullivan, 2009). Responses were recorded and transcribed, and then used to frame the themes and structure of this chapter and the case study itself.

Place Activation, Breathing and Lockdown: The *Wanna Dance* Response to Place-based Collective Trauma

Hard lockdowns were enforced in response to COVID-19 in all Australian states across 2020–21, and particularly in the major cities of Melbourne and Sydney. There were extended periods when outdoor spaces were closed, and authorities monitored public places to enforce social distancing. Spontaneous, cathartic, and joyful experiences of arts and music, including dance, were impacted by the lockdowns, as spending time in public space for anything other than utilitarian purposes became a social norm. These losses were experienced in uneven ways across the city, with policing of breaches of lockdown rules later revealed to have been differentially enforced along socio-economic, class and racial stereotypes (Johnson, 2021; Faruqi, 2020).

City People is a Sydney-based, creative placemaking company providing arts and cultural strategy, culture-led placemaking, arts programming, production, and project implementation. From their professional understanding of the relationship between place, health and wellness, City People team members recognised the losses that the public domain and public life of the city were suffering during COVID-19 lockdowns in Sydney, identifying the personal impact or loss for individuals resulting from lockdowns. Additionally, they noticed the city became non-functional and lifeless with local businesses, employees and routine economic activities disrupted. The absence of business activity affected the sense of connection to place and community, a role that local businesses usually provide as social infrastructures. City People also noted the devastating impact of lockdowns on the arts sector, and particularly the livelihoods and communal belonging to place of artists and creative placemakers like themselves.

In August 2020, City People hosted an *Accelerator*, or temporary 'think tank', for Sydney-based arts, creative professionals and city governance and policy sector. Understanding that the distress caused by COVID lockdowns in urban spaces was related to the separation of people from each other and special places for social connection, the *Accelerator*'s intention was to 'generate creative ideas about how arts and culture can continue to make places people love, [and how to create] places where people [can] connect with each other while being at a safe physical distance' (Cohen et al., 2020, p. 4). The participants in the *Accelerator* discussed feelings of 'collective loss' and mourning for experiences 'of movement and connection with others in a place, which can't be replicated indoors or online' (Cohen et al., 2020, p. 5). Before the pandemic the goal of much public domain arts and culture was to encourage 'freedom of association and expression' (ibid.) of as diverse an audience as possible. The Accelerator participants asked whether 'we now need new art practices and new kinds of places to make art and culture that connects us all?' (ibid.)

In response to these provocations, during late 2020 and early 2021, and as part of the Sydney Solstice Festival (2021), *Wanna Dance* was developed and installed in the southern end of the targeted Central Business District (CBD) precinct. Intentions for the project included activation of a disused city laneway to address the needs of community after intensive lockdown periods, and restoring opportunities for group celebration and connection, through public gatherings and collective movement, following a period when these had both been prohibited.

Wanna Dance was a deceptively simple project, in which three industrial-looking vents were constructed as sensor-operated light and sound boxes designed by sculptor and architect Chris Fox, who had participated in the *Accelerator*. These interactive vents were installed on the walls of Parker Lane, a one-way inner-city laneway used primarily for deliveries and rubbish collection, amongst graffiti and service pipes along the side of the fledgling Museum of Chinese Australia, which was awaiting renovation. While the lane can be used as a pedestrian thoroughfare, before *Wanna Dance* this laneway was not inviting to foot traffic. However, Parker Lane is very close to public transport on the main thoroughfare of George Street, providing easy access to *Wanna Dance*. Operating from 5pm to 3am, as the darkness of winter afternoons set in, the motion sensors on *Wanna Dance*'s vents detected passers-by, illuminating a small sign spelling out 'Wanna Dance?' enticing passers-by into the lane and inviting interaction. Once a person stepped underneath a vent, dance music sound swelled out, rainbow disco lights flashed, and the groove was on. Located in part of Sydney which has longstanding ownership and association with Sydney's Asian population and businesses, particularly Chinese-owned businesses, the *Wanna Dance* sound designer, Tyson Koh, was briefed to include this heritage in the soundtrack design, including a range of genres and languages.

Wanna Dance was intentionally created to provide both a physical and emotional release from individual and collective trauma associated with the losses of public social connections. Its project team, including dance artist and choreographer Emma Saunders alongside Cohen and Fox, were aware of dance as a vehicle for creating emotional experience: recognising dance as a very tangible art form that connects the dancer to the world, or to place, often within a social context.

Responding to the *Accelerator*'s finding that arts and culture practitioners were optimistic and flexible in their responses to COVID-related restrictions, the project built on 'a willingness to consider how we can reconfigure artforms, not just audiences, to respond to the regulations… aimed at restricting the transmission of COVID-19 in the community' (Cohen et al., 2020, p. 4). By bringing the city's dance cultures out of nightclubs and performance venues into public space, the project engaged with the losses felt by performers and local sub-cultures during restrictions on public gatherings whilst promoting social diversity by crossing boundaries between performance cultures and music genres.

Supporting the existing local dance culture and relating local arts to local place, local dancers were recruited and over the full 27 days of activation other curated dance performances and demonstrations, including contemporary, hip-hop, and street dance, were simultaneously held. The resultant pumping lights, tunes and movement brought a 'vibrant', 'fun' and 'party' feel to the city spot, according to some of the estimated 3000 actively dancing public participants, from different community groups, ages, and backgrounds (City People, 2021).

Dancing in public for the *Wanna Dance* project returned spontaneity to the city streets – a fundamental element that had gone missing since pandemic-triggered lockdowns began. Cohen attributed the success of the project to the emotional liberation that resulted from the intersection of spontaneity, culture, location, and dance: providing physical and emotional release for participants and professional dancers alike made and restored connections with culture and place. Joy was evident among the sudden transformation of a collection of passers-by into a ballooning of music, lights and dancing, triggered by the quiet invitation: 'Wanna Dance?'

Solastalgia and Creative Places

Wanna Dance is one example of the desire to use creative placemaking to reinvent people–place attachment following a place-related trauma, and by extension negotiate the sense of loss and the 'homesickness one experiences when one is still at home' (Albrecht, 2006, p. 35). Recent work in urban planning and creative arts has investigated the deep relationship between place and emotion by looking at the intersections of trauma, place and placemaking. In the urban design context, such as our case study site, urbanist and architect Jenny Donovan defines place-trauma as a major disruptive event in a location, which requires external resourcing and intervention to promote community recovery (2013). Maria Tumarkin names the 'physical places constituted by experiences of particular events and their aftermath' (2019, p. 5) as 'traumascapes'. Traumascapes belong to both individual people and their families directly affected by events, as well as collectives such as witnesses, journalists, and artists. Much of the discussion on place-trauma focuses on the disruption of people–place connections following a major traumatic event at scale, otherwise described as solastalgia. As defined by Glenn Albrecht, 'Solastalgia exists when there is recognition that the beloved place in which one resides is under assault' (2006, p. 35). Although penned prior to the COVID-19 pandemic, Albrecht describes the destructive impacts of disease and ecological disaster on our attachments to place and the planet as a whole (Albrecht, 2020). Scholars investigating experiences of solastalgia seek to highlight the importance of people–place relationships to wellbeing and community – hence the root word 'solace' as being intrinsic to Albrecht's conceptualisation of a 'sense of lost home while still at home' (Askland & Bunn, 2018, p. 18).

When thinking about trauma-informed placemaking initiatives, it is useful to consider how art-focused projects may help communities recover following a traumatic event or situation, particularly alongside longer-term physical place improvements, such as renovating damaged public and private structures or generally making the public domain safe and accessible (Madden, 2018). Creative projects can assist communities with their recovery from disasters, particularly by promoting social, economic, and environmental recovery (Donovan, 2013). However, Donovan emphasises that creative placemaking initiatives only positively impact place-trauma when projects are focused on reimagining a place and genuinely involve local or otherwise invested communities (2013, p. 240).

Valuing Creative Placemaking in the Context of Trauma

To better understand how *Wanna Dance* demonstrates how creative placemaking can be used to address solastalgia, we used our *VCPM Toolkit* with its range of indicators of placemaking success under three main value indicator impacts: Social, Environmental and Economic. Using the *VCPM Toolkit*, elements of *Wanna Dance* were categorised according to these known indicators of value in placemaking projects (Table 21.1).

Wanna Dance addressed a lockdown-related economic problem stemming from difficulties finding employment that many creative industries workers worldwide were facing (Sherwood, 2022) – which could be argued as another place-related trauma in its own right. Nonetheless, temporary placemaking projects like *Wanna Dance* primarily aim to affect and amplify people's social and emotional experiences of places, thereby overcoming solastagia. However, because these social dimensions are often less tangible, their legacy is difficult to quantify without

TABLE 21.1 Analysis of *Wanna Dance* using the *Valuing Creative Placemaking Toolkit* indicators

Placemaking value aspects	Indicators of value	Unique value indicators	Examples of evidence of value in Wanna Dance
Social	Civic engagement	Number of attendees Diversity of participants Community-assisted events Repeat visits	Approximately 3000 in-person and 50,000 virtual participants from different community groups, ages, and backgrounds 25 creative sector community volunteers from diverse backgrounds Culturally diverse music selection
	Safety/Crime reduction	Reducing vandalism Addressing urban problems Integration into urban planning Reduction of crime	The intent of *Wanna Dance* was to address the urban problem of a city that had become very quiet and still due to COVID-19 lockdowns
	Place attachment	Attracting new audiences Changing perceptions of community Storytelling	Cultural, gender and age-diverse audiences attended Instagram campaign shared the story to 52,754 people with 92,293 views
Environmental	Facility provision	Improving public facilities	Temporary free dance space, lights and music for 27 days
	Diversity support	Amenable/accessible environment Catering to diverse needs	Located in an accessible location with close proximity to public transport Dance demonstrators from diverse backgrounds
Economic	Real estate	Uplift in property prices Investment in housing	Can only be verified by future real estate data
	Education and employment	Local business growth Increased business activity Increased career development/ education	Provided employment and experience to the arts sector: engaging 30 creative sector workers and 25 creative sector volunteers Businesses reported that *Wanna Dance* was an excellent attractor to crowds, contributing to vibrancy of the area and increasing trade

Source: Authors (2023).

the help of value indicators that specifically name them (Table 21.1). Social value indicators for creative placemaking:

> *include acts of community participation during temporary, event-based and tactical initiatives which celebrate community process, enable deliberative discussion and collaboration amongst community, and foster a sense of ownership, civic pride and community spirit of a place.*
>
> *(Vaughan et al., 2021, p. 435)*

The *VCPM Toolkit* provides value indicators identifying where a project is achieving these social values. The case study of *Wanna Dance* demonstrates actions and undertakings that can be identified as indicators of social value, not only in a creative placemaking context, but also in the intersection of addressing solastalgia trauma. While there were several indicators of social value that can be identified in *Wanna Dance*, the social value indicators relating to expanding participant diversity and addressing urban problems in its underused site were demonstrated consistently in different ways throughout the project duration.

Diversity, including cultural and linguistic diversity, is a strong indicator of social value when different diverse groups interact. This form of public and urban interaction happens 'where participants learn to interact with others beyond and within their immediate familial and social groups and [are] welcomed and welcoming to strangers' (Till & McArdle, 2015, p. 59). In the recorded conversation between the research team, Cohen remembers one evening during the installation when 'You could tell that we had touched a nerve with the project':

> *[It] had this kind of magic about it … it wasn't like it was just pumping the whole time. It was a thing that you walk past and experience, and then you'll keep going because you're off to whatever, and so it would go completely dormant.*
>
> *And then you'd see somebody discover it. And then you'd see some other people looking at those people and wondering what they're doing, and then somebody would encourage them to get involved, and it would kind of grow… you know there's this one moment where this whole family of Middle Eastern men and women came past and it had been really quiet, and a couple of the street dancers were there. And anyway… just this kind of crazy coincidence, a piece of music came on that … had some kind of, you know, Middle Eastern kind of rhythm or something to it.*
>
> *Suddenly the laneway just ballooned, and you had all these men with their arms folded doing this kind of, you know, kind of traditional dance, and then [dancer] was doing this kind of hip hop moves to it. So, this is kind of crazy, you know, meeting of things. So yeah, I think very much they were the moments of observation and of discussion that kind of showed us that it was working.*

As part of managing the COVID-19 pandemic whilst ensuring safe and equitable access to public space, city governments, planners and communities trialed innovative and inclusive practices. The *Wanna Dance* project provided a solution to both, through a temporary intervention to bring the city's dance and performance cultures outdoors and occupy 'leftover' urban space. The project, and the thinking behind it, signals the key role of arts and culture in leading creative thinking around shifts in access to public space and has highlighted the potential for similar projects to continue, post-lockdowns, demonstrating the ongoing social, environmental and economic value of creative placemaking. By instigating, in a small way, a momentary collective ownership of the city and reappropriation of the urban commons through movement and music, the project shows a possible way forward for communities experiencing place-based trauma and distress. Perhaps most importantly, using the *VCPM Toolkit* to evaluate creative placemaking projects like *Wanna Dance* helps to evidence their underlying value to stakeholders. To quote from one of the participants of the Accelerator: 'Let's not Band-Aid the old normal but make use of this opportunity – a creative pause – to rethink what comes out [for arts and culture] at the end of COVID-19' (Cohen et al. 2020, p. 13). Given that environmental distress is likely to increase with the impacts of human-induced climate change, the lessons of *Wanna Dance* are likely to resonate for some time to come.

Conclusion and Discussion: Creative Placemaking, Trauma and Healing Legacies

This chapter presents *Wanna Dance* as a case study on how trauma-informed creative placemaking might not only address solastalgia in COVID-lockdown affected urban spaces but how to articulate and evaluate creative placemaking's underlying intangible social impacts. Potential intersections between trauma-informed placemaking and creative placemaking have been identified through the ways that both surface and address experiences of solastalgia. A trauma-informed lens on creative placemaking has been used to understand a range of impacts of a specific project, *Wanna Dance*. A common thread when observing the *Wanna Dance* case study through the *VCPM Toolkit* value indicators is the focus of this project on addressing an urban problem and providing inclusion for diversity from the very inception of the idea through the entire project. These insights are offered to help placemakers better respond to negative or traumatic experiences of place. Finally, we invited further discussion and thinking about the benefits and challenges of creative placemaking projects as ways of encouraging healing and improving people–place relationships from the trauma of loss of familiar places and their contributions to urban social life during times of environmental distress.

References

Albrecht, G. A. (2006). 'Solastalgia', *Alternatives Journal*, 32(4/5).

Albrecht, G. A. (2020). 'Negating Solastalgia: An Emotional Revolution from the Anthropocene to the Symbiocene', *American Imago*, 77(1), Spring.

Askland, H. H. & Bunn, M. (2018). 'Lived experiences of environmental change: Solastalgia, power and place', *Emotion, Space and Society*, 27.

City People (2021). 'Wanna Dance Vox Pop Final Edition Compressed' [online], *YouTube*. Available: www.youtube.com/watch?v=OxR4ktQSlts. [Accessed: 10 March 2023].

Cohen, M., Mondy, C. & Spinks, K. (2020). 'Accelerator: How arts and culture can connect people in public spaces during COVID-19 and beyond', *City People*. Available: https://citypeople.com.au/portfolio/covid-accelerator/. [Accessed: 31 August 2021].

Cohen, M., Gajendran, T., Lloyd, J., Maund, K., Smith, C. &Vaughan, J. (2023). 'Valuing Creative Placemaking: A survey toolkit for public and private stakeholders', Landcom, NSW, Sydney. Available: https://landcom.com.au/approach/research-and-learning/landc oms-new-approach-to-research/valuing-creative-placemaking-toolkit/. [Accessed: 1 April 2023].

Donovan, J. (2013). *Designing to Heal: Planning and Urban Design Response to Disaster and Conflict*. Melbourne: CSIRO Publishing.

Faruqi, O. (2020). 'Compliance fines under the microscope', *The Saturday Paper*, April 18–24 [online]. Available: www.thesaturdaypaper.com.au/news/health/2020/04/18/com pliance-fines-under-the-microscope. [Accessed: 10 January 2023].

Galway, L. P., Beery, T., Jones-Casey, K. & Tasala, K. (2019). 'Mapping the Solastalgia Literature: A Scoping Review Study', *International Journal of Environmental Research and Public Health*, 16(15).

Johnson, P. (2021). 'Western Sydney's COVID lockdown raises questions on Q+A about racism and classism', *ABC News* [online], 30 July 2021. Available: www.abc.net.au/news/2021-07-30/western-sydney-covid-lockdown-racism-claims-qa/100335718. [Accessed: 20 March 2023].

Madden, K. (2018). *How to Turn a Place around: A Placemaking Handbook*, Project for Public Spaces, New York, NY.

Sherwood, H. (2022). 'Unesco warns of crisis in creative sector with 10m jobs lost due to pandemic', *The Guardian*, 8 February 2022 [online]. Available: www.theguardian.com/culture/2022/feb/08/unesco-warns-of-crisis-in-creative-sector-with-10m-jobs-lost-due-to-pandemic. [Accessed: 10 February 2023].

Smith, H., & Dean, R. T. (eds.) (2009). *Practice-led research, research-led practice in the creative arts.* Edinburgh: Edinburgh University Press.

Sullivan, G. (2009). 'Making space: The purpose and place of practice-led research', *Practice-led research, research-led practice in the creative arts*, 2.

Sydney Solstice 2021, Wanna Dance for Sydney Solstice. Available: www.sydneysolstice. com/wanna-dance-sydney-solstice. [Accessed: 20 January 2023].

Till, K. E., & McArdle, R. (2015). 'The improvisional city: Valuing urbanity beyond the chimera of permanence', *Irish Geography*, 48(1).

Tumarkin, M. (2019). 'Twenty Years of Thinking about Traumascapes', *Fabrications*, 29(1).

Vaughan, J., Maund, K., Gajendran, T., Lloyd, J., Smith, C. & Cohen, M. (2021). 'Determining and representing value in creative placemaking', *Journal of Place Management and Development*, 14(4).

Zitcer, A. (2018). 'Making Up Creative Placemaking', *Journal of Planning Education and Research*, 40(5).

22

HEALING FROM TRAUMA IN POST-DISASTER PLACES?

Placemaking, *Machizukuri* and the Role of Cultural Events in Post-Disaster Recovery

Moéna Fujimoto-Verdier and Annaclaudia Martini

Introduction

Using cultural events to reshape narratives and imaginaries in places hit by disasters is not uncommon. In addition to their economic benefits, cultural events can be used as an essential tool for placemaking, offering possibilities for local communities to make sense of traumatic pasts, reimagine their future, and generate a sense of place through connections between people, whether locals or non-locals. This chapter investigates how and in what ways the locals involved in the Reborn Art Festival (RAF) in Ishinomaki, Japan, directly or indirectly utilize the festival to negotiate trauma and placemaking processes after the triple disaster which took place on 11 March 2011. On that day a seism of 9.0 magnitude was followed by a tsunami, and by a meltdown in the Fukushima Daiichi nuclear power plant, causing the evacuation of 409,146 residents from the Fukushima, Miyagi and Iwate prefectures and 19,630 casualties (Tagore-Erwin, 2018). Ishinomaki, the second-largest city in Miyagi prefecture with a population just over 139,000 inhabitants (Ishinomaki no jinko oyobi menseki, 2021), is one of the places hit the hardest: 3,553 people were lost, 423 are still missing, and 50,758 people were evacuated (Tagore-Erwin, 2018).

The affected cities and towns are still in the process of recovery and grassroots initiatives and cultural events are flourishing to aid the physical reconstruction and the process of re-imagining the cities' identity. This chapter is based on research collected by the first author, mostly through archival and media material as well as 16 semi-structured interviews with locals, volunteers and Reborn Art Festival organisers, and a focus group with volunteers, with a focus on the relationship between trauma and placemaking.

DOI: 10.4324/9781003371533-26

Placemaking to *Machizukuri*

Definitions of placemaking are numerous. Placemaking has often referred both to the top-down approach used to control or shape the perception of a certain place and the associated efforts to influence people's identities, behaviours and experiences (Smith, 2002), or, conversely, as the organic, bottom-up approach and process shaped by individuals in their daily social practices (Lems, 2016). This chapter adopts the latter, as it is concerned with the tensions between the institutional side and the local communities who are trying to have their voices heard (Vainio, 2020) through a set of tools, which give communities agency in determining how their place should look like, just as much if not more than architects, planners, designers, developers, politicians and governmental officials (Courage, 2023).

Rather than placemaking, in this chapter we adopt the autochthonous term *machizukuri* which translates into 'town-making'. *Machizukuri* denotes the use of grassroot initiatives which aim to bring change through the engagement and participation of local citizens in building a physical, social, and emotional environment that reflects their values (Kusakabe, 2013), and hence has a broader role in generating a sense of community, place, and resilience. *Machizukuri* is closely related to the notion of placemaking but is a word that more closely attunes to a non-Western-centric approach to place specific to the context of Japan (Tagore-Erwin, 2018). *Machizukuri* planning is often described as an alternative to the conventional top-down urban planning approach led by the government (Kusakabe, 2013; Posio, 2019). However, as experienced by some localities in their post-disaster reconstruction planning, such as in Kobe in 1995, there has been criticism around the fact that *machizukuri* is often instrumentalised by governmental institution to gain consent but does not embrace residents' perspectives enough. Similarly, in Tōhoku, despite efforts from authorities in taking into account past criticism, current reconstruction plans seem to be following the same pattern, with excessive control of government authorities in the reconstruction process (Edgington, 2011).

Cultural Events, Disaster, and Trauma-healing

Trauma, in its basic meaning of severe and lasting physical or emotional shock and pain (Pain, 2019), can modify the sense of place, placemaking processes, and sense of belonging, and it is a force that actively shapes place for communities. Cultural events can be a tool for post-disaster community economic and physical recovery, as well as in the recovery from mental and emotional trauma (Itoh and Konno, 2019) by giving meaning to and offering future prospects of life – a production of value (Orr, 2016) – to rural populations with the compound experiences of depopulation, aging, outmigration of local industries, and loss of jobs and education opportunities. Festivals can be considered as a transient but essential placemaking tool, as a site where displaced communities can meet and through active or passive participation reimagine place through activities and storytelling, making sense of the past and delineating a trajectory for the future. Also, in places impacted by traumatic events, the most impactful consequence long-term is at times the 'slow trauma' that continues to permeate places and bodies post-crisis (Coddington & Micieli-Voutsinas, 2017). Festivals are key sites for the ultimate re-working of grief (Fullilove et al., 2004), utilizing a larger context of social-cultural engagement as a space of healing. Festivals can be utilized as placemaking for communities to 'renew their connection to where they live, to ask questions, to empower themselves, to intervene in planning and policy, to change infrastructure and impact the cultural, social and economic life of their neighbourhoods' (Courage, 2023).

Festivals can revitalize rural places as they offer tools for the community to reimagine their city, and they attract not only external individuals such as tourists, but also volunteers, often connected with non-profit organizations (NPOs). Some of these eventually become future permanent settlers as they are fascinated by the situation and decide to stay. The fact that there are new migrants, in turn, can increase the attractiveness and liveability of those places due to the new influx of people in a place where other displaced citizens do not necessarily want to come back (Sanders et al., 2015; Tagore-Erwin, 2018). In a post-disaster context, Solnit (2010, p. 22) describes the 'extraordinarily generative character of disasters' in which affected places have the power to bring different people together, forming new networks and innovative ideas while imagining and aspiring to new values and quality of life. It is worth pointing out that, on the other hand, when external elements such as new residents, external NPOs and influences from the government enter a context, especially a post-disaster one, it is crucial for locals' wellbeing that *machizukuri* remains mostly under the local community's control and agency (Sorensen & Funck, 2007).

Using festivals as an instrument for revitalization can be a problem, as these events are often aimed mainly at economic revitalization (McLeod, 2019). Festivals, when facing pressure for their economic viability, usually intensify their commercialization, leaving out community resilience and environmental sustainability. This trend of using festivals as an instrument for creative placemaking does not come without criticism as it can lead to potential risks such as gentrification and serial-reproduction (Richards, 2020). Moreover, while events such as festivals can be a powerful tool to recover a sense of agency over narratives and individual and social management of trauma (Martini & Minca, 2021), there is an emotional distance perceived by locals towards governmental institutions at different scales, which produces a feeling of injustice. Unevenness in access, social services, networks, opportunities, intensify not just experiences, but also feelings and perceptions of exclusion (Cook & Butz, 2016).

Our findings reveal that while the Reborn Art Festival can be considered as a successful event, which has potential to aid processes of trauma-healing, locals' engagement is ambivalent, resulting in *machizukuri* as well as healing processes that are unequally enacted. In the case of Reborn Art Festival, and more generally in post-disaster recovery processes, trauma is complex and multifaceted, apprehended and negotiated by different actors in different ways. However, trauma-management for communities is much more complex.

Ishinomaki City and the Reborn Art Festival

Ishinomaki city in Miyagi prefecture is located on the coast of Tōhoku, less than 100km from the epicentre of the earthquake. Historically the Tōhoku region had been subjected to a government place-branding (Vuignier, 2017) using nostalgia to funnel revenue as a response to economic migration in urban areas. The broader aim was to reaffirm values such as tradition, as the roots of Japanese identity are identified with the place 'Where the heart, the rice and the grandmother's house is' (Hopson, 2017 p. 8).

After 2011, people were subject to profound trauma during and after the disaster, having experienced loss and displacement. Many were sent to live for years in temporary housing complexes far from their original neighbourhoods. In these circumstances, the notion of community and people's sense of community were both radically challenged and became an essential tool mobilized by both residents and institutions to recompose and make sense of the traumatic experience moving forward. Like many other towns, the reconstruction process is still ongoing even after a decade (Tagore-Erwin, 2018). As discussed by Vainio (2020, p. 7), while the recovery process is advertised by the government as being 'community-focused', in reality, the locals often feel left out 'voiceless' and 'unheard'. The clashing imaginaries concerning the future envisioned by the government focuses on reconstruction plans, reports and policies, which are not always well-received by the local communities and have become a major source of tension (Vainio, 2020), as they constrained the freedom of local governance in pursuing *machizukuri* and creating sustainable and livable environments (Sorensen & Funck, 2007). The Japanese government and prefectural institutions focused on erasing and overcoming trauma completely, creating slogans such as '*Smile, Japan!*' or '*Gambarimasu!*', a non-translatable word which implies the idea of 'holding on', 'moving forward', or working hard to get to an objective. These slogans have an emphasis on future-oriented thinking, and implicitly presume that, after a certain amount of time for mourning, residents should think positively and move on. This has unfortunately much to do with the fact that in Japan seeking mental health services is still heavily stigmatized (Ando et al., 2013) and, accordingly, long-term suffering and residual trauma are often deemed socially inconvenient.

In such a situation, community-led initiatives have the potential to play a significant role in maintaining social relations and communication, opening possibilities for community resilience in contrast to top-down recovery plans that do not consider the social aspect and realities that the local communities are facing. As one resident explains: 'The landscape of the city is progressively being restored and although there are some improvements, the reality is that terrains and views that we want to keep are being scraped' (Interviewee, Ishinomaki).

While the government and local authorities had mainly focused on physical reconstruction, according to one local from Ishinomaki, new locals coming from outside the prefecture to volunteer after the disaster have been a tremendous support for some of the local communities. These volunteers were not only actively invested in rebuilding the city but also provided support for the locals' emotional and psychological trauma through interpersonal relationship building via taking care of children that could not go to school anymore, or creating space and giving time to listen to the stories of those who lost their relatives. They were confronting the realities that local communities were facing, leaving a stronger impression than governmental authorities, which mostly focused on providing temporary housing.

It is in this context that The Reborn Art Festival was devised and organised in different parts of Ishinomaki, in 2017. Its aspiration is to 'reconstruct from the inside [of the region] and to create a new cycle' (Reborn Art Festival 2021–22, 2021). A contemporary art festival with a tourism development aim, Reborn Art Festival involves Japanese and international artists and showcases art installations, food and music events that encompass aspects of the local culture. It takes place during 51 days over the summer months of July and September and it differs from other festivals in that it is not organised by local authorities but by a well-known musician, Takeshi Kobayashi, who also serves as executive committee chairman of AP Bank, a non-profit private organisation that is known for their active engagement in supporting the reconstruction of the Tōhoku region (Itoh & Konno, 2019) and for their environmental advocacy. Initially, AP Bank started a three-day pre-event in Ishinomaki in 2016, called the AP Bank Festival. It mainly focused on music events. This was also a way to attract people from all over Japan and have an idea of what it could be to set up a festival in Ishinomaki. While Reborn Art Festival is an initiative from a private organisation, it has collaborated with local authorities from the municipalities of Ishinomaki and Matsushima as well as the prefectural authority of Miyagi and other local organisations throughout its festival editions. As Ishinomaki is the main stage for the Reborn Art Festival, the festival delegated the secretariat to Gota Matsumura, the representative of the organisation Ishinomaki 2.0.

Making Place, Healing Trauma?

The mission of The Reborn Art Festival heavily implied the will to 'produce' placemaking for the community, through the festival form to offer a time and place for groups to celebrate and remember their history and culture through exhibits and displays, food, arts, music, dance, and performance (Peyton, 2021). However, when agency over the construction of the festival event is not in the hands of the communities themselves, results can be mixed. The urban-centred institutionalised approach to placemaking through the festival has served only a part of the stakeholders, mainly non-locals, exacerbating a cultural conflict between the local and the global, the urban and the rural (Qu, 2019). Possibilities to build placemaking processes upon endogenous capabilities exist, except priority is given to something that does not appeal to a large part of the local community (Turok, 2009), such as urban-centred artworks. As demonstrated in previous studies that concern art festivals organised in Japan like the *Echigo-Tsumari*

Art Triennial or the *Setouchi* International Art Festival (Klien, 2010; Qu, 2019), 'artists implant their art–which connotes the elite, global, and urban–into the rural context' (Qu, 2019, p. 23).

The Reborn Art Festival mirrors the very complexity of placemaking and *machizukuri* processes. Places are always contested and constantly changing rather than fixed. The festival generated a sense of place, but alienated the majority of locals as it mostly united urban stakeholders that share common practices (Aldrich & Meyer, 2015). Placemaking processes too were influenced by a dominant narrative, creating a parallel imaginary that is detached from the needs and values of a large part of the local community. *Machizukuri* is supposed to be a bottom-up process that reflects local citizen's values (Kusakabe, 2013) but the one taking place through the medium of the Reborn Art Festival is not perceived as genuine because it omits local communities from the decision-making process (Qu & Cheer, 2020). Discourses and festival representations of the disaster, as proposed by The Reborn Art Festival organisers, as well as correlated institutional narratives around post-disaster recovery, posit Japan as a resilient country, which has mourned, and can now stop and move on. However, processes of trauma healing are not linear: trauma is embedded within particular times and spaces, representing a social wound grounded in the built environment (Coddington, 2017).

Yet, at the same time, trauma is timeless, placeless, contagious, permeable, a site of continuous potentiality. Caruth (1995, p. 153) argues that trauma represents 'a history that literally has no place, neither in the past, in which it was not fully experienced, nor in the present, in which its precise images and enactments are not fully understood'. This means that the site of trauma cannot be precisely located, as it transpires across times and spaces, like an unmappable haunting. A volunteer that participated in the Reborn Art Festival, in describing their experience, highlights how trauma has a contagious element, which resembles 'a process of connecting several and often unrelated traumas, a process whereby trauma adheres, spreads, and expands' (Coddington, 2017): 'How can I say… after all, I wasn't affected by the disaster, so it's absolutely impossible to understand. But after bringing back home what I have felt while being there, I started to think about it on a daily basis, things like what are natural disasters' (Interviewee, Tokyo).

The placemaking process happening with the Reborn Art Festival is not community-based, but closer to the one defined by Smith (2002) which is a top-down way to shape the image of a place. The Reborn Art Festival's narrative revolves around their aim to help revitalisation from the 'inside', which is supposed to help locals through endogenous development. But the reality is quite the opposite, as the Reborn Art Festival implemented extraneous elements which were embracing the values of non-locals more than locals (Lew, 2017). The festival had the potential to reflect local communities' values, desires, and imaginaries but did not embrace them enough, omitting local's voices while advertising their activity as the opposite (Vainio, 2020). It ended up contributing to an increased fracture of the city with, on the one hand, the inflow of new locals (mostly coming from Tokyo) actively engaged in the reconstruction process and creating new networks, and on the other locals that have been living there for a long time (Klien, 2016).

The Reborn Art Festival shows an aspiration, a process of placemaking where the post-disaster countryside is reimagined by urban dwellers as rural fantasy for the use of outsiders. This could be said to have reinforced the urban-rural dichotomy (Klien, 2010) as the Reborn Art Festival prioritised the modern over the traditional, the extraneous over the endogenous. Within the festival, trauma takes on a life of its own, forging its own connective tissue through encounters in which individual emotions find expression (Mountz, 2017). The dichotomy is also partly sustained by locals themselves. From the locals' perspective, minimising external influences while maintaining their values as people from *inaka* (countryside) can be a way to reinforce their cultural identity, which reflects the conservative mindset described by some locals themselves and as shown in past studies about festivals in the region (Klien, 2010). While they did not feel connected to or interested in that event, they still appreciated parts of it, as using art or music to remember the catastrophe and having people coming to the place is something that is very meaningful for the locals (Figure 22.1).

Some art projects, however, were highly contested and received complaints from local communities during the 2017 edition of the Reborn Art Festival. *Inside out* consisted of printing photographs of visitors in black and white on a large-scale poster. Local communities showed their discomfort towards the cultural

FIGURE 22.1   Kohei Nawa, *White Deer* (*Oshika*) (2017). Image credit: Roger Smith (2017).

insensitiveness of the project, as black-and-white photos are usually printed for funerals in Japan. Similarly, another project required the use of public loudspeakers, which are part of the everyday soundscape of Japanese people. The artist used the equipment to transmit a chime with their singing voice within the venue of the festival, which made some locals feel uneasy. Both art installations were removed within one week due to a number of complaints. An interviewee also shared their thoughts about this:

> *No matter how much they try to stand close to us, it's a place where many people died with the disaster and only those people know where their home used to be. Even if people say, 'I used the debris to make art!' it was once part of someone's house or life. 'A famous photographer takes pictures of everyone's smiles and decorates the embankment with black and white photos!' This was also removed because everyone perceived it as portraits of deceased persons and disliked it (…) It's not the artist's fault, but I think that with the context behind the place, it makes it difficult.*
>
> *(Interviewee, Ishinomaki)*

Ultimately, festivals have the potential to articulate trauma as 'haunting' (Orr, 2016), a transmission of feeling that wants to perform. Throughout the artworks and their spatialization trauma erupts in bursts, 'revealing moments wherein past erupts into the present, rendering more visible the haunting' (Mountz, 2017, p. 75). It is when trauma is made visible, that it is then possible to access its transformative power through posttraumatic growth, 'a positive change that occurs as a result of the struggle with highly challenging life crises' (Tedeschi & Calhoun, 2004, p. 1). If engaged with sensitively, representations of traumatic pasts can be used for the emotional benefit of communities, and not just to pursue superficial forms of government-mandated *machizukuri*. For the locals of Ishinomaki, the Reborn Art Festival has been a contested and at times disengaging experiment, which has not contributed to *machizukuri* in the city. One important aspect of *machizukuri* is the utilisation of diverse, community-proposed strategies to plan an environment that reflects the residents' values and lifestyles, and to gain greater involvement and legitimacy for local organisations (Sorensen & Funck, 2007). The festival, ultimately, was not perceived by locals as contributing to *machizukuri*, as it involved too many external stakeholders who controlled too much of the decision-making process. However, despite this, or partly because of this, as Orr (2016, p. 274) states, 'trauma may also produce unexpected openings, the effects of which will also circulate'. Far from being a completely negative experience, the Reborn Art Festival also contributed to the creation of new movement in the city, new dynamics between locals, tourists, and new locals, and ultimately, by agreeing with it or opposing it, offered a platform for discussions on trauma and its consequences.

References

Aldrich, D. P. & Meyer, M. A. (2015). 'Social capital and community resilience', *American behavioral scientist*, 59(2).

Ando, S., Yamaguchi, S., Aoki, Y. & Thornicroft, G. (2013). 'Review of mental-health-related stigma in Japan', *Psychiatry and clinical neurosciences*, 67(7).

Caruth, C. (1995). *Explorations in memory*. Baltimore: The Johns Hopkins University Press.

Coddington, K. (2017). 'Contagious trauma: Reframing the spatial mobility of trauma within advocacy work', *Emotion, Space and Society*, 24.

Coddington, K. & Micieli-Voutsinas, J. (2017). 'On trauma, geography, and mobility: towards geographies of trauma', *Emotion, space and society*, 24.

Cook, N. & Butz, D. (2016). 'Mobility justice in the context of disaster', *Mobilities*, 11(3).

Courage, C. (2023). 'The art of placemaking' in *RSA Journal*, 1.

Edgington, D. W. (2011). *Reconstructing Kobe: The geography of crisis and opportunity*. Vancouver: UBC Press.

Fullilove, M. T., Hernandez-Cordero, L., Madoff, J. S., & FULLILOVE III, R. E. (2004). Promoting collective recovery through organizational mobilization: the post-9/11 disaster relief work of NYC RECOVERS. *Journal of biosocial science*, 36(4), 479–489.

Hopson, N. (2017). *Ennobling Japan's Savage Northeast: Tōhoku as Postwar Thought, 1945–2011*. Leien: Brill.

Ishinomaki-shi no jinko oyobi menseki (2021), *Ishinomaki City*. Available: www.city.ish inomaki.lg.jp/cont/10102000/0040/2204/2204.html. [Accessed: 17 August 2023].

Itoh, Y. & Konno, Y. (2019). 'A Study on the Role of Art Festivals on Disaster Recovery: A Case Study of the Great East Japan Earthquake and the Reborn-Art Festival', *Yokohama Keiei Kenkyuu*, 40(2).

Klien, S. (2010). 'Contemporary art and regional revitalisation: selected artworks in the Echigo-Tsumari Art Triennial 2000–6', *Japan Forum*, 22(3).

Klien, S. (2016). 'Reinventing Ishinomaki, Reinventing Japan? Creative Networks, Alternative Lifestyles and the Search for Quality of Life in Post-growth Japan', *Japanese Studies*, 36(1).

Kusakabe, E. (2013). 'Advancing sustainable development at the local level: The case of machizukuri in Japanese cities', *Progress in Planning*, 80.

Lems, A. (2016). 'Placing displacement: Placemaking in a world of movement', *Ethnos*, 81(2).

Lew, A. A. (2017). 'Scale, change and resilience in community tourism planning', in A. Lew (ed.) *New Research Paradigms in Tourism Geography*. Abingdon: Routledge.

Martini, A. & Minca, C. (2021). 'Affective dark tourism encounters: Rikuzentakata after the 2011 great East Japan disaster', *Social & Cultural Geography*, 22(1).

McLeod, K. (2019). 'Festivals, Place, and Placelessness', *Canadian Theatre Review*, 179.

Mountz, A. (2017). 'Island detention: Affective eruption as trauma's disruption', *Emotion, Space and Society*, 24.

Orr, J. (2016). 'Body Animations (or, Lullaby for Fallujah): A Performance', *Critical trauma studies* (pp. 157–176). New York: New York University Press.

Pain, R. (2019). 'Chronic urban trauma: The slow violence of housing dispossession', *Urban Studies*, 56(2).

Peyton, S. I. (2021). ' "Showing Out": African American Cultural Festivals as Sites of Reworking Grief from Historical Trauma' (Doctoral dissertation, The University of North Carolina at Chapel Hill).

Posio, P. (2019). 'Reconstruction machizukuri and negotiating safety in post-3.11 community recovery in Yamamoto', *Contemporary Japan*, 31(1).

Qu, M. (2019). 'Art interventions on Japanese islands: The promise and pitfalls of artistic interpretations of community', T*he International Journal of Social, Political and Community Agendas in the Arts*, 14(3).

Qu, M. & Cheer, J. M. (2020). 'Community art festivals and sustainable rural revitalisation', *Journal of Sustainable Tourism*, pp. 1–20.

Richards, G. (2020). Can Placemaking in Canadian Public Greenspaces Bring Suburban Communities Together? Case Studies of City Park Community Gardens in Mississauga, Ontario and Surrey, British Columbia.

Sanders, D., Laing, J., & Frost, W. (2015). Exploring the role and importance of post-disaster events in rural communities. *Journal of rural studies*, *41*, 82–94.

Smith, N. (2002). 'New globalism, new urbanism: gentrification as global urban strategy', *Antipode*, 34(3).

Solnit, R. (2010). *A Paradise Built in Hell: The Extraordinary Communities that Arise in Disaster*. London: Penguin Books.

Sorensen, A. & Funck, C. (eds.) (2007). *Living cities in Japan: citizens' movements, machizukuri and local environments*. Abingdon: Routledge.

Tagore-Erwin, E. (2018). *Post-Disaster Recovery Through Art: A Case Study of Reborn-Art Festival in Ishinomaki, Japan*.

Tedeschi, R. G., & Calhoun, L. G. (2004). 'Posttraumatic growth: conceptual foundations and empirical evidence'. *Psychological Inquiry*, 15(1), 1–18.

Turok, I. (2009). 'The Distinctive City: Pitfalls in the Pursuit of Differential Advantage', *Environment and planning A,* 41(1).

Vainio, A. (2020). '*They don't understand how we feel': An affective approach to improving the 'best practice' of community-based post-disaster recovery.* The University of Sheffield.

Vuignier, R. (2017). Place branding & place marketing 1976–2016: A multidisciplinary literature review. *International Review on Public and Nonprofit Marketing*, 14(4), 447–473.

23

PLACEMAKING, PERFORMANCE AND INFRASTRUCTURES OF BELONGING

The Role of Ritual Healing and Mass Cultural Gatherings in the Wake of Trauma

Anna Marazuela Kim and Jacek Ludwig Scarso

Prologue

In the spirit of this project to develop a dynamic community of praxis, the authors have decided to frame their contribution to this volume as a dialogue rather than a univocal essay, keeping the commitment to posing theory against practice, and bringing a new voice, that of an artistic researcher-practitioner, into the discussion. In what follows, Dr Anna Marazuela Kim (AMK) provides a wide-ranging introduction to explore the potential of mass cultural gatherings in the healing of societal traumas, specifically as they invoke placemaking in the *absence* of place – a dynamic which forms a new contribution to the literature on this topic. This is followed by Dr Jacek Ludwig Scarso's (JLS) presentation of a case study, Artichoke's *Sanctuary* (2022) (Figure 23.1). This is intended as an example to apply our considerations, as it both reflects and challenges notions of placemaking in trauma-informed practice.

Lastly, an exchange of ideas between the two authors provides a series of reflections highlighting both the poignancy and complexity of addressing trauma through ephemeral, artistic placemaking.

AMK

While the premise of this volume – *trauma-informed placemaking* – might seem new, the project of placemaking has long been predicated upon the value of understanding the history of the places it seeks to remake, in order to make visible the complex set of structures and forces that have shaped and continue to inform their trajectories. The US-based *Thriving Cities Project* (now *Lab* and

DOI: 10.4324/9781003371533-27

Group) with which I have been involved pioneered a paradigm of the city as a complex ecosystem, with the understanding that the basis for engaging any given city was not limited to identifying its present set of challenges and circumstances, but the deep structures of the history that formed it. Often it is only by means of this socio-political archaeology that one uncovers the potential of what was once a richly layered, complex and diverse community of interests living and working together, the interrelation between what Richard Sennett describes in terms of the '*cité*' and '*ville*' (Sennett, 2019). Historical investigation is also crucial in revealing the hidden, or nearly invisible, fractures and traumas in a landscape which are later obscured by newer layers of sedimentation: whether the grim detritus of post-industrial abandonment; the shiny patina of gentrification covering wounds of displacement and a flattening of layers of cultural diversity; or the newly demarcated borders instituted by war, political force, or the forces of global capitalism, further dividing societies and communities. That is to say, placemaking often begins in the aftermath of histories of trauma.

Places – even when they have suffered such radical changes as to render them unrecognisable to the communities who once made them, for better or for worse – are nonetheless a material repository of histories which preserve the *potential* for their remaking, even if limited to remembering the traumas which have formatively shaped them. But what if trauma takes place not in a specific place or spaces, but throughout the body politic of a particular community, a society, or society at large? How might placemaking address these non-site specific, collective wounds, particularly when they are rendered less visible by the mandate to carry on despite the injury, such as in the aftermath of the devastation of COVID-19? Given the cumulative effect of these wounds, how might placemaking heal the deeper fractures that cripple the social infrastructure foundational to a healthy, inclusive civic life?

The multiplication and intersection of societal traumas – from pandemic isolation to race riots, to the deterioration of infrastructures of societal support at the level of community, democratic processes at the national level, and indeed our shared, planetary conditions – make these questions of increasingly urgent concern. Beyond the debilitating effect on individual and collective wellbeing of more recent events, there is the growing awareness of the toll that living in the aftermath of long histories of structural violence and injustice brings to the equation. Anne Cvetkovic's description of depression as a cultural, social and political phenomenon or 'public feeling', rather than an individual or chemical malady, illuminates the psychological burden we all carry by virtue of a shared inheritance of trauma, regardless of whether we are conscious of it (Cvetkovic, 2012, pp. 1–2). There is little doubt that more recent traumas are doubled by a sense of fatigue in the face of long historical chains of social violence – intractable and seemingly without easy solution – further draining our capacity to address them. Yet as Cvetkovic explains, such an investigation, which is part of a larger 'affective turn' across disciplines that recognises the importance of the life of feeling, presents the possibility to regenerate the hopeful foundations for political agency (ibid., pp. 2–3).

In searching for a means to undertake the healing of these invisible, collective wounds, we might also turn to the sociological perspective of Jeffrey Alexander in his exploration of 'cultural trauma', when 'members of a collectivity [sic] feel they have been subjected to a horrendous event that leaves indelible marks upon their group consciousness, marking their memories forever and changing their future identity in fundamental and irrevocable ways.' Alexander notes that social scientists stress the importance of 'public acts of commemoration, cultural representation, and public political struggle – some collective means for undoing repression and allowing the pent-up emotions of loss and mourning to be expressed', as methods of amelioration and healing (Alexander, 2004, pp. 1–30).

In describing the experiential operation of an artwork dedicated to recently exhumed remains from the Spanish Civil War by Hrair Sarkissian, whose family history was shaped by the Armenian genocide, critic Vali Mahlouji gives richer definition to this otherwise generic set of sociological principles, which also brings it into the affective realm. The artwork under discussion is Sarkissian's *Deathscape* (2021), described by the artist as 'a temporal sound installation that enables the audience to experience transmitted soundscapes made by forensic archaeologists during the exhumation of mass graves that hold the bodies of those executed by the authoritarian regime headed by Francisco Franco from the time of the 1936–1939 Spanish Civil War until his death in 1975' (Mahlouji, 2023, p. 244):

Moving from materiality to immateriality, and from evidentiary truth to experience, [Sarkissian's] artwork shifts the focus away from the absent-present object-to-be-salvaged into an immaterial sensorial field, and then into a field of mnemonic and affective experience.

(ibid.)

In order to mitigate the damage of traumatic experience transmitted intergenerationally and inherited through history, against those who caution against reopening closed wounds, Mahlouji emphasises the importance of the act of witnessing: 'There is no possibility of healing without the truth, and no possibility of healthy reconciliation with history in the absence of the collective acknowledgement of truth' (ibid., p. 245). What is made possible by the physical artwork in question is an archaeological re-presentation of trauma that is effectively a double operation: first of witness or recognition, aiding the recovery of the repressed, or in this case, suppressed, memory, followed by an active form of recovery through 'somaesthetics': what Richard Shusterman has described as 'thinking through the body' (Shusterman, 2012, p. X).

Such approaches explicitly reference an understanding of trauma and theories that have their origins in psychoanalytic theory. When trauma studies emerged in the 1990s, it was initially primarily shaped by Freudian thought, particularly regarding the mechanism of repression and the deleterious effects of traumatic events inaccessible to the individual patient or *analysand*. While the sociological and social constructivist methodologies with a basis in psychoanalysis described above do emphasise the material and experiential, they are also focused to a large degree on the retrieval of memory, and therefore to processes strongly allied with conscious thought, even if through somaesthetics or 'thinking through the body.' But there is another strand of origin for the term 'trauma', literally 'wound', which offers a different point of entry for consideration; this lies in Ancient Greece. Turning to that linguistic and cultural source, I would suggest that beyond the deictic and experiential dimension of psychoanalytic approaches to trauma, the importance of communal ritual and performance, as they present the potential for catharsis, offer a valuable contribution as well. Within this frame, we might note the interest of contemporary artists in Ancient Greek tragedy as the foundation for addressing histories of trauma, such as Grada Kilomba's *Illusions* Vol. II (2018) and Vol. III (2019), which re-stages the stories of Oedipus and Antigone to unpack the violence of a colonial past, or Edmund Clark's *Oresteia*, which documents his work as artist-in-residence in Britain's only therapeutic prison, through a retelling of the play by inmates.

Extending these approaches more broadly, we begin to see the potential of cultural gatherings that draw upon performative rituals of witness and catharsis to heal the collective wounds which we have earlier described. The positive capacity of festivals to bring people into a shared experience of wonder and joy through the arts, and rituals of healing, capture a further dimension of these gatherings (Alvarez, 2020). Joining together these methodologies, we can now return to the initial question of how placemaking might address collective trauma that is not specific to place, but affects the larger body of society, trauma which is furthermore left to the individual to address in relative isolation, as in the case of COVID-19 and the continuing pain and loss left in its wake (Schupbach, 2020). The case study we have chosen as exemplary, *Sanctuary* (2022) commissioned by Artichoke, adds a further dimension to specifically address the absence of a site as historic *locus* for these processes. This is the intentional creation of a temporal architectural structure and space that presents the possibility of publicly sharing and memorialising the trauma of COVID-19 and its aftermath: a building co-constructed by artist and community, providing both a process and an infrastructure for communal healing (see also Beinart and Lugalia-Hollon & Quintara, this volume). In further articulating the operation of such an artwork, we might invoke what theorist Mechtild Widrich defines as a 'performative monument', one which is constituted in relation to its interactive audiences, temporal rather than permanent, processual rather than fixed (Widrich, 2014, p. 6). However, the temporal nature of the intervention, its constitution of being led by an artist rather than co-designed with a community, and its seeming disassociation with the particularity of its chosen site, all raise questions of potential importance for not only this volume but for placemaking as a whole, which the following section seeks to articulate and to address.

FIGURE 23.1   Top, the built structure of *Sanctuary*; middle, participatory engagement in *Sanctuary*; bottom: The burning of the structure of *Sanctuary*, an Artichoke project in association with Imagineer. Created by David Best and the Temple Crew, 2021. All images courtesy of Artichoke.

JLS

Produced by Artichoke in collaboration with Imagineer and Nuneaton & Bedworth Borough Council, Warwickshire, UK, in May 2022, *Sanctuary* (Figure 23.1) was a participatory work created by the US artist David Best in Miners' Welfare Park in close collaboration with members of the local Warwickshire community. A temporary memorial to COVID-19 victims, a wooden architectural structure of monumental proportions (65ft/19.8m high) was erected to stand for the duration of a week. Throughout this period, visitors were invited to inscribe wooden pieces with messages dedicated to loved ones lost during the pandemic and attach these to the structure. As an act of collective catharsis, the week-long duration culminated in the spectacular act of setting the structure on fire, intended as a way of closure in the traumatic experience of grief. An estimated 16,700 people visited during the week and 10,000 the burning event, with further dissemination achieved through features on the national press and the BBC.

My research with Artichoke began in 2021, looking at the impact of COVID-19 on the changing perception of mass gatherings and their cultural value in reclaiming public space, post-pandemic. In an article co-authored with Kirsten Jeske Thompson (Scarso & Thompson, 2022), we looked at three case studies from Artichoke's body of work since its debut with the itinerant performance *The Sultan's Elephant* in 2006. Alongside the latter, our research explored the recurring light festival *Lumiere* (biennially, 2009–) and the participatory production *Processions* (2018), commemorating one hundred years of women obtaining the right to vote in Britain. The aim of the publication was to analyse the unique potential and resilience of these case studies in involving large audiences in mass-scale acts of performance. These acts, we argued, would at once need to adapt to the locations used, responding to the complexities of multiple sites and the countless logistical restrictions inherent to these; at the same time, they would creatively disrupt, for a temporary period, such contexts, their daily routines and social conventions, in the creation of spectacular experiences that would radically shift the perception of the city and of the individual's role within it. The argument here was that, in a moment of deep social change and unprecedented restrictions as those caused by global lockdowns, the process of gradually reclaiming public space could be informed by the learnings developed over the years by companies like Artichoke, that have made public space their stage and art gallery.

Across this research, with Kirsten Jeske Thompson (Scarso & Thompson, 2022), Artichoke posed an interesting example of what could be defined as 'ephemeral urbanism': a way of relating to the built environment by highlighting what happens in between the buildings, in the 'soft city', that is in the intangible ways in which we may creatively 'mis-use' a city environment. To follow the 2022 study, which used the pandemic as a starting point, it is particularly poignant to now look at the way this artistic approach has been used as a memorial gesture to directly address COVID-19. Somehow significantly, the piece analysed here is set apart from the urban context that provided the backdrop to so many projects by Artichoke: an almost symbolic decision of moving away from the hustle and bustle of the city and to look for a contemplative space that transcends the meaning of a specific locality, while directly engaging with its local community: a place to ponder on the human loss of COVID-19 and, in doing so, to meditate on the fragility and on the value of life.

The relationship between Artichoke's work and the concept of placemaking is not without the need for problematisation. As an organisation commissioning and producing public art events, placemaking may not be straightforwardly applied to Artichoke's projects (nor indeed does Artichoke consciously reference placemaking in its work). Yet, placemaking does provide a lens through which to appreciate their potential and impact further. If we use Courage's definition of placemaking as putting a community 'front and centre' in deciding how a place looks and functions (2020, p. 2), we may see Artichoke's approach as sitting at odds within this framework: here, the public is most certainly a participant in the project, but the commissioning process used by Artichoke, actively selecting artists, curating the artistic production of the work and the modes in which participation takes place, could be regarded as only partially co-creative. This is a deliberate choice by Artichoke: when we interviewed its CEO, Helen Marriage (Scarso & Thompson, 2022), she emphasised that co-creation should not be at the expense of the artist's expression: that the apparent default application of co-creative strategies (Marriage was referencing current trends in public funding in this respect) risks inadvertently contributing to an implicit dismissal of what an artist can do and bring. Thus, Artichoke operates as a facilitator of a dialogue between creative professionals and a community, drawing on the expertise of the former and the experience and insight of the latter, to generate new possibilities of artistic exploration.

It is also important, in this respect, to consider the scale of Artichoke's projects. With *Sanctuary* creating a temporary place for multiple thousands of participants over the space of a week, there is a clear need for the organisation to supervise operations in a highly structured way, whilst also ensuring active engagement by the public in the creative process. Discussing this with Beth King, Head of Participation and Learning at Artichoke (interviewed on 25 November 2022), projects like *Sanctuary* require long-term planning in order to be logistically feasible. The project had to be postponed several times due to COVID restrictions and, in addition to the complex practicalities of its installation, it also entailed a range of preparatory activities engaging the public in its making: for instance, local schools were invited to create designs for some of the wooden panels that would be eventually incorporated in the structure, and this took place months before its realisation. Furthermore, Artichoke actively engaged local job centres to provide paid employment for members of the community to work in the creation of the main structure, in addition to training volunteers to take part in the hosting of the week-long installation.

Inevitably, the large-scale format discussed here, in relation to Artichoke and more broadly across this chapter, leads to the question of whether such scale, combined with the ephemerality of projects like these, can indeed be conducive to actual placemaking. King conveyed that there was some initial resistance amongst the community of Nuneaton & Bedworth when originally consulted on the making of *Sanctuary*, with many questioning why public money should be used for an artistic intervention of this scale that was designed to be only temporary. This was in direct contrast with the overwhelmingly positive response that the project received once realised, including by many of those who had opposed it initially. This leads to complex questions with regards to the process of consultation in relation to creative risk taking. Artichoke's ethos is about creating 'extraordinary and ambitious ephemeral events' that 'invade our public spaces' (Artichoke website); their approach champions innovative artistic ideas that may be difficult to fathom until they are materialised. Public consultation should of course take place, but to what extent should an artistic vision be trusted regardless, informed but not potentially blocked by this?

According to these complexities, it may be worth considering the different phases of co-creation in a project like *Sanctuary*. In the first phase, a site is created through the commissioning process, hence positioning the producing organisation and the artist involved as leading such creation in a structured way – undoubtedly, here notions of hierarchy are at play, notwithstanding the collaborative ethos that has always been championed by Artichoke. Indeed, it could be argued that the work is created *for* participants, as opposed to *with* participants at this stage (Page, 2020, p. 157). This however changes once the situation is created: participants join in and begin to take ownership of the location, 'making' this a 'place' through the act of sharing and commemorating their lived experience, giving meaning and collective purpose to what up until that point can be seen as a spectacular artwork.

King communicated the team's surprise at just how deep the engagement by the public was in *Sanctuary*. Thousands of participants visited the site, sharing very personal stories of loss, bringing photographs, memorabilia and in some cases even the ashes of dear ones, to the point that the piece quickly transcended the specific context of COVID-19 and became a powerful vehicle for a collective expression of grief. While this testified the sheer impact of the project, it came with complications too. King remembers that the team began to refer to the project as the 'temple of tears' and many would feel overwhelmed by the participants' sharing of so many stories of emotional pain. While NHS counsellors were present at the site for part of the project and useful contacts were actively disseminated for further psychological support where needed, it was clear that additional aid would be necessary, and Artichoke followed this up by providing counselling for the team involved in the aftermath of the project's conclusion.

The healing of traumatic lived experience is central to projects like *Sanctuary*, and this process necessitates that such experience is remembered and possibly relived. This implies further responsibility for its creators as to the safeguarding of participants. While, inevitably, a key point we may want to interrogate is whether further measures in this respect would have been necessary, including more specialist support in mental health or more training for staff present at the art event, it is important to consider that a project like this is about nurturing mutual support in a community. It is by encouraging this communal support, made of a genuine exchange of personal experiences, that a project like this can be seen within the frame of placemaking. Indeed, we may ask whether, should the presence of mental health specialists have been more visible, the project may have been perceived differently and its immediacy and community spirit may have been somewhat lost. This is an important question in trauma-informed creative practice: to what extent should strong emotions, which may be key to artistic engagement and active dialogue, be mitigated by the need to safeguard the participant from the very feelings that the project is designed to address?

Numerous citations from participants as well as quantifiable data of engagement with the project (Artichoke, 2022) testify to *Sanctuary*'s overall success. Significantly, while any initial reluctance to this piece appeared to be overcome in its positive realisation, more controversy arose in the actual act of burning the Sanctuary structure, as the culmination of the project. King recalls the disappointment that this caused for many participants, though the burning event was indeed powerfully symbolic in relation to the cathartic concept of the work. Aside from this event being always intended as the climax of the piece according to David Best's vision, King observes that making *Sanctuary* a permanent memorial would not only have had substantially different logistical implications, but, crucially, may have created a very different relationship with its participants. We may indeed contemplate whether the temporary nature of this work was itself integral to the enabling of a deep engagement with its community. That said, this does raise the issue of community legacy for ephemeral projects, one that we will return to later in this chapter.

Reflections

AMK

As Cara Courage notes in her introduction, 'Placemaking represents a paradigm shift in thinking about planning and urban design, from a primary focus on buildings and macro urban form to a focus on public space and human activity – what happens in these spaces, why, how, and with and by whom, and not: this is all the stuff of placemaking' (Courage, 2021, p. 3). Within this expanded framework, we might understand temporal artistic interventions such as *Sanctuary* not only within this category of activity in the public realm that aims to reconfigure it: the 'soft city', as Thompson and Scarso have described with regard to other creative works by Artichoke, such as processions and festivals. More radically, it might also be viewed as a form of tactical urbanism, which seizes upon the opportunity presented by temporal incursions by its inhabitants into public space, with the potential for more permanent reclamation and agency in that sphere. In an earlier volume, architect and commons practitioner Torange Khonsari explicates the operation of temporary spatial objects and architectures as a tactic in citizen-led placemaking: the role of 'disobedient objects' as a means to resist the increasing diminution of public space from 'the waves of city-development, supported by capital interests moving at high-speed' (Khonsari, in Courage & McKeown, 2019, p. 127). Such tactics, according to Khonsari, can have the effect of encouraging the chance encounters that build a sense of community, which I have argued is a crucial first step in fostering the civic agency that empowers individuals and communities to effect change (Kim & Yates, in Courage & McKeown, 2019).

Placemaking's role in creating 'the commons', in the many ways it is defined, is clearly important. But historically examples tend to focus on the urban realm, for a variety of reasons: because of the focus on cities as the primary, and increasingly growing, site of inhabitation for the majority of the population, a concern of UN Habitat and other organisations, from grassroots to government; cities as the traditional seat of political power and governance and the greater impact on communities and society; and the rapid change (ranging from gentrification to dereliction) exerted by global capitalism in the urban realm. But what of the extra-urban spaces and landscapes of our countries, the countryside, which at times has truly been forgotten in these discussions? A shift in thinking in urban design from centre to periphery was clearly signalled by AMO and Rem Koolhaas' 2022 *Countryside. A Report*, a title that might be juxtaposed with the ground-breaking work that established the architect as a leading intellect and visionary in the field, *Delirious New York: A Retroactive Manifesto for Manhattan* (1994). Since then, and even before, if we take *Countryside* as a bellwether of significant cultural shifts in the landscape, there has been an increasing interest in the towns and rural communities outside the hubs of our cities. This is particularly true in Great Britain, which has shifted its priorities from continued cultural investment by government in powerhouses such as London to those which have been left behind: at attempt to 'level up' opportunities for creative placemaking and the creative industries across the country. Similarly in the United States, there are signs of a shift in government investment, from the nation's prize cities to areas outside of them in the reinvigoration of the manufacturing industries. As the example of *Sanctuary* makes clear, the role of temporary, creative placemaking without further tangible infrastructure in areas of relative deprivation, such as the mining town which was its chosen site, raised serious concerns. How placemaking will be engaged in and take on meaning for the communities in these areas, in the wake of economic and other traumas, remains to be seen.

JLS

One of the key challenges in the case study explored in this chapter is measuring its impact in relation to the concerns of our discussion. This is particularly so, due to the ephemerality of the events and, in the case of *Sanctuary*, their one-off nature. As often is the case with festivals and large-scale events, we may get trapped in the tendency to measure impact only in terms of quantifiable data (audience numbers, statistics, financial return, etc.), but these are precisely what trauma-informed practice should not be reduced to. When researching the project featured here, we have come across a wide range of audience comments affirming the success of these events and the importance of these in their lived experience. Of course, how do we measure personal responses other than simply citing these anecdotally? Notions of trauma and healing are inherently difficult to define, and we must remember that the events studied here are, first and foremost, artistic projects: they are not intended to 'solve' problems, but, arguably, to provide a creative opportunity for dealing with such problems through respite, contemplation, thought provocation or even simple escapism.

If we cannot approach the impact in case studies such as *Sanctuary* in terms of how much the 'problem' they explore is 'resolved' as a result, how do we make a case as to the necessity of these? The answer may have to stay open-ended: like the ephemeral nature of such projects, their impact must be understood as fluid, changeable and indeed subjective. Most importantly, such impact may be noticeable beyond their actual duration. As Alvarez states in relation to festivals, which, by extension, may be applied here too:

> *instead of thinking of the event as the final product, the festival inspires a horizon of impact that includes the multiple ways artists and communities might 'use' the festival as fuel for activities of cultural autonomy and renewal. Most of those activities will be after-effects, carried out beyond public view and statistical reach, or outside the curatorial guidance of event organizers.*
>
> *(in Courage (2021), p. 476)*

Conclusion

We see through our chosen case study, *Sanctuary*, the potential to heal collective and individual trauma through ephemeral, performative, placemaking strategies, whereby new creative structures, spaces and ritual invite communities to participate in a process of public witness and emotional sharing, with the ultimate aim of catharsis. While such an intervention at first seems to fall outside principles of placemaking as traditionally construed, whether because of its ephemeral nature or beginning from traumas that are not specific to place, we have sought to show how this might contribute to a further development of the field by its methodology of creating *place*, not only in the sense of the temporary structure co-constructed by artist and a particular community, but by providing, through their narratives, a 'supplementary layer' (Schipper, 2014, p. 24) through which familiar spaces are re-imagined and re-discovered by their inhabitants under a different light. Here, the locations used are symbolically transformed in temporary sanctuaries, an idea that our case study directly references, where personal and societal traumas can be addressed through artistic practice.

As we have seen, projects like *Sanctuary* entail mass-scale operations, involving large numbers of facilitators and attendees (in both cases, exceeding the tens of thousands.) Such scale necessitates careful planning, which at face value may seem to contradict the notions of co-creation and anti-hierarchy that are inherent to an understanding of placemaking; on the other hand, our case study encourages a level of spontaneity and active dialogue, according to which the public functions as more than spectators, becoming participants, whose personal stories and experiences help shape the events. As we navigate the aftermath of the COVID-19 pandemic, this mass scale becomes all the more poignant, reminding us of the power of large gatherings in creating an alternative social dimension that is activated through creative expression.

In this sense, projects like *Sanctuary* may be seen in relation to the possibility of healing traumas, but such healing should not be understood in an art-therapy context: the art experiences entailed here are not to be confused with those that may be designed specifically with a therapeutic aim, thus based on the technical methodologies of application that are inherent to these. Rather than aimed at providing healing to the individual, it is in the communal dimension they promote through artistic expression, that, using placemaking as a reference point, these projects can provide an opportunity to collectively acknowledge trauma and transcend this through the power of art and participatory performance.

Hence, ephemerality should not be seen as an obstacle to legacy and impact. In fact, it is in the impermanent nature of our case study that its significance is amplified. Returning to King's observation, a project like *Sanctuary* may be perceived as all the more meaningful, precisely as it is only temporary, making its duration and presence precious. This lack of permanence and the changeability of these events reflects the nature of the traumas they address, which are also never monolithic experiences, but are just as variable, from person to person, from place to place, from time to time. Impermanence allows for new ideas to follow, highlighting a sense of dialogical community as opposed to static institutionalisation. And particularly in discussing trauma, impermanence is indeed a poignant reminder of the idea, sobering and reassuring in equal measure, that 'this too will pass.'

References

Alexander, J. C., Eyerman, R., Giesen, B., Smelser, N. J., & Sztompka, P. (2004). *Cultural Trauma and Collective Identity* (1st ed.). Berkeley: University of California Press.

Alvarez, M. (2021). 'Rituals of regard: On festivals, folks, and findings of social impact', in Cara Courage (ed.), *The Routledge Handbook of Placemaking*. Abingdon: Routledge.

AMO & Koolhaas, R. (2022). *Countryside, A Report*. Cologne: Guggenheim and Taschen.

Artichoke. (2022) *Sanctuary Evaluation Report* Available: www.artichoke.uk.com/ accessed 29/11/2022. [Accessed: 12 August 2023].

Clark, E. (2018) *Oresteia*. Available: www.edmundclark.com/works/oresteia/#1. [Accessed: 12 August 2023].

Courage, C. (2021). 'Introduction 'What really matters: moving placemaking into a new epoch', in Courage, C. (ed.). in *The Routledge Handbook of Placemaking*. Abingdon: Routledge.

Cvetkovic, A. (2012). *Depression: A Public Feeling*. Durham: Duke University Press.

Khonsari, T (2019). 'Temporary spatial object/architecture as a typology for placemaking', in Courage, C. & McKeown, A. (eds.) (2018) *Creative Placemaking: Research, Theory and Practice*. Abingdon: Routledge.

Kilomba, G. Illusions Vol. II, Oedipus (2018); Vol. III, Antigone (2019) Available: www.tate.org.uk/art/artworks/kilomba-illusions-vol-ii-oedipus-t15691. [Accessed: 12 August 2023].

Kim, A. M & Yates, J. J. (2019). 'Towards Beauty and a Civics of Place: Notes from the Thriving Cities Project', in Courage, C. and McKeown, A. (eds.) *Creative Placemaking: Research, Theory and Practice*. Abingdon: Routledge.

Koolhaas, R. (1994). *Delirious New York: A Retroactive Manifesto for Manhattan*. New York: Monacelli Press.

Mahlouji, V. (2023). 'Trauma and Trace in the Photography of Hrair Sarkissian' in Hrair Sarkissian, *The Other Side of Silence*. Kholeif, O and Ringborg, T. (eds.) Exh. cat.

Page, T. (2020). *Placemaking A New Materialist Theory of Pedagogy*. Edinburgh: Edinburgh University Press.

Scarso, J. L. & Jeske Thompson, K. (2022). 'Safety in Numbers: Reflecting on the work of Artichoke as 'Adaptor-Disruptor' in Reclaiming Public Space', *The Journal of Public Space*, 7(3).

Schipper, I. (2014). 'City as Performance', *TDR* (1988-), 58(3) *Performing the City: Special Issue*.

Schupbach, J. (2021). 'Placemaking in the Age of Covid and Protest', in Courage, C. et al. (2021) *The Routledge Handbook of Placemaking*. Routledge.

Sennett, R. (2019). *Building and Dwelling:Ethics for the City*. London: Penguin.

Shusterman, R. (2012). *Thinking through the Body: Essays in Somaesthetics*. New York: Cambridge University Press.

Widrich, M. (2014). *Perfomative monuments: The rematerialisation of public art*. Manchester: Manchester University Press.

24

RETHINKING PLACEMAKING IN URBAN PLANNING THROUGH THE LENS OF TRAUMA

Gordon C. C. Douglas

Planning and placemaking are crucial aspects of urban development that have the potential to create significant social and economic benefits for communities. However, it is essential to recognize that the very history of urban planning is in no small part a history of trauma. From its earliest incarnations in the religious-ceremonial foundations of some early cities to its professional origins in response to social problems in the 19th century, on up through 20th- and 21st-century schemes of urban renewal, megadevelopment, and gentrification, planning has both responded to and created traumatic events.

It is sadly predictable then – if nonetheless ironic – that the same can be said even of the contemporary planning trend known as placemaking. It too can be a cause of trauma. As practiced in contemporary planning and economic development, placemaking is the process of creating or enhancing public spaces that promote social interaction and a sense of community. Placemaking can be used to address a variety of challenges, including general disinvestment, crime, and cultural displacement. Yet, as an approach to urban design and local economic development that emphasizes the creation of new public spaces and local character, placemaking can also be place-threatening where it causes displacement, overwrites places that already exist, or creates 'placeless' places.

Fortunately, understanding the relationship between planning and trauma can help us to do better planning and placemaking. In particular, placemaking efforts informed by historical trauma (or the threat of new violence) can work to rebuild and heal, reclaiming place and community for marginalized groups. Drawing on student experiences and ideas from unique urban design studio courses and other cases from planning's past and present, this chapter emphasizes the value of considering trauma in planning pedagogy and practice, highlighting the power of

DOI: 10.4324/9781003371533-28

placemaking efforts that are not only mindful of past traumas, but perhaps even work to heal and prevent them.

The Historic Violence of Urban Planning

Urban planning is a complex process that involves creating and transforming the physical spaces of neighborhoods, cities, and regions. While the profession of urban planning as we know it today has its origins in responding to the social ills of rapid urbanization and planning is often seen as helping to create safe, livable, and sustainable communities, it can and has often had negative impacts as well. Urban planning can cause trauma through the overt destruction and displacement caused by large-scale redevelopment projects or through more subtle place-threatening actions, including gentrification and cultural changes, which may accompany placemaking efforts.

One of the most significant traumas associated with planning is the forced destruction and expulsion of communities of color in the name of 'urban renewal.' Tied to efforts at improving – or just as often simply removing – low-income areas, urban renewal programs were implemented in many cities across the United States throughout the 20th century. These programs were intended to revitalize urban areas by demolishing old buildings and replacing them with new housing and commercial developments, as well as freeways, universities, stadiums, and other large developments. However, these programs often resulted in the displacement of low-income communities, particularly communities of color.

The construction of interstate highways through urban areas may have been especially motivated by racist policies, as transportation officials, desperate to maintain connections between suburban housing and downtown commercial districts, plowed new highways through disenfranchised communities with little political power to fight them (Ebeling, 2013; Henderson, 2013; Loukaitou-Sideris et al., 2023). For instance, the demolition of the historic Rondo neighborhood in St. Paul, Minnesota, in the 1960s displaced thousands of Black residents and destroyed the social fabric of a vital African American community (Avila, 2014; Bills, 2023). We see similar patterns from Atlanta to Chicago to San Francisco. These projects had a devastating and traumatic impact, as the demolition of homes, businesses, and cultural institutions forced people to relocate to other areas, resulting in the loss of social ties, the disruption of cultural practice, and even impeding access to basic services (Fullilove, 2001; Carter, 2011; Ebeling, 2013).

In the last few decades, reinvestment in long-neglected urban neighborhoods has generally taken different forms, most prominently the gentrification caused by the arrival of new residents, businesses, and real estate capital, but still also megaprojects such as sports venues and office or entertainment districts. This 'new urban renewal' continues to displace long-time residents and change the character of communities in potentially devastating ways (Hyra, 2008; Elliott-Cooper, Hubbard & Lees, 2020).

Placemaking and Placetaking

In urban planning, design, and economic development, the term placemaking typically describes efforts to create public spaces, streetscapes, or whole districts that appeal to people at the human scale, often featuring locally-specific design elements and signage, creative lighting, decorative plantings, and interactive furniture (Fleming, 2007; Madden, 2011; Thomas, 2016; Douglas, 2023). Like good planning generally, the practice is often associated with positive outcomes, such as pedestrian activity, place identity and visibility, vibrant public spaces or shopping districts, and generally bringing a particular place 'to life'. However, as an established planning and real estate development tool, mainstream placemaking efforts often prioritize mainstream values, aesthetics, and consumption opportunities, discouraging other uses (political expression, cultural transgression) and users (especially the poor). Indeed, the planning, design, and regulation of public space is often highly concerned with prescribing acceptable uses and excluding 'undesirables' in order to produce a safe and orderly landscape for economic activity (Whyte, 1980; Davis, 1990; Sorkin, 1992; Mitchell, 2003).

For all these reasons, placemaking has also been recognized as potentially culpable for many of the same known impacts of urban renewal and gentrification: exclusivity on class lines, cultural displacement, and the transformation or even erasure of already-existing places (Starowitz & Cole, 2015; Douglas, 2023). For example, curb-side parklets and pedestrianized plazas have been associated with gentrification in New York, Los Angeles, and San Francisco (Douglas, 2018). The primarily African American Shaw-U St. neighborhood in Washington, DC, which has been the target of developers seeking to profit from the area's proximity to other affluent and trendy central districts, has been gentrified in part through (repetition of trendy) commercially oriented placemaking efforts. Many long-time residents have been forced out of their homes and businesses, and the cultural significance of the neighborhood has been threatened (Hyra, 2017). Such outcomes, like the violence of 'slum clearance' before them, can be deeply traumatic.

The critique here is not that placemaking is necessarily harmful, but that it must be informed by local character and created with sensitivity to what place may already be there. When done right, placemaking can be just the opposite – incorporating meaningful cultural symbolism, supporting and strengthening local identity, and even responding to past trauma. For planners, this can be a prime example of what Knapp et al. (2022) have called 'reparative planning praxis' in the profession.

Placemaking as a Response to Trauma

Trauma-informed placemaking offers a design tool for planners, developers, and community members to respond to historic trauma by addressing past wrongs while working to preserve local identity and embolden the disempowered.

First, the conscious community-led 'making of places' has potential to foster a sense of belonging, community, even power – especially perhaps for subaltern, underprivileged, excluded, or displaced people. Indeed, researchers have looked at 'sense of place' among disadvantaged communities, including with regard to community empowerment, improvement, and place attachment (Bennett, 2000; Tester et al., 2011; Fullilove, 2013; Manzo, 2014).

Chicanx muralism is one example of placemaking that can be said to be trauma-informed and perhaps trauma-healing. These murals often address issues of social justice, community identity, and resistance to oppression in Mexican American neighborhoods. Campaigns to return land to indigenous stewardship are also examples of trauma-informed placemaking that recognize the history and heritage of (de)colonized places. Chinatowns and other ethnic or cultural enclaves, which are often spatially the direct outcome of discriminatory policies directed at these communities, work to promote and defend cultural identity in the face of the racism that helped shape them.

Consider the legacy of San José's Japantown. Founded by Japanese immigrants in 1900, *Nihonmachi* had become a vibrant neighborhood and one of the largest Japantowns in the United States by the start of World War II. When Japanese Americans were forcibly imprisoned in internment camps in 1942, many Japantowns disappeared. In San José, while there was no shortage of anti-Japanese sentiment and even vandalism, many Japantown businesses were actually protected, not only by the Japanese community's own organization and security, but by non-Japanese residents and by the police (perhaps due in part to the economic importance of the large Japanese population to the city). A group of non-Japanese San Joseans formed a 'Japanese Protection Committee' to help defend businesses and institutions from vandalism and violence. Although San José's Japantown was still slow to recover after internment, its persistence as one of just three remaining Japantowns in the United States and, today, a vibrant urban neighborhood filled with popular businesses, cultural institutions, and visible monuments to the challenges of racism and internment, can be understood as a living process of explicitly trauma-informed placemaking (Saito et al., 2021). It is a place in which 'Memories of generational traumas and recoveries are passed through generations' (Saito et al., 2021, p. 21, after McAlister, 2008).

We might consider any number of efforts at place-based identity expression or promotion in ethnic or cultural enclaves to be a form of trauma-informed placemaking. For instance, analogous placemaking efforts can be found in Black communities like Chicago's Bronzeville. The Crenshaw and Leimert Park areas in Los Angeles, long the cultural center of that city's Black community, are today engaged in a powerful placemaking effort called Destination Crenshaw intended explicitly to celebrate cultural heritage and resist displacement in the face of major new transit-oriented development investments. These projects are also explicitly working to resist displacement, helping to ensure that communities are not forced out of their homes and businesses in the name of development.

As placemaking can work to address past wrongs and embolden the disempowered, it has the potential to play a role in advancing racial equity as well. In recent years, protests following the murder of Black people by police in the United States, and the Black Lives Matter movement more generally, have sparked a wave of new placemaking efforts that are trauma-informed and responsive to injustices. For example, the intersection where Minneapolis resident George Floyd was killed has become a site of public mourning, celebration, and resistance. Known as George Floyd Square, this place was created by the people in response to the murder to literally transform the street using artwork, memorials, and community events into a new place where people could come together to grieve, to protest, and to demand justice. Similarly, the renaming of Madison Square Park in Oakland for Wilma Chan after her death and in the context of anti-Asian hate crimes more broadly, was an important recognition of anti-Asian violence as well. Murals have likewise been a significant part of placemaking for racial justice in cities around the world as a way for people to demand change and express their anger, their grief, and their hope for the future right on the pavement and buildings of the built environment. Again, in the context of the Black Lives Matter movement, murals created during protests have remained on the streets as a reminder of the ongoing struggle.

Trauma-Informed Placemaking in the Design Classroom

The principles of trauma-informed placemaking can be tremendously powerful in the teaching of urban planning and design. Of course, planning's history of traumatic impacts, especially for communities of color, must be central to an introduction to the discipline and curriculum on planning history, land use and zoning, transportation, and housing. But these ideas can be equally impactful in thinking about community participation and engagement, urban design, and the future of cities. At San José State University in California, several urban design studio courses have taken this approach.

In 2018, professors Gordon Douglas and Virginia San Fratello, working with a unique university-community partnership called *CommUniverCity*, led a two-semester design practicum focused on how to do community-driven 'tactical placemaking'. The site was centered on an intersection in San José's Northside/Luna Park neighborhood, an area with rich history as a destination for immigrants and people of color but also facing the challenges of increasing unaffordability and displacement for those groups. Unlike some studio courses, the students' work began not with precedent studies or physical site analyses, but with ethnography and pavement-pounding neighborhood outreach. Only after multiple community meetings, charettes, even design-through-play activities with some local third-graders, and a lot of walks, conversations, and observations, did the student groups begin to attach design ideas to what residents said they needed. Students realized that in order to really hear from everyone who lived in the neighborhood– not just those who own homes, speak English as a first language, and have plenty of time to attend meetings –they needed to go to them, hanging out in the park and chatting in Spanish and Vietnamese with people doing laundry at the laundromat. In the end, this informed the entire project, called *Mi Sala Tu Sala*, in which students built an 'urban living room' outside of the laundromat for everyday San Joseans to hang out in while doing laundry, complete with a massive mural of San José's agricultural and immigrant history. The proposal was realized and installed with support from two local foundations.

Another design studio focused explicitly on the traumatic events of 2020. With Downtown Oakland as the study site and the theme of 'Urbanisms of Crisis and Hope', students were tasked with developing proposals addressing the traumatic events of the COVID-19 pandemic, police violence and the murder of George Floyd, and flooding, fire, and extreme weather. The resulting designs drew on precedents of trauma-informed placemaking ranging from pop-up urbanism to post-disaster recovery and climate-change mitigation. One group looked to the history of the civil rights movement and urban design for protest as well as the murals, graffiti, and other physical signs of struggle marking the streets of Oakland. They proposed a pedestrianized street preserving and memorializing artworks of resistance leading into a civic plaza renamed for police violence victim Oscar Grant and redesigned to better support political demonstrations (Figure 24.1).

In addition to teaching the importance of the sensitive, trauma-informed designs themselves, this pedagogical approach demonstrated the power of thoughtful placemaking in the *process* of community engagement and participatory planning practice. Indeed, successful community engagement may rely on successful placemaking to create spaces for participation that feel safe, inclusive, and welcoming. When people feel safe and welcome, they are more likely to connect with others and feel comfortable talking about their experiences. And, as already emphasized, involving communities in placemaking itself is essential not only to making sure the places being created are meaningful, but in empowering the communities in the process while doing so. When communities have a voice in the design and development of their neighborhoods, they may feel more invested and perhaps even more likely to heal from trauma.

A Future for Trauma-informed Planning and Placemaking

Placemaking can be a powerful tool for addressing trauma. When done right, placemaking can also be a way to uplift and strengthen local identity, to commemorate past events, to address ongoing injustices, and to promote healing. In these ways, placemaking can be a part of 'trauma-informed care' – an approach to care that recognizes the impact of trauma on individuals and communities and seeks to create safe and supportive environments where people can heal. And it can help to heal through the literal repair of the built environment itself. According to Knapp et al. (2022), reparative planning practice requires 'radical honesty,

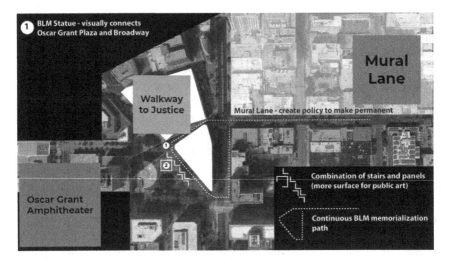

FIGURE 24.1   A site plan for a proposed redesign of part of downtown Oakland by SJSU urban planning graduate students Adriana Coto, Kevin Lee, and Branka Tatarevic in 2021. The design aims to embrace the existing usage of Frank Ogawa Plaza / Oscar Grant Plaza and surrounding streets near Oakland City Hall as spaces of resistance by preserving protest-oriented street art, installing a new Black Lives Matter memorial, and reshaping the plaza itself to better support demonstrations.

confronting whiteness, and radical imagination'. All of the examples above do just this.

Perhaps the most exciting conversations in the realm of trauma-informed placemaking in urban planning in the United States right now are those at the intersection of reparations and freeway removal. Freeway removal can be understood as a direct response to the violent trauma of urban renewal and a way to reconnect communities that were divided by highway construction. This approach seeks to create a more just and equitable city by acknowledging the damage caused by planning in the past and seeking to address it in a direct and meaningful way. For example, in Syracuse, New York, the city is exploring the possibility of removing a section of Interstate 81 that bisects the predominantly Black and low-income neighborhood of Southside. This would be accompanied by significant investments in the neighborhood, including affordable housing, small business support, and job training.

Because freeways are essentially non-places, their removal and replacement with urban streets and amenities is literal placemaking where there was no place before. Still, a priority of these freeway removal and neighborhood restoration projects must be to consider the damage that urban renewal has done in the past and to make sure that the redevelopment that comes in place of freeways is not just a new sort of displacement through reinvestment. Freeway removal is one of the clearest opportunities for trauma-informed placemaking in planning.

Of course, a fundamental takeaway must be that placemaking is not a one-size-fits-all solution. The specific histories, identities, and needs of each community vary widely, and almost by definition placemaking must be tailored to those differences. Practically speaking, however, we might articulate some general principles that can be applied to placemaking in communities that have experienced trauma.

First, efforts should be as much about preserving or restoring place identities, not necessarily making new ones. At a minimum, the result should be spaces where all people feel safe and welcome. Also, with a goal of creating more inclusive and equitable public spaces that reflect the needs and desires of communities, the process of creating these spaces should involve residents by giving them a voice in the design and development of their neighborhoods. Planners and designers need to engage in ongoing conversations with communities, listen to their needs and desires, and work together to create spaces that are truly reflective of the people who live in them. This can help to build a sense of shared place and belonging, and ultimately of trust and resilience, which can be healing for people who have experienced trauma. By creating safe, inclusive, and empowering spaces, placemaking can help to heal the wounds of trauma and build a more resilient and just society.

References

Avila, E. (2014). *The Folklore of the Freeway: Race and Revolt in the Modernist City.* Minneapolis: University of Minnesota Press.

Bennett, S. (2000). Possibility of a Beloved Place: Residents and Placemaking in Public Housing Communities, *St. Louis University Public Law Review*, 19(2), 259–307.

Bills, T. (2023). 'A Contemporary Path to Transportation Justice in Rondo' in R. Reft, A. K. Phillips de Lucas & R. C. Retzlaff (eds.) *Justice and the Interstates: The Racist Truth About Urban Highways*. Washington: Island Press.

Carter III, G. (2011). 'From Exclusion to Destitution: Race, Affordable Housing, and Homelessness', *Cityscape*, 13(1): 33–70.

Davis, M. (1990). *City of Quartz: Excavating the Future in Los Angeles*. London: Verso.

Douglas, G. C. C. (2018). *The Help-Yourself City: Legitimacy and Inequality in DIY Urbanism*. Oxford: Oxford University Press.

Douglas, G. C. C. (2023). 'Reclaiming Placemaking for an Alternative Politics of Legitimacy and Community in Homelessness', *The Journal of Politics, Culture, and Society*, 36: 35–56.

Ebeling, M. (2013). 'Rethinking the Urban Freeway'. *Mayors Innovation Project.* Available: https://cows.org/wp-content/uploads/sites/1368/2020/04/2013-Rethinking-the-Urban-Freeway.pdf. [Accessed: 6 July 2023].

Elliott-Cooper, A., Hubbard, P. & Lees, L. (2020). 'Moving Beyond Marcuse: Gentrification, Displacement, and the Violence of Un-Homing', *Progress in Human Geography*, 44(3): 492–509.

Fleming, R. L. (2007). *The Art of Placemaking: Interpreting Community Through Public Art and Urban Design*. London: Merrell.

Fullilove, M. T. (2001). 'Root Shock: The Consequences of African American Dispossession', *Journal of Urban Health*, 78: pp. 72–80.

Fullilove, M. T. (2013). *Urban Alchemy: Restoring Joy in America's Sorted Out Cities.* New York: New Village Press.

Henderson, J. (2013). *Street Fight: The Politics of Mobility in San Francisco.* Amherst: University of Massachusetts Press.

Hyra, D. S. (2008). *The New Urban Renewal: The Economic Transformation of Harlem and Bronzeville.* Chicago: University of Chicago Press.

Hyra, D. S. (2017). *Race, Class, and Politics in the Cappuccino City.* Chicago: University of Chicago Press.

Knapp, C., Poe, J. & Forester, J. (eds.). (2022). 'Repair and Healing in Planning', *Interface: Planning Theory & Practice*, 23(3), 425–458, DOI: 10.1080/14649357.2022.2082710. Available: www.tandfonline.com/doi/full/10.1080/14649357.2022.2082710. [Accessed: 6 July 2023].

Loukaitou-Sideris, A., Handy, S. L., Ong, P. M., Barajas, J. M. et al. (2023). 'The Implications of Freeway Siting in California: Four Case Studies on the Effects of Freeways on Neighborhoods of Color', *Pacific Southwest Region University Transportation Center* research report. Available: www.metrans.org/assets/research/the%20implications%20of%20freeway%20siting%20in%20california.pdf. [Accessed: 6 July 2023].

Madden, K. (2011). 'Placemaking in Urban Design', in T. Banejee, A. and Loukaitou-Sideris (eds.) *Companion to Urban Design.* Abingdon: Routledge.

Manzo, L. C. (2014). On Uncertain Ground: Being at Home in the Context of Public Housing Redevelopment. *International Journal of Housing Policy*, 14(4), 389–410.

Mitchell, D. (2003). The Right to the City: Social Justice and the Fight for Public Space. Guilford Press.

Saito, R., Faas, A. J. & McClure, J. (2021). 'This Must be the Place: Partnerships for Disaster Preparedness in San Jose's Historic Japantown', *Practicing Anthropology* 43(4): 23–30. Available: https://doi.org/10.17730/0888-4552.43.4.23. [Accessed: 6 July 2023].

Sorkin, M. (ed.) (1992). *Variations on a Theme Park: The New American City and the End of Public Space.* New York: Hill and Wang.

Starowitz, S. & Cole, J. (2015). Thoughts on Creative Placemaking. *Lumpen Magazine.* Retrieved from www.lumpenmaga zine.org/thoughts-on-creativeplacetaking/

Tester, G., Ruel, E., Anderson, A., Reitzes, D. C., & Oakley, D. (2011). Sense of Place among Atlanta Public Housing Residents. *Journal of Urban Health*, 88, 436–453.

Thomas, D. (2016). *Placemaking: An Urban Design Methodology.* Abingdon: Routledge.

Whyte, W. H. (1980). *The Social Life of Small Urban Spaces.* New York: Project for Public Spaces.

Our Call to Action

Nurturing Healing Through Action

25

THE PLACE HEALING MANIFESTO

Charles R. (Chuck) Wolfe

Not so long ago, during Summer 2020, we were all grappling with the necessary retrofitting and recalibration of our urban landscapes to curb the spread of COVID-19. This was soon compounded by civil unrest in the USA in cities such as Minneapolis, Atlanta, Seattle, Portland. An insidious racial injustice exposed itself, an ominous addition to the pandemic crisis – anecdotally termed by some as the 'second pandemic', which became more ominous and complex as the summer progressed.

The United Kingdom echoed similar responses from across the Atlantic. Statues of slave traders from a bygone era were re-evaluated in a new light; Edward Colston was suddenly submerged in Bristol Harbour, and at the time of writing his podium of many years is still vacant. As the pandemic and protests played out concurrently, it became glaringly apparent that calls for healing and reconciliation were universal, necessitating responses that evoke the deeper essence of our physical locales.

With this in mind, and drawing upon recent research and experience, let us delve into the prospect of bridging the gap between our past and our future. Inspired by commonplace examples that lie within our immediate purview, we have now seen three years of the adaptability of cities and their tales. I have chosen to label this phenomenon as 'place-healing', a term that seems apt for our times.

So, what does place-healing entail?

Place-healing manifests through civic engagement. In the United States, we witnessed voluntary and spontaneous clean-up and repair efforts after violent protests in June 2020. Meanwhile, in London, protestors demonstrated solidarity and provided aid across ideological boundaries. This form of place-healing is rooted in empathy and interconnectedness, illustrating the human potential to restore community cohesion, reminiscent of the early days of the now-dismantled

DOI: 10.4324/9781003371533-30

Capitol Hill Autonomous Zone in Seattle (several blocks left police-free for three weeks during Black Lives Matter protests in Summer, 2020).

Adaptation Examples

Further, unintentional place-healing adaptations, however commonplace, have exhibited how our urban spaces will evolve in appearance and experience. Cities like London, steeped in history and rich diversity, showed this transformation. Continuity with the past provided a strange solace and merger of the 'before times' with what followed.

Consider the photograph above, in Richmond upon Thames, London – a retro red phone (Figure 25.1) booth merged with a pandemic-induced sidewalk modification. This vibrant icon (an artifact of a bygone era) stood alongside an

FIGURE 25.1   From phone booth to artifact.

FIGURE 25.2   An adapted gate at Kew Gardens, London, during the pandemic.

emergency public health response. It might seem insignificant, but this coexistence of the new and familiar encapsulates the essence of place-healing. More deliberate examples included impromptu entrances (Figure 25.2), alfresco dining, vintage furniture framing take-out windows, and modified subway cars to ensure social distancing.

Moreover, institutions, including private businesses, may endure in evolved forms – even after the pandemic – upholding an ethic crucial to place-healing: adaptation and respect for the people involved in various contextual settings. Greater London's Bobtail Fruit, a customer-focused and community-sensitive enterprise that transitioned from a fruit stall to five physical outlets and now an online delivery service, embodies this ethos.

Based on my interviews with business development staff, the enduring success of Bobtail isn't merely about tangible aspects of place-healing but rather a respect for customers and dedication to community service (such as the delivery of gifts at holiday time in neighborhoods where its stalls once stood). Notably, during the COVID-19 pandemic, Bobtail expanded its delivery area to accommodate more residential customers as its traditional office clientele worked remotely (Figure 25.3).

In the Bobtail example, the physical location no longer confines the selling process. However, the customer service ethos, harking back to the original owner's treatment of fruit stall patrons in Old Covent Garden days in Central London, remains integral to the business, which has adjusted to meet customer needs in a new form. In sum, urban images showcasing adaptation and the people-centered

FIGURE 25.3 The Bobtail Fruit adaptation at work.

approach of Bobtail Fruit indicate how to convert anxiety into action. Adaptation and amalgamation will shape the next steps towards something novel. However, we still require a pertinent roadmap to get there.

The Place-Healing Manifesto

If place-healing is the catalyst for more stories of adaptation and transition, we should devise ways to highlight more stories and pathways to significant change. To meet this challenge, I propose the following place-healing manifesto:

1. **Emphasize communication**: Municipalities should continually refine their communication with citizens, particularly given the lengthy periods of isolation we endured.
2. **Acknowledge differences and limitations**: Greater focus should be placed on language, age, technological access, and the potential of facilitated dialogue.

3. **Revive the American-championed 'front porch'**: Global placemaking networks have recently revitalized attention towards 'front porch' projects, allowing communication from dwellings to passers-by compatible with social distancing measures (and see, Rusk, this volume, for a UK-housing-design-specific similar initiative).

4. **Strive for mutual understanding and education**: We must all commit to a deeper understanding of each other, sensitive to issues of privilege and class.

5. **Honor diverse narratives of minority populations**: We need to genuinely understand the cultural history of others and its impact on their everyday experience.

6. **Explore new place metrics**: Alongside popular metrics like *WalkScore* in the United States, why not consider a non-fiscal *EquityScore*?

7. **Expand *pro bono* work**: What if all professions committed to *pro bono* obligations, offering more free services to those in need?

8. **Encourage social networks to serve the public**: Imagine if social networks routinely highlighted opportunities to help combat ongoing discriminatory practices in a user's area.

9. **Advocate co-creation**: In shaping our urban condition, there is a growing preference for collaborative creation and empowering local knowledge over top-down imposition.

10. **Question the necessity of consultants**: In the post-pandemic era, professionals adapted their expertise to prescribe solutions, but the role of facilitator, one who can help curate a collective strategy for place-healing, may be more essential.

This manifesto represents a roadmap for healing and reconciliation in our cities, providing a means to navigate the complexities of our post-pandemic reality.

Reference

Walkscore. Available at: www.walkscore.com/. [Accessed on: 14 August 2023].

26

LEADERSHIP HORIZONS IN CULTURE FUTURISM AND CREATIVE PLACEHEALING

Theo Edmonds, Josh Miller and Hannah Drake

Preface

This learning resource provides a deep dive into two interconnected concepts: creative placehealing and culture futurism, focusing on the practical application of these theories in real-world contexts. It has been designed with the following reader learning goals in mind:

1. Understand and articulate the concepts of creative placehealing and culture futurism: by the end of this resource, the reader should be able to define these terms, understand their theoretical underpinnings, and appreciate their significance in contemporary community development and population health discourse.
2. Analyze the role of leadership in creative placehealing and culture futurism: the reader should be able to distinguish between different types of leadership – exploratory, combinatory, and transformational – and understand how these leadership types can drive initiatives that promote creative placehealing and culture futurism.
3. Apply theoretical concepts to real-world scenarios: using *(Un)Known Project* as a case study, the reader will learn how theory translates into practice. They will examine how creative placehealing and culture futurism were applied to promote communal healing and reconciliation in the context of systemic racism and the legacy of enslavement.
4. Develop a practical approach to community development and health promotion: through the *MOTIF* model, the reader will learn a step-by-step process to guide the design and implementation of culture-centered health promotion initiatives in their communities.

DOI: 10.4324/9781003371533-31

5. Reflect on their role as future practitioners or leaders: the reader will be encouraged to think critically about their role and responsibility in promoting health equity, social justice, and cultural wellbeing in their communities.

I. Introduction to Creative Placehealing and Culture Futurism

For over a decade, the three authors of this chapter have worked to bring clarity to the concept of *creative placehealing*. We define it as an evolutionary approach to population health research and development that integrates the principles of creativity, culture, and place (physical and virtual) into a cohesive framework for innovating through a lens of cultural wellbeing (Lister, 2021; Edmonds, 2021). This process underscores the importance of the social-ecological approach (McLeroy et al., 1988) to communal healing as the key driver of functional design across physical and virtual spaces where individuals live, learn, work, heal, and explore.

Creative placehealing is a multidimensional process that goes beyond spatial and architectural elements to encompass the spiritual, emotional, cultural, and social aspects of a community. It recognizes that places are not merely physical locations but living, breathing entities that shape and are shaped by the collective imagination and wisdom of community members and stakeholders. Through this perspective, place is understood as a holistic entity, interconnected with individuals and the community, deeply embedded in its historical, cultural, and social context.

Culture futurism is a progressive approach that seeks to anticipate and shape meaningful futures through culture, an intentional action by a group to activate greater agency in devising their culture rather than just a cultural approach. It involves envisioning futures that are divergent from the present and exploring new ideas, technologies, and possibilities that encourage collaboration across differences of all kinds. It challenges existing paradigms and constructs an inspiring, inclusive, and sustainable vision for multi-generational awareness across time, past, present, and future. Storytelling is a crucial aspect of culture futurism, serving as an act of individual agency to speak into existence and explore alternative futures and imagined worlds that might not exist yet or even seem possible.

Culture futurism is interdisciplinary, encompassing art, literature, music, film, design, generative AI, and media technology. It can also involve social and political movements and scientific and technological advancements. As practice, it entails a profound understanding and appreciation of history, culture, and tradition, recognizing that culture is not static but evolves over time. This evolution is influenced by various factors, including social, political, economic, technological, environmental, and creative forces. It seeks to shape culture in a way that is responsive to these forces and promotes positive change.

Because it seeks to anticipate and shape culturally responsive futures of human flourishing, culture futurism is a critical but often overlooked driver of creative placehealing. Both are deeply interconnected and synergistic, with creative placehealing providing the physical and emotional space for the envisioning and manifestation of cultural futurism. This chapter presents three types of futurist leadership horizons – exploratory, combinatory, and transformational – for their contribution to culture futurism and creative placehealing.

II. Futurist Leadership Development in Creative Placehealing

Understanding how creativity interacts with creative placehealing and culture futurism provides a pathway to identify emergent leadership phenotypes, particularly in cultural and communal contexts. Margaret Boden, a renowned scholar in the field of creativity research, postulates three primary forms of creativity: exploratory, combinatory, and transformational (Boden, 2010). While traditionally used to describe the artistic process, these forms can also be repurposed to elucidate the evolving characteristics of leadership in the context of creative placehealing and culture futurism.

Exploratory Leadership

Exploratory leadership navigates the existing paradigms of a given system, delving into established traditions, routines, and patterns to discover uncharted territories. In creative placehealing, exploratory leaders identify latent potentials within communities, fostering inclusivity and equity through uncovering and embracing indigenous wisdom, communal narratives, and cultural heritage. These leaders are akin to explorers, setting out on a journey into the unknown with an open mind and heart, ready to learn and absorb new knowledge, experiences, and perspectives. They see the potential in the existing elements of a culture or place and seek to elevate these to new levels of expression and manifestation. Exploratory leaders are innovative and intuitive, embracing ambiguity and complexity as sources of inspiration and growth.

Combinatory Leadership

Combinatory leaders blend diverse elements to generate innovative approaches. They draw from multidisciplinary knowledge, juxtaposing disparate ideas to birth fresh perspectives and solutions. This approach is particularly relevant to culture futurism, where leaders must interweave historical, cultural, and technological strands to visualize and manifest compelling and sustainable futures. Combinatory leaders possess a unique ability to see connections and synergies where others see separateness and dissonance. They are akin to alchemists, transforming raw materials into gold through their innovative thinking and creative problem-solving skills. These leaders are inclusive and collaborative, valuing diversity and multiplicity as catalysts for creativity and innovation.

Transformational Leadership

Transformational leaders defy and rewrite the rules of the game. They revolutionize systems, institutions, and thought patterns, fostering paradigm shifts that nurture communal healing and positive societal transformation. This form of leadership is crucial to culture futurism, as it necessitates challenging entrenched narratives and norms to shape responsive, inclusive, and inspiring visions of the future. Transformational leaders are change agents, visionary thinkers who are unafraid to disrupt the status quo and push boundaries to bring about meaningful and lasting change. They are courageous and resilient, willing to face resistance and adversity to pursue a better, more equitable, and inclusive future.

The intersection of these leadership types – exploratory, combinatory, and transformational – with creative placehealing and culture futurism can be elucidated through real-life projects. The following case studies, drawn from *(Un)Known Project*, illustrate these leadership phenotypes, providing insightful examples of how they can contribute to leaders seeking to lead as culture futurists in creative placehealing.

III. Case Studies: Exploratory, Combinatory, and Transformational Leadership in Action

Case Study One: Exploratory Creativity – Making the Invisible, Visible

The *(Un)Known Project (2023)*, co-founded by Hannah Drake and Josh Miller, exemplifies Margaret Boden's concept of exploratory creativity. Their initiative re-evaluates existing cultural and historical narratives to give voice to the unheard stories of enslaved Black men, women, and children in Kentucky and beyond. Collaborating with partners including the Frazier History Museum, Roots 101 African American Museum, and Louisville Metro Government, *(Un)Known Project* endeavors to unearth hidden histories while promoting racial healing and reconciliation. Exploratory creativity in this context involves a deep dive into the historical and cultural fabric of a place. It requires a commitment to truth-telling and a readiness to confront uncomfortable realities of the past. In this process, leaders like Drake and Miller serve as guides, helping communities navigate the complexities of their shared history, illuminating hidden narratives, and facilitating conversations that foster understanding and reconciliation.

The *(Un)Known Project* emerged in a socio-political backdrop marked by racial tensions and heightened awareness of systemic racism. The murder of Breonna Taylor by the Louisville Metro Police Department and the subsequent civic unrest intensified the community's focus on its racial history. In this atmosphere, the city of Louisville confronted its history of racism by removing the Confederate Major John B. Castleman statue. Mayor Greg Fischer declared racism a public health crisis, acknowledging the deep-seated and pervasive nature of this social toxin.

In response to this heightened awareness and urgency, *(Un)Known Project* used the tools of public art and crowd-sourced information to breathe life into the untold stories of enslaved Black people. In doing so, it enabled the community to connect with their history and foster a sense of identity and belonging. The project used exploratory creativity to disrupt the narrative of erasure, challenging the community to question the dominant narrative and explore alternative histories. It urged the community to see beyond the official histories and understand the human toll of enslavement. The leaders of *(Un)Known Project* served as exploratory leaders, mapping a route to collective healing through the discovery and acknowledgment of these previously invisible stories.

Case Study Two: Combinatory Creativity – Reclaiming Stories through Creativity

The *(Un)Known Project* employs various art forms – from public art installations to performance poetry – to highlight enslavement narratives. This multimodal storytelling technique exemplifies combinatory creativity in action, where disparate elements are blended to create something entirely new. In this context, combinatory leadership involves curating and orchestrating diverse art forms, voices, and perspectives to stimulate dialogues and inspire actions that shift societal relationships and perceptions of a shared history. This leadership style enables *(Un) Known Project* to navigate the complexities of cultural memory and communal healing, merging the personal and the collective, the past and the present, the local and the global.

The *(Un)Known Project* used sculpture, spoken word poetry, dance, music, and augmented reality to bring the stories of enslaved Black people to life. The team collaborated with local and international artists, including Ed Hamilton, a renowned sculptor, and Sidney Monroe Williams, a professor and theatre artist. The project demonstrated how combinatory creativity can drive culture futurism. It merged different art forms, technologies, and narratives to challenge existing representations of history and present a more nuanced, multi-dimensional understanding of the past. It stimulated conversations about racial justice and systemic racism, engaging the community in critical dialogues about the legacy of enslavement and its contemporary ramifications.

Case Study Three: Transformational Creativity – New Relationships with Place and History

The *(Un)Known Project* seeks to redefine the relationship between people, their history, and the spaces they inhabit. This exemplifies transformational creativity, where established paradigms are disrupted and reimagined. In the context of creative placehealing, transformational leadership fosters an understanding of our shared history and its impact on the health of individuals and communities. It involves reimagining the narratives embedded in our environments, reframing how we relate to space, and redefining our collective identity.

The *(Un)Known Project* challenged the community to reconsider how the history of enslavement is remembered and commemorated in public spaces. By installing a public art monument (Figure 26.1) on the Ohio River waterfront – a significant site in the history of the slave trade – the project transformed the city's landscape and its relationship with its past.

Moreover, *(Un)Known Project* inspired transformational change by mobilizing the community to participate in the narrative process. The project asked the community to submit names of enslaved individuals, which were then inscribed on the monument. This interactive and inclusive approach fostered a sense of ownership and agency in the community, empowering them to contribute to the rewriting of their city's history. In these ways, *(Un)Known Project* used transformational creativity to reimagine the city's relationship with its past, fostering an understanding and acknowledgment of the impact of enslavement on contemporary disparities. It prompted a shift in the community's perception of its shared history and identity, facilitating a path toward communal healing and reconciliation.

FIGURE 26.1   *On the Banks of Freedom* public art installation overlooking the Ohio River. Image credit: Josh Miller.

IV. Practice Model: *MOTIF Framework*

From 2017 to 2021, we developed and published the *MOTIF Framework* as a culture futurist practice framework for creative placehealing, designed to guide practitioners in the process of transforming places and fostering cultural wellbeing (Edmonds et al., 2017; Edmonds et al., 2021). The five-step framework incorporates self-determination, hope, and a sense of belonging, recognizing culture as a key community asset (Table 26.1).

Map: The first step in the framework involves mapping the social toxins in a given community. Social toxins are policies and practices contributing to inequity and injustice, impacting community health and wellbeing. This first step underscores the necessity of understanding and addressing these structural barriers for effective health promotion.

Orient: The second step emphasizes the importance of cultural responsiveness. It recognizes the power of the arts as a tool for understanding, appreciating, and respecting the diverse cultures within a community. It encourages practitioners to approach their work with humility and curiosity, fostering a climate of mutual learning and respect.

Translate into Cultural Wellbeing Science: The third step involves building a transdisciplinary team of local cultural workers and bearers, artists with radically diverse practices, and health professionals and researchers. This multi-dimensional approach facilitates the blending of diverse perspectives and expertise, fostering innovation and responsiveness to community needs and aspirations.

TABLE 26.1 MOTIF Innovation Framework

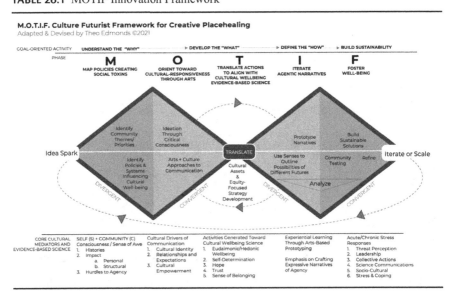

Note: Adapted and developed by Theo Edmonds © 2020.

Innovate: The fourth step involves the design and implementation of culture-centered interventions that promote community health and wellbeing. These interventions can take various forms, from art installations and performances to policy advocacy and community organizing.

Foster: The final step involves fostering community ownership and leadership. It recognizes the importance of nurturing local capacity and building sustainable structures of support to ensure the long-term viability of the initiatives.

Exploring (Un)Known Project through the MOTIF Framework

The *(Un)Known Project*, developed through IDEAS xLab (2023), provides an ideal case study for understanding the *MOTIF Framework*, as a five-step theoretical framework rooted in artistic devices for understanding recurring, narrative elements with symbolic significance. In applying the *MOTIF Framework*, the steps are cumulative, albeit not necessarily linear. This case study examines how *(Un)Known Project* utilized this framework for creative placehealing and culture futurism.

Step One: Map Policies Creating Social Toxins

At the core of *(Un)Known Project* was the mapping of policies that created social toxins – harmful influences on societal wellbeing – through the lens of enslavement's social and economic implications. These influences have enduring impacts on current policies and practices, ranging from access to healthcare to loan eligibility for housing. The project focuses on how enslavement casts a long, persistent shadow over the modern American landscape. *(Un)Known Project* made a conscious effort to examine and spotlight historical injustices, underlining the systemic racism deeply entrenched in current policy formation and implementation. The result was a thought-provoking exploration of the social toxins created by discriminatory policies and their influence on the lived experiences of Black communities.

Step Two: Orient toward Cultural Responsiveness through Arts

The *(Un)Known Project* embraced the transformative power of art as a culturally responsive medium to engage with the public. Many artistic modalities were deployed, from public art installations to immersive experiences, storytelling, poetry, and photography, creating diverse avenues for discourse around enslavement and promoting racial healing and reconciliation. Narratives were generated that represented the experiences of Black women, queer people, and more, challenging and expanding the public's understanding of the full human complexity of the enslaved. These narratives opened the door for community members to contribute and participate in the project, fostering a more inclusive and nuanced discourse on enslavement's legacy.

Step Three: Translate Actions to Align with Cultural Wellbeing Science

Actions were translated to align with cultural wellbeing science through a call-to-action to memorialize sites associated with the lives of enslaved people. This included sharing historical information and encouraging the public to contribute knowledge and personal experiences to this collective memorial. For instance, Hannah Drake, with the help of a genealogist, traced her ancestry back to a plantation in South Carolina where her ancestors were enslaved. This knowledge imbued her with a renewed sense of belonging, connection to community, and personal agency, highlighting the power of cultural wellbeing science in fostering individual and communal healing.

Step Four: Iterate Agentic Narratives

The project iteratively created and amplified agentic narratives, using its various components – from art to *(Un)Known Project Trail* website – to engage the community continuously. It identified and collaborated with partners, shared new and existing stories, and disseminated the narratives more broadly, further fueling the process of community engagement and healing.

Step Five: Foster Continued Wellbeing

The ultimate goal of *(Un)Known Project* was to foster continued wellbeing through the transformative power of art. The team is now working to develop a series of web accessible tool kits and support resources that expand the project into other states beyond Kentucky by enabling other communities to execute their own versions of the project.

Ultimately, the art created and shared served as a catalyst for understanding, action, and connection within and between each step of the *MOTIF Framework*. The project leveraged the power of art to reimagine and reshape relationships within the community – how people relate to the history of their city or state and how they relate to each other. By forging new relationships, building bridges across different community groups, and keeping history and stories at the forefront, the project fostered an atmosphere of increased communal wellbeing.

Through *(Un)Known Project*, the *MOTIF Framework* not only facilitated the process of community healing and reconciliation but also proved instrumental in shaping a more inclusive, understanding, and equitable societal narrative. This case study underscores the transformative potential of creative placehealing and culture futurism and the vital role they play in promoting health equity, social justice, and cultural wellbeing in our communities.

Applying this *MOTIF Framework* in creative placehealing initiatives, like *(Un)Known Project*, ensures that the innovation process is participant-driven and systematic, synthesizing core ideas from the literature into a practical leadership approach for culture futurism and creative placehealing.

Conclusion: An Intimate Gaze at the Future of Culturally Responsive Leadership

As we, the authors, come to the end of this chapter, we find ourselves reflecting on the rich tapestry of experiences, perspectives, and insights we've shared. Through the lens of our collective lived experiences as different combinations of Black, female, queer, or neurodiverse individuals, we daily traverse a complex American landscape where creativity and mental wellbeing are seemingly in decline (Graham et al., 2021). Yet, we've also discovered that even in this tumultuous terrain, there exist pockets of resilience and wellsprings of transformation waiting to be activated. Stories we tell through the arts serve as catalysts, signaling shifts below the surface, waiting for their time to break forth. These 'wonder points' – moments where assumptions unravel – can be anticipated by stepping out of our proverbial comfort zones and engaging with diverse perspectives. The seemingly invisible from one viewpoint might be a striking revelation from another.

The journey of an artist is one of resistance, a declaration of presence, and the importance of voice (Jones, 2023). Artists shape and influence our cultural, political, and economic environments. We ignite the spark of wonder, making new possibilities visible and feasible. Through our work, we have seen how art can help people recognize and make sense of past trauma, an undeniable part of our human experience. By finding opportunities for creative placehealing, we overcome the disconnection from our senses and the places and spaces we navigate daily. Wonder is the fuel that powers our creativity, allowing us to navigate the unfamiliar, and traverse between awe and curiosity. It's the beacon when we stand on the precipice of what we know and what is yet to be discovered. Maintaining this sense of wonder can be challenging, especially in the face of ongoing trauma. But as culturally responsive leaders, we must find ways to keep this sense of wonder alive, to transform trauma through art, and to change narratives.

In conclusion, the future of culturally responsive leadership resides within us. As we confront grand societal challenges, we nurture wonder, inspire and lead by example. Using our unique perspectives, resilience, and creativity, we stand ready to shape a future where all voices are valued. We invite you to craft a culture futurist leadership journey of your own.

Postlude: Practice and Reflection

To consolidate your understanding of the concepts discussed in this resource and reflect on their applicability to your context, engage with the following activities:

Reflective Journaling: Write a reflection on how the concepts of creative placehealing and culture futurism resonate with your understanding of community health and wellbeing. How does the notion of place as a 'living, breathing entity' change your perspective on community development and health promotion?

Case Study Analysis: Revisit *(Un)Known Project* and consider how each type of leadership – exploratory, combinatory, and transformational – was manifested in the project. Can you identify specific actions or strategies that exemplify these leadership types?

Community Mapping: Consider your own community or a community you are familiar with. Map out the social toxins that might be present and think about how creative placehealing and culture futurism might be applied to address these issues.

Design a Culture-Centered Intervention: Using the *MOTIF Framework* as a guide, design a hypothetical culture-centered intervention that promotes community health and wellbeing. Be sure to consider how you would engage local cultural leaders, artists, and health professionals in this initiative.

Leadership Reflection: Reflect on your own leadership style. Which type of leadership – exploratory, combinatory, transformational – do you gravitate towards? How can you cultivate these leadership skills to promote creative placehealing and culture futurism in your work?

By engaging in these activities, you will gain a deeper understanding of creative placehealing and culture futurism, enhance your analytical skills, and develop a more nuanced perspective on community development and health promotion. Remember, the journey to community health and wellbeing is a collective endeavor – and as a future practitioner or leader, you have a crucial role to play in this journey.

References

(Un)Known Project (2023). Available: https://unknownprojecttrail.com/. [Accessed 1 July 2023].

Boden, M. A. (2010). *Creativity and Art: Three Roads to Surprise*. Oxford: Oxford University Press.

Edmonds, T., Drake, H., & Miller, J., et al. (2021). 'A Framework for Integrating Arts, Science, and Social Justice Into Culturally Responsive Public Health Communication and Innovation Designs', *Health Promotion Practice*, 22(1_suppl): 70S–82S.

Edmonds, T., Persad, P. & Wendel, M. (2017). Project HEAL (Health.Equity.Art.Learning). *PEW. Health Impact Assessment*. Available: www.pewtrusts.org/en/research-and-analysis/data-visualizations/2015/hia-map/state/kentucky/project-heal. [Accessed: 21 August 2023].

Graham, C., Allen, J. R., Bera, R., Chandra, A., de Neve, J., Edmonds, T., Eyre, H. A., Hey, N., Hill, F. & Keller, A. (2021). Addressing America's Crisis of Despair and Economic Recovery A Call for a Coordinated Effort. *RAND Corporation*. Available: www.rand.org/pubs/external_publications/EP68697.html. [Accessed: 21 August 2023].

IDEAS xLab (2023). Available: https://ideasxlab.com/. [Accessed: 1 July 2023].

Jones, B. T. (2023). [podcast] *Spark and Fire*. Available: https://sparkandfire.com/billtjones/. [Accessed: 1 July 2023].

Lister, C., Salunkhe, S. S., O'Keefe, M., Payne, H. E. & Edmonds, T. (2021). 'Cultural Wellbeing Index: A Dynamic Cultural Analytics Process for Measuring and Managing Organizational Inclusion as an Antecedent Condition of Employee Wellbeing and Innovation Capacity', *Journal of Organizational Psychology*, 21(4).

McLeroy, K. R., Bibeau, D., Steckler, A. & Glanz, K. (1988). 'An ecological perspective on health promotion programs', *Health Education Quarterly*, 15(4).

27

WHERE HEALING HAPPENS

A Working Theory on Body, Relationship, and Intentional Structure for Restorative Placemaking

Elena Quintana and Ryan Lugalia-Hollon

Our bodies are the literal forces that take us places. They are also a place unto themselves, a community of systems that support us everywhere we go, holding our histories and our hopes. They are always on, at least to some degree, mediating our experiences of safety or threat, connection or isolation, functionality and/or demobilization (Van der Kolk, 2014). When we walk into a room, shake a hand, give a hug, sit down in a chair, hear another person's story, or tell one of our own, our body is there, mediating our relationship to relationships themselves. Likewise, our bonds with others, or lack thereof, also shape our experience of our bodies, affecting our breath, feelings, and entire nervous system (Porges, 2011).

The quality of interactions we have with others is experienced through sounds and tones, physical touch, nonverbal cues, words selected, and agreements made. Through these dynamic exchanges, our bodies can become vehicles for connection, which is why we believe examples of 'relational placemaking' have emerged across the world (Pierce et al., 2010). Another major driver of experience is the physical structures that shape space, such as architecture and the material arrangement of objects in a space. Working together, they mediate and set the conditions in which all our interactions occur.

Restorative Peace Circles as Placemaking

Trauma-informed care, which supports treating all people with dignity, equity, and respect, is the result of a growing suite of Neurobiology, Epigenetics, Adverse childhood experiences, and Resilience sciences, or the 'NEAR sciences'. These NEAR sciences emphasize the fact that we heal in community and that community capacity shapes individual health outcomes. A 2012 comparative study of communities within Washington State made this clear. The study demarcated 'high

DOI: 10.4324/9781003371533-32

capacity' communities, those who had received high amounts of training around Adverse Childhood Experiences (ACE), and 'low capacity' communities that, while economically better off, had no or very little training in ACE. It found that an outcome of building community capacity in the NEAR sciences is a direct increase in positive mental health and wellness (Longhi, 2012).

The practice of restorative peace circles is one way to build community capacity for trauma-informed care (Pranis, 2005). These are literal circles, typically arranged through chairs all facing a common centerpiece that holds various 'talking pieces' used for facilitated conversation, that provide the space for deep inter-personal connections across a group. The number and type of talking pieces can vary greatly, as can the number of participants, the level of decoration of the centerpiece, and the number of other objects the centerpiece holds. The rooms where these circles are formed also vary greatly but they must always offer enough separation from the outside world to enable calm, deliberate, one-person-at-a-time conversation around the circle. Rounds of discussion are guided, question by question, by a specially trained circle keeper or pair of keepers. Whoever holds the talking piece is invited to speak, while all other participants are invited to listen.

This replicable structure, rooted in highly connected power sharing, offers all participants a sense of reliability and belonging. Whether bare bones or elaborate, the physical arrangements help to ensure that people can come as they are and participants often find a sense of being understood and a willingness by others to extend care in the form of a listening ear, connections to resources they need, a meal, or just a general sense of hope created by positive human connection.

As the NEAR sciences grow, it becomes clear that what we live through, as well as what our parents and grandparents have lived through, affects our gene expression – our own perceptions, and experiences of our body (Yehuda, et al., 2016). It also becomes increasingly clear that when the connection to our own bodies is not relaxed, it is difficult to feel completely relaxed anywhere. However, even for those who feel uncomfortable in their own body, setting the scene for a fully inclusive space can improve upon the relationship each of us has with ourselves, each other, and our communities.

The power of inclusion works across many scales. A two-year study in 2004 (Hatzenbeuhler et al., 2012) in Massachusetts of 1200 men sought to understand more deeply why gay men were more depressed, saw doctors more frequently, and took more medication than other members of the general public. In the middle of the survey period, Massachusetts became the first state to pass marriage equality. Twelve months after this policy change, gay men in the study experienced significant declines in symptoms of mental illness, in numbers of visits to the doctor, and of medication consumption, suggesting equity and inclusion are healing, and that marginalization sickens.

As we offer here, 'restorative placemaking' applies this inclusion principle in targeted, room-bound ways, providing participants opportunities for truthful self-expression, deepened interpersonal bonds, and connection to the non-clinical medicine that is enabled by loving support and respect. This can be true even when they occur in close proximity to exclusive practices and institutionalized violence.

Peace Rooms – Healing at School

School is the first experience that many of us have of life in an institution. It may be where we first navigated the traumatic experiences of feeling like we did not belong, of bullying or being bullied, or of enforcing what is right and normal within a rule-bound, proscribed social setting. These parameters aim to create a managed and manageable learning environment. What happens, however, when the immediate or ongoing trauma endured by students in their lives outside of school disrupts the managed environment as symptoms bubble up and over?

We have helped to build and facilitate *Peace Rooms* across more than 35 school settings in Chicago and San Antonio, designated to the processing of these disruptive events, places where props, policies, room design, and human training, can all prepare the way for healing and transformation, working in support of more equitable school experiences for those enduring harms from the outside world.

Peace Rooms are designed to be intentional, trauma-informed places that support restorative, rather than punitive, responses to harm. They are inviting and accountable to all members of a community, with deep roots in Indigenous cultures and Tribal communities. When activated well, *Peace Rooms* become sites for advancing trauma-informed restorative justice at the level of both individual school campuses and big citywide systems. Formerly dull physical spaces can be transformed to help reground relationships in trust, reconciliation, and accountability, thereby enabling interconnectedness to ourselves, each other, and nature. While *Peace Rooms* themselves cannot eliminate harm, they can be key anchors in larger community efforts to reduce and address harm, vital touchstones that can help to sustain healing culture. As one participant said, 'To me, circle is life because it helped me become the person I am today. It helped me grow' (Middle school student, 2022. Interview conducted by StriveTogether in March 2022 in San Antonio).

In stark contrast, school policies traditionally focus on exclusionary, punitive practices such as out-of-school suspension, which disproportionately impact students of color, especially young Black men in many school districts and networks across the urban United States. Across Texas, for example, a recent study showed that 83% of Black males experienced an exclusionary punishment between grades 7 and 12. A huge part of what is absent from these processes is the examination of what happened and a subsequent resolution. Dr Paula Johnson, the Chief Student Advocacy Officer for Judson Independent School District in San Antonio, shares in a 2022 StriveTogether interview:

We used to have physical incidents in high schools where the child involved would disappear for a few days, then just helicopter-drop back in. But no one was actually looking at what happened. So, the student would come back to school without any resolution.

In its first year of restorative practices, the Judson district saw significant results, particularly in disciplinary incidents. One Elementary School experienced an immediate decline in one month, with incidents measuring at 29% of the recorded incidents from the year prior. That trend continued throughout the year, holding steady at a 70% decline five months later. In the words of Dr Jeanette Ball, the district saw 'more students staying in school, fewer students being suspended and diverse students achieving more' (StriveTogether Interview, 2022).

Healing in Prison – Community Anti-violence Education

In the mid-1990s, Sly (a fabricated nickname) and two friends set out to collect on a street debt from a man who was with his wife and their new-born baby in her arms: the young men thought he had money, and he had not paid them what they believed he owed. One of the group of young friends produced a gun and killed the man in front of his wife. Moments later, the three accomplices were on the run, and a woman would be left to raise a baby whose father had just been murdered.

Sly found himself facing a 55-year prison sentence for this murder, although he was not the shooter. Over the years, Sly busied himself by going to school and preparing himself to contribute to a better world in whatever ways he could. Sly wrote a proposal to form a violence prevention study group with men in the University of Illinois college prison program, the *Education Justice Project*. Like Sly, most of the men in attendance were serving long sentences for murder and they had committed to study the NEAR sciences and to improving themselves and those around them. With the support of outside facilitators, a peer-led group was created called *Community Anti-Violence Education*, or *CAVE*. Over a decade since its founding, this group of incarcerated men continue to work to prevent violence through learning, unpacking their own trauma, and participating in outreach and dialogue with their incarcerated peers.

Rooted in circle-based conversations, this work is guided using a curriculum to support the Sanctuary Model, a model of care designed to address trauma, by Dr Sandra Bloom. Part of the ritual of the group includes making weekly *Commitments to Sanctuary* (Andrus, 2022). These commitments include the following:

1. Social Learning: Being open to learning from other people, perspectives, and ideas.
2. Nonviolence: Doing your part to effect, protect and preserve physical, psychological, social and moral safety within a group and within yourself.
3. Emotional Intelligence: Understanding the connection between trauma and managing our emotions; making choices that support positive relationships.

4. Shared Governance: Active participation from every member of the group to include their voice and listen to others to successfully build and share a sacred space.
5. Social Responsibility: Supporting the idea that we belong to each other, and we need to do right by each other in community.
6. Open Communication: Choosing to speak clearly and fully about the realities of our lives, needs, and perspectives allows for fuller interactions with others, and allows us to identify and express a full range of emotions and needs.
7. Growth & Change: The science of trauma allows us to put our experiences in proper perspective and challenge ourselves to better our responses to the people and world around us in a way that will help us live better and more functional lives.

Critically, being in community in prison is a high contact sport that pays dividends quickly and effectively through the healing magic of uninterrupted emotional connection, and interpersonal accountability. In *CAVE,* men facilitate discussion about topics requiring great vulnerability such as loss, victimization – both experienced and perpetrated – losing control emotionally, physically, and mentally, self-loathing and more. They provide a safe place for each other to speak openly and honor the ground rule of confidentiality. Men in prison, while more likely to have been victims of all types of abuse in childhood, are less likely to share these realities with others, especially true of sexual assault survivors. The original ACE study found that 16% of men in the general population stated they had been a victim of sexual abuse; however, Wolff and Shi (2012) found that only 4.5% of incarcerated males admitted to having been the victim of sexual abuse. This disparity could be a sign that incarcerated men think that being seen or portrayed as a sexual victim may increase the likelihood that they will be victimized in prison. Therefore, it is imperative that the *CAVE* participants honor each other even in the cases that they don't particularly like each other.

CAVE has proved to be a powerful force for transformation. Sly has earned so much good time for the work he has done that he was able to get out of prison after 26.5 years. He grew into a man with an ability to control impulses, manage his anger, and understand the long-term effects of his actions. Through his tutoring of men within the prison, he also understood how healing it is to support others on their path to better themselves and their lives. Of the scores of similar men that founded, facilitated or attended *CAVE* and that left prison, none have returned. Many have become civic leaders working for peaceful communities, strengthening their own families, and supporting programming and policy that addresses rather than recreates trauma in prisons and communities.

Sadly, the baby Sly left at home, Sly Jr., grew up with many of the same challenges Sly faced; no father, poverty, surrounded by violence. Sly Jr. was locked up as a young man for a shooting. This raises the need for healing spaces at a systemic level. How do we commit to create spaces of healing for the child of the man who was killed, or for the son of the man who was jailed? In questioning our deeper beliefs about who deserves healing, we can better design the present and future spaces we need to create safety.

Healing in Community – Circle Room

In the early 1980s, Fr. Dave Kelly was a young priest in Chicago dedicating himself to prison chaplaincy and working at Cook County Jail and the Cook County Juvenile Temporary Detention Center, also referred to as the JTDC, or the 'Audy Home'. Over 40 years later, Fr. Kelly is still dedicated to his work in the JTDC, and after watching the same young people cycle through the JTDC to one of the 26 prisons in Illinois, he wondered if he could do more to prevent this path, rather than just pray for the people sucked into the prison pipeline. In 2002 he and three other Missionaries of the Precious Blood established a community center called the Precious Blood Ministry of Reconciliation (PBMR) in the Back of the Yards neighborhood of Chicago.

PBMR began in a modest way as a community center on the second floor of an old school building that was owned by the Chicago Archdiocese. At the center of it was the Circle Room, a space utilized for restorative justice circle training, private conversations about personal transformation, or expressing grief. Immediately outside the doors sits a large dining room off from a busy kitchen, filled with tables, chairs and couches that allow people to come rest, eat, and gather in community. Down the hall is another large room used for different meetings and activities. In the early days, a small corner room housed a makeshift gym, and a utility closet housed a washer and dryer for anyone who needed it.

The PBMR was created to do 'the harder work', as Fr. Kelly referred to it, of keeping young people in the community, and out of the prison pipeline, established on the cornerstone of restorative justice, a philosophy of belonging to each other, and doing right by each other. PBMR began to work more deeply with whole kinship networks of the young people who had been locked up to attempt to go farther upstream to keep the young from getting permanently sucked back into the churn of the prison system. Part of the reason Fr. Kelly found that people kept going back to crime, and ergo, prison, is because there was no kind and safe place for young people who had become gang involved. Indeed, although PBMR borders a lush, spacious park, many of the Chicago Park District sites do not welcome groups of gang-involved youth. These policies, created to deter violence, create a self-fulfilling prophecy of cutting off the most traumatized and vulnerable youth from community mentors who may be able to support a different path forward.

Creating a sacred space in community means challenging all of the rules created to 'weed out the bad element', as in the Park District example. Instead, at PBMR, the youth who were the constant target of police shakedowns and searches were seen as the most promising, the most desirable, full of potential and light.

Restorative placemaking at PBMR started with the premise that you have a place where people you respect find you worthy of their investment.

In 2011 PBMR began the process of working with a number of other organizations that also prioritized work with system-involved youth to create *Community Restorative Justice Hubs (2023)*. These RJ Hubs, joined in a set of shared practices undergirded by restorative justice. Founders of the *RJ Hubs* created a list of shared practices called the 'Five Pillars (Restorative Justice Hubs)':

1. Radical Hospitality: Meeting people where they are, and providing them a safe and hospitable place to gather in community.
2. Accompaniment: Walking with youth and adults in support of their goals and positive development.
3. Building Relationships: To support strong, interconnected communities that work in right relationship with one another.
4. Relentless Engagement of Services and Supports: Creating ongoing relationships with service providers, and advocacy and connection to link these supports to community members who most need them.
5. Participation in a Learning Community: Leadership from all of the RJ Hubs meet regularly to challenge and learn from one another to deeply practice restorative justice in a way that is meaningful to community building.

The Five Pillars act as a philosophical guidepost and throughline, offering essential non-negotiables of creating a safe space to support trauma-responsive work for both clients and workers. In these spaces, it is deceptively simple to put relationship building first; however, the ongoing tugs to bend to the whim of funders, evaluators, or other interest groups are constant challenges to staying centered in work that helps and heals according to the science. Indeed, the fifth pillar, participation in a learning community, has been a vital part of the strengthening of each of the participating organizations to continue to walk a path that avoids missteps that may knock any one of the hubs away from their shared mission. In part because of this, the *RJ Hubs* have become synonymous with high quality, effective work that walks with community and has stabilized all of the hubs over the course of the last decade that the hubs have been in existence.

Guided by the needs of the community, PBMR has grown from a modest space to an expansive campus that houses a basketball court, a huge field where a decrepit church and moribund building once stood, an expansive vegetable garden that houses chickens and bees, a media center, three re-entry homes, a transitional home for women and children, an art center with a wood shop and a t-shirt making shop, a healing center, a mothers center, a small gym, a food pantry, a program that helps community members buy their own homes, and more. They employ community members as case workers, grief counselors, farmers, crafts workers, and more.

Spaces that Heal

Working across settings, practitioners of restorative justice can build and multiply zones for healing, inviting more and more people into community-centered healing, where the innate desire for love, belonging, and attention can be fulfilled. These trained practitioners intentionally connect the threads between our bodies, our bonds, and the spaces we inhabit. Through restorative practices, we see people as people – even when they do wrong – and we seek to understand the basis of their behavior.

For Ramon D. Vasquez, a leader at the American Indians in Texas at the Spanish Colonial Missions (AIT-SCM), restorative justice is an opportunity to reconnect with culture, honor values and recognize the trajectory of the systems that were created and forced upon populations throughout the years. He states:

Healing circles, called 'circulos' by my people, are part of the way we conduct ourselves, and they can apply to any situation. They create space in a way that is central to who we are, create healing when it's needed and create opportunities to deepen the relationship to our culture on an indigenous level.

(StriveTogether Inc. 2022)

One of the restorative properties of these practices is to return power to those who may feel they have none, a shift that requires participants to change their mental models. That is particularly true when restorative justice is applied in schools – and especially true when dealing with young populations of color. It is also true in the otherwise toxic space of prison where – through this sacred practice – participants are able to be honest with each other about the difficulties they've faced, providing a rare space where they can talk about their true feelings and the difficult realities of their lives. It is true in neighborhoods as well, where dedicated community spaces can be used to welcome and help people who have historically been barred from schools and/or targeted for arrest and detention to re-integrate.

Of course, it is not enough to use restorative placemaking to create scattered sites of inclusion and safety. We must also uproot the systems of oppression, repression, and marginalization that feed the trauma pipeline. As we learn to do so, restorative practices can help makers of places to expand their focus beyond single events and momentary symptoms, and to engage the larger relational context and the underlying challenges at play.

References

Andrus (2022). The Sanctuary Model. Available: www.thesanctuaryinstitute.org/?s=sanctu ary+model. [Accessed: 28

Hatzenbuehler, M., O'Cleirigh, C., Grasso, C., Mayer, K., Safren, S., & Bradford, J. (2012). 'Effect of Same-Sex Marriage Laws on Health Care Use and Expenditures in Sexual Minority Men: A Quasi-Natural Experiment' in *American Journal of Public Health*, 102(2): 285–291.

Longhi, D. and Porter, L. (2012). Stress, Strength, Work, Hope (Technical Appendix). Family Policy Council, Olympia, WA.

Pierce, J., Martin, D. G., & Murphy, J. T. (2010). 'Relational placemaking: the networked politics of place' in *Graduate School of Geography, Clark University*.

Porges, S. W. (2011). *The polyvagal theory: Neurophysiological foundations of emotions, attachment, communication, and self-regulation*. Washington, D.C. W. W. Norton & Company

Pranis, K. (2005). *The Little Book of Circle Processes: A New/Old Approach to Peacemaking* (The Little Books of Justice and Peacebuilding Series) Intercourse, Pennsylvania, Good Books

Restorative Justice Hubs (2023). *Five Pillars*. Available: rjhubs.org. [Accessed: 30 June 2023].

StriveTogether Inc., UP Partnership (2022). *Changing school disciplinary practices to improve academic results*. Available: www.strivetogether.org/wp-content/uploads/2022/05/ST-UP-Partnership-Case-Study-ChangingDisciplinary_Final.pdf. [Accessed: 30 June 2023].

Van der Kolk, B. A. (2014). *The body keeps the score: Brain, mind, and body in the healing of trauma*. New York: Viking.

Wolff, N. & Shi, J. (2012) 'Childhood and Adult Trauma Experiences of Incarcerated Persons and Their Relationship to Adult Behavioral Health Problems and Treatment' in *International Journal of Environmentally Responsive Public Health*, 9(5).

Yehuda, R., Daskalakis, N. P., Bierer, L. M. Bader, N., Klengel, T., Holsboer, F., & Binder, E.B. (2016). 'Holocaust Exposure Induced Intergenerational effects on FKBP5 Methylation' in *Biological Psychiatry*, 80(5).

28

THE ART OF PLACE

Daria Dorosh

Introduction

Once upon a time, there were wandering storytellers in every culture. They were the guardians who passed on the memory of place to future generations. In the Celtic tradition, for instance, the *seanchai* traveled from village to village, entertaining, informing, and keeping history alive through the spoken art of poetry, legend, and tales of adventure. For many of us, home is no longer rooted in a specific location or its cultural memory. Home was transformed in the 1800s by the telephone when it merged public and private space and made non-local networks possible. Since then, communication devices and machines continue to expand our roaming range in every way.

The relationship of art to place has been affected by the exponential speed of change occurring in the last two centuries. Questions have surfaced about the role of art in a mobile culture and how it fits in the many places we inhabit over a lifetime, either by choice or necessity. If home once signified ownership, stability, and history, then placemaking is always in a process of change. It has led artists to investigate new forms of art and to ask, where does art belong? How can our individual narratives unfold in the ongoing process of placemaking? If we integrate our physical location with the global meet-up place of the internet, can we nurture what we care about and build meaningful relationships with each other in this new hybrid space?

It has been proposed that our text-based culture is transitioning to an image based one. Over time, writing has changed art from symbol to story, and in turn, art has shaped writing by emancipating it from accounting. This 5000-year-old relationship is shifting again in the data age of technical images at our fingertips.

DOI: 10.4324/9781003371533-33

Perhaps the ubiquitous cell phone that lives on our body is the opportunity for creative expression by everyone – a stage where image, text, community, and commerce intersect. The examples and interactions I offer reflect a way of integrating the once dominant product world with the new world of information culture. In this hybrid network of product and process, difference is the springboard in which new patterns wait for discovery.

Illustration One: Damage as Springboard for Self- discovery

The persistent stress of climate change and several years of political upheaval and growing economic uncertainty were followed by a pandemic we were not prepared for. By mid-2020, 'sheltering in place' became a global survival strategy. A sense of danger in the workplace took its toll emotionally and led to decisions about the future way of working for many. When our social spaces closed down, deeply personal and imaginary realms had room to surface in the silence.

The emotional gravity of these years led me to core questions about my existence. I began to integrate images with text mapping, technology, and textiles in my art, and considered formerly unimaginable sites for my art (Figure 28.1). Trauma helped me discover new resources and integrate them with older strategies. I believe that there are no rules to follow in the new normal. The personal space of being is where the action is. I believe our public lives before the pandemic had too much 'noise' that blocked out the uniqueness of being. Hours of daily travel, dinners and events that were not particularly satisfying, work we did to further the goals of others that did not resonate with our own all seemed expendable. Being homebound led to finding new options for interaction.

The governmental strategy of sheltering in place was an opportunity to look inward and find resources to work with. A visual mapping of one's questions, desires, and assets is a way to take charge of the unique possibilities inherent in each of us. The gift that is our body, the generosity of our inherited DNA from scores of unknown ancestors, the history of our life's activity and richness of relationships with others are all material for ongoing creative self-fulfilment, but only if we dare to question the rules we bring from a past that no longer works. I have found text mapping to be a non-linear way to examine any question. The safety in making lists no longer informs me. To discover relationships and dynamic connections, a visual method closer to art making may be the key.

Process to Create a Mind Map

1. draw a central branch-like curve
2. write your question
3. pull out branches of associations
4. add sub-branches when inspired

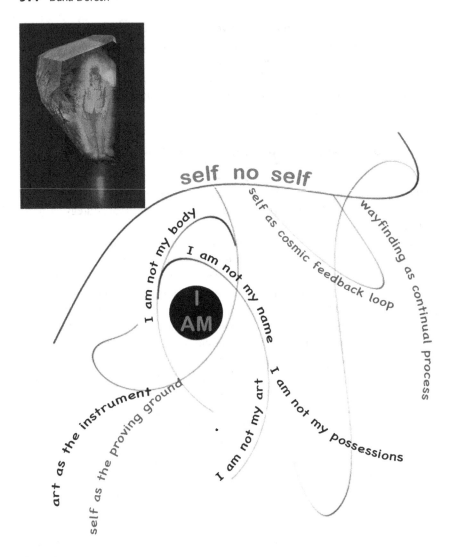

FIGURE 28.1 Dorosh, D. (2020). *Damage as springboard for self-discovery*. Image credit: Dorosh (2020).

Illustration Two – Discovering the Digital Self

As the reality of the pandemic settled in, I noticed a growing sense of confusion in apparel. Daywear and sleepwear lost their dividing line and we started to dress in dissonant layers of anything at hand. Without our public spaces, the self-image we present to each other through clothing lost its function.

Zoom, a virtual communication technology that gained usage during the pandemic, offered a digital landscape where we could be together – if only from the waist up. Our computer screen became a window through which some of us looked into the home of others while they looked back into ours. Our computer camera and home environment did not anticipate sheltering in place and did not represent us well. Most of us did not understand that we were on a world stage or how to take charge of this public/private way of being. Technology managed to keep us connected virtually during our isolation and for some of us, led to the hybrid social space we choose to inhabit today.

For me, technology opened a door to unexpected opportunities for making art. Personalizing this public/private arena became an interesting challenge. When I realized it was possible to 'wear' my own background on Zoom, I created hundreds of patterned backgrounds from my photographs for meetings (Figure 28.2).

FIGURE 28.2 Dorosh. D. (2020). *Discovering the digital self.* Image credit: Dorosh (2020).

I presented myself to others in a context of my artwork and found my personal way of 'dressing' for a virtual stage.

Another surprise was exploring the green screen, which made it possible to see my image in infinite patterns. Zoom selfies became a new form of art for me, and a rebellion against the pressure so many of us feel when being photographed. Looking 'good' has oppressive cultural baggage attached to it set by others, especially for women. I discovered my criteria for self-representation by ignoring the rules built into the software. The freedom to present myself through infinite patterns was exciting and provocative and truer to the complexity of identity.

Exploring unlimited pattern variations in image and sound is open to anyone with a cell phone, computer, or iPad. Some very basic apps and features built into our devices can be a creative playland of discovery. The process of looking inward with everyday digital tools is an antidote to the trauma of confinement. If we do not recognize the digital power at our fingertips, we impose confinement on ourselves, which intensifies the trauma of outside events over which we have no control.

Process to Create a Zoom Background

1. Take a landscape format photo.
2. Format it to Zoom dimensions of 1920 x1080 pixels at 72dpi in Adobe Photoshop.
3. Download Zoom, click on 'new meeting' and see your image on screen
4. on the lower-left side menu bar. Click on the 'up' arrow over the video icon.
5. Click on 'choose virtual background'.
6. See yourself on the small screen.
7. Click the plus sign on the left under it.
8. Choose 'add image'.

Illustration Three – Infinite Spaces of the Mind

'Once upon a time…' are magical words that invite adventure. Storytelling is an event in which the story takes on a unique sensory form in the mind of each listener. It is a co-creation between an individual and a culture traversing time and space. Any image can be a stage for storytelling. A vast sea of images wait for completion by anyone who is reading this chapter and willing to see a story in an image. When I photograph the natural world, it becomes a base for further transformation. Using digital software, I rearrange the light, patterns, and shapes in an image to feel a new connection to my subject.

There are many ways to tell a story using imagination and the mood of the moment. I weave different stories from the same image because I am different every day. I can alter the image and see how it affects my story. I share the image with others and listen to their story. The games that can be played are many if we remember that we once inhabited images as children and can take them to more complex levels as adults. In our storytelling, trauma is cloaked in metaphor created by the psyche so that we can safely face our anxiety to expel it.

The trauma of confinement in the pandemic became an incentive for breaking out by looking at what is close at hand that had been previously overlooked (Figure 28.3). For instance, most of us live with plants and yet know little about

FIGURE 28.3   Dorosh. D. (2020). *Infinite spaces of the mind.* Image credit: Dorosh (2020).

them. The search engine at our fingertips has the power to change how we see life, forever. Our cell phone camera can take us deep into the heart of a flower where a silent universe fulfils its mission unobserved. It is no less spectacular than looking at the stars with a telescope and sensing a connection to a process we are part of but did not create. The macro and micro universe is accessible to us wherever we are to experience a personal sense of place.

Process to Tell a Story from An Image

1. Take a photo or find an image that interests you.
2. Find a listener or two of any age.
3. Share the image and then start with 'once upon a time...'
4. Free associate to the picture and make up a fantastical story.
5. Indulge in a little drama in your presentation style.
6. Let the listener tell you a story from the same image.

Illustration Four – Be the Museum

This substantial project grew out of conversations in the first year I was sheltering in place in my studio in upstate New York. Through my art network online, I met artist Yvonne Shortt who in 2020 became a new member of A.I.R. Gallery, an institution where I have been an artist and cofounder since 1972. With some 30-years difference in our timeline, we found that both of us were interested in a similar range of issues that impact our art practice. Yvonne defines herself as a question-based process artist who works with materials and is active in digital social spaces; and I am a born analogue artist working with analogue and digital tools to find a visual form for my questions. Because of the lockdown, we engaged in a weekly conversation about art on Zoom for a year before we met in real space. During that time, we put in place a Research and Development program for our arts organization as a collaborative artist-led exploration for ourselves and we shared our findings online with others. Some of the issues we addressed were: how many spaces are there for art? Where does art belong? What could an artist-led model of legacy be? Can it be less transactional and more like stewardship? Each question led to more questions because our art practice is permeated with social structures from the past that no longer work. Basic issues such as self-selection, hierarchy, and transparency in the arts had to be integrated into our practice because they reflected the present data age of information.

Be the Museum (Figure 28.4) is the theme of our collaborative artist-led project in which each of us established our own version of *The Museum for Contemporary Artists* (MfCA), in which the artist's studio, with all its history, documents, artifacts, lifestyle, and activity is a museum. It is a perceptual shift in which one's skills, resources, physical and virtual assets are the center from which the artist participates in a living cultural stewardship and legacy. It is creative placemaking

FIGURE 28.4 Dorosh, D. (2020). *Be the Museum*. Image credit: Dorosh (2020).

against the trauma of traditional art practice in which there is no stewardship and legacy plan for the art we spend a lifetime creating. MfCA is a response to the realization that an artist's physical location combined with a digital presence is a rich springboard, where innovative concepts can be explored and new attitudes put into place. It is a sustainable model created with our own resources.

My generation of artists who were born analogue in the last century matured in a world of slow art, slow food, and the art market as the only destination for their work. Their expectations were based on the prevailing economics of scarcity that fueled competition and gatekeeping in which art was a commodity. Product ruled over process and handmade work was valued over work made from code. In the artists' live/work lofts that populated lower Manhattan during a real estate slump in the 1960s and 1970s, artists combined their personal and professional activity. Their compact lifestyle was driven by the need for raw working space, little financial resources, and time for making art. The artist's live/work loft is being revisited by scholars to better understand how the real estate sector commodified and benefited from the unrecognized labor of artists and other creative workers in neighborhoods on the edge of commercial collapse. The prevailing myth that sustained us was that good art will find its way to institutions that will care for it. Within the generation that straddled the analogue/digital paradigm shift, the impact of it is embedded in their work. However, the increasing number of artists in that generation continued to multiply as we moved toward the digital present, which makes it unlikely that public institutions as we know them could accommodate this very important history. An archive for the future is at risk because there is no institutional legacy plan in place for the direction art has taken in the paradigm shift.

I believe that the data age shows abundance and that the universe manifests generosity, as in the plant world of seeds all around us. Perhaps code, like DNA, replicates data so that each end product is an original, which implies that there should be enough for all. In contrast, reproductions in the last century produced limited copies of an original, fueling scarcity, competition, ownership, and hierarchy. Artists who straddled both analogue and digital centuries are presented with the challenge of finding a new model for art practice that respects the value of the artifact while expanding its playing field in the digital realm. However, we were not prepared for art making that included making a place for art as well.

The empowerment of analogue and digital options has changed how I conduct my art practice. It has led me to test new strategies in sustainability, stewardship, and placemaking for art. *Be the Museum (2021)* is a challenge to shift my viewpoint from scarcity to abundance, and from waiting to be selected to self-selection, so that I can create the change I want to see in the world.

Process to Be the Museum – *Implementation Prototype*

1. Artist Yvonne Shortt decided to become a museum.

   As a way to break from a scarcity mindset, Yvonne Shortt selected her own space to exhibit her work. Deciding a space and determining how her work would be viewed meant that her museum was bound by her own terms, rules, and timelines. Ms. Shortt focuses on process. When people come over they go in the creek, sometimes harvest clay, have a meal, play with the art, and engage in conversation.

2. She created a logo and made two signs: one to show the logo and another sign to explain what the museum is about.

   Every institution has a brand and a recognizable mark that identifies it to the public. Yvonne decided for her museum she would create a logo so those visiting would know where they were. The logo was printed along with a description for visitors to learn about the museum and what it meant to her.

3. A website was created to RSVP to visit and learn more about the project.

   With a visible sign and a website link, the museum was almost ready.

4. The museum location became public through a Google map entry.

To make the museum accessible to visitors, they need to know how to find it. A few clicks on Google Maps allows anyone to add a 'missing place' and helps art seekers find The Museum for Contemporary Artists. A link on social media helps the public know of the new museum.

5. A staff ID card was created to test reciprocal access at other museums
   Every art museum has an ID card for their staff. Artist Yvonne Shortt created a staff card to test reciprocal access at other museums.
6. A stewardship program and gift shop were established.
   Artist Yvonne Shortt was approached by a non-profit to purchase her headwear after seeing it at her museum. The artist decided to create a stewardship program where the non-profit became a steward of the artwork for twenty years. This stewardship program came with funding that will help the artist continue to grow her museum ideas and share what she learns with others.

Concepts to Explore

- Decide the space.
- Decide what will be in your space.
- Will the space be indoors, outdoors, static, or changing?
- Will the space be process based or offer workshops?
- What are your hours? Do people have to RSVP?
- Will there be a collection that you lend out?
- Will you have a gift shop? Will you charge or give freely?
- Will you share a meal with your visitor?
- Will you upload your location to Google Maps?

Conclusion/Dream

I believe that our tools have changed the definition of home to be anywhere we are because we carry our home inside us. We nurture it with information we select from our communal digital home, the internet, and combine it with the richness of our physical spot on the planet. The vast new library of digital images and data belonging to all of us waits to be repurposed, played with, and re-shared in the global community. The entire planet is home now, and trauma ought not to be self-inflicted by fear to move forward and explore this new expanse of home. We have come out of a history of scarcity into a world of abundance and have work to do to learn how to manage our new self and our new home. Most of us have little experience managing abundance so that it does not become as toxic as the scarcity we have lived by until now.

If we were to consider the planet as a communal home for all life – people, plants, and animals – how would we care for this home? How would we share resources equitably so all life could thrive? Our former concept of home was a protective possession that kept us safe in a dangerous environment of competition with others for resources. Our beliefs fueled hoarding, and our insecurity justified taking more than we need from all other life.

Today, a re-evaluation is underway, notably in women-led projects across all disciplines. One example is the work on biomimicry in designing the built environment in which design is a process not a product. Another is the science-led work by visionaries who are sharing their deep knowledge about the planet we all call home. It is a shift from asking what can we learn about other life forms to what can we learn from them?

The strategy of dominance we have lived by until this century is not compatible with a data culture of replication. It produces monstrous excess for some and valleys of deprivation for others. And so, we must be the ones to change ourselves if we are to have a future.

Reference

Be The Museum (2021) [online]. Available: www.airgallery.org/research-and-development. [Accessed: 8 July 2023].

29

UNRAVELLING MEMORIES

The Metaphor as a Possibility of Resilience

Pablo Gershanik

Introduction

I do theatre. My artistic work is divided in three main fields: as an actor-clown, as a director and as a professor at the National San Martin's University's in Buenos Aires. I work on the link between personal and collective memory. After the creation in 2017 of *Eighty Bullets in the Wing* (a personal creation in the Argentinian ESMA, the biggest extermination camp in the Americas used by the Argentinian Junta between 1976 and 1983, formerly the Navy's Mechanic School), in 2019, I was invited by French cultural institutions to Paris, where I established and created the *Intimate Model's Lab*. This Lab has been a meeting point for artists, people having experienced traumatic events, therapists, and academics from more than 20 nationalities. Throughout the Labs, theatre companies, museums, universities and cultural spaces from Argentina, Mexico, France, Belgium, Switzerland, the United States and Canada have mixed in interdisciplinary groups to explore this method. In a full immersion, five-day experience the participants work resiliently in group and on their own through different aesthetics tools:

1) Physical play through movement and dance.
2) Objects and image theatre, writing and drawing through personal and collective memory.
3) A plastic work to create the models, a transposed representation of a lived experience.

The Intimate Model's Lab has been awarded with creation residencies at the Cent-Quatre and the Cité Internationale des Arts of Paris and the Argentinian and French branches of the French Institute. This chapter narrates the creation and development

DOI: 10.4324/9781003371533-34

of intimate models as a tool for approximation and aesthetic elaboration of a traumatic event experienced.

At the end of 2016, I, along with a small group of multidisciplinary artists, was invited by the Visual Arts Team of the Haroldo Conti Cultural Centre of Memory in Buenos Aires to do a two-month creative residency called *This Is Not an Exhibition*. Situated in a middle-class neighbourhood in the city, the centre was created in 2008 by the Argentinian government and human rights organisations *Madres de la Plaza de Mayo*, *Abuelas de la Plaza de Mayo* and *HIJOS*. The aim of the residency was that each one of us would come up with a piece that would be the result of the meeting between our initial proposals and the space we were working in.

It was an interesting challenge for me, not coming from the visual arts, but from acting and theatre disciplines. It allowed me an indirect entry into an extremely intimate and close subject through visual arts, and an opportunity to leave my daily work and language. I started to work – with fewer resources and at the same time with less automaticity – with a subject that is part of my personal history and my artistic search: the murder, in 1975 in the city of La Plata, of my father, Dr Mario Gershanik, a young Jewish paediatrician, rugby player and political militant, at the hands of the Argentine Anti-Communist Alliance (AAA), a fascist parapolice group created by the Peronist right wing in power to exterminate sectors of the Peronist and non-Peronist militancy.

I immersed myself in the project, driven by the following reflection:

I wonder what it means to reconstruct a tragedy… Is it going back to the ground zero of the nightmare to organize the shards of that pain? Or will it be to retell the story (to others and to ourselves), to knead it, reinvent it and share it until the poison dissolves?

I think of the absences, the lack, the lead and gunpowder, and all the joy and life frozen by the green Falcons, the cars that had become an infamous symbol, used by the fascist gangs to do their raids. A city on the dissection table. La Plata in 1975, the setting for this journey in search of Mario, my dad, and the story that binds us forever like a misty and fragmented mirror of other stories, so many, indelibly marked by that time of bullets and wings.

II.

The result of that residency was *Ochenta Balas sobre el Ala* (*Eighty Bullets in the Wing*) (2016) – so named as Mario used to play as wing for the University's rugby team and he was murdered with 80 bullets. This was a plastic and audio-visual work that consisted of a 'model-installation': the city of La Plata as it could have been that day, with its geography, architecture, characters, ordinary and extraordinary events, all co-existing with another half-real, half-fictionalized city where images and sounds of the social and political reality of that moment implanted by me were intermingled. The work included images of our family's life and a filming-projection in real time from a kinetic device of an electric train that ran across the piece filming and projecting the resultant scenes in the former concentration camp walls, integrating the audience to the projected landscapes: a crime scene, and a *ricochet* of situations, possibly happening at the same precise moment in the same block, in the neighbourhood, in the city, the country and the planet – tiny little everyday-life situation and relevant historic events. All that at my father's height: 1.84m, the scale of a father – life as it could have been, and for that the idea of a model.

I wanted to create a device that worked autonomously, that would allow me not to be on stage as I use to do, but rather to take a step back so that another body could become present in the play: the body of Mario, 'the Wing'. Every night I went to the museum to activate the objects in the piece and that's how I started talking to the people who came by. Visitors to the exhibition, people who had lived years before, within those same walls, the experience of confinement, disappearance, torture, and extermination of their comrades and who found in the model elements of their own lives that allowed them to share fragments of the experience traversed. I discovered different layers in the model: the real facts, the archives, and the fable that sometimes is necessary to talk about some issues. And the metaphoric step between reality and fiction embroidering between what had happened and what I could do with that story.

It was like throwing a bottle into the sea. Through word-of-mouth different people familiar with the subject or our history began to approach and say: 'I was your father's nurse'; 'Mario was my paediatrician and brought me into the world'; 'We played rugby together' – very moving encounters with people that I did not know but that thanks to the model came to share fragments of our common stories and allowed me to understand that every personal story is, at the same time, a collective story. This experience made me wonder if making a model could help other people to 'revisit' their story and the way it is told in a similar way to how it was helping me.

During a period of two months new spectators came every night to decipher what was happening in that miniature world, leading me to understand the need for a narrator: someone that could receive the audience in the first person, look them in the eyes and tell them: 'Here is a story that I would share with you'. Thus, the installation began to propose a new theatrical layer: a present body pointing little objects and lights closer to put focus on some elements. Each night the story evolved, influenced by the presence of the spectators and what happened when sharing that moment with them.

Witnessing this modification of the narration gave me the clue as to how my condition as a theatre artist should address this new stage: the narration is modified because memory is changing. But also, because we change positions, the individual's point of view (and in this sense the model is an invitation to look at a reality with 360 degrees of different perspectives) necessarily changes what we say: to modify the context is to modify the text. We can interpret it in very different ways and this exercise of approaching and stepping back from a lived experience allows us to create space, and that creative space lets arise veils, metaphors, non-linear or direct ways of embodying a story.

The story acquires a breath, and a distance between the viewer and the performance, between the actor and the character, between the biography and the story, that allow both the narrator and the audience to have a place in the story. Each person completes the empty spaces with their own imagination, memory, and emotion. The combination of documentary elements from the intimate and collective archives coexists with totally 'reconstructed' objects, half true, half modified, to turn them into illusory objects, riding between reality and fiction.

However – and here the magic of theatre lies – they become a plausible story in the eyes of the audience. The quest of the missing elements that I believed had to be in the model but that were not necessarily at my disposal, including documents, bullets, photos that were part of the scene of my father which I had never had or would have access to, led me to understand that it was important to create them with my own hands. This became, somehow, the process of creating objects from my past in the present.

Through this process of devising signs for the public, I found a constitutive element of my work that later became a fundamental aspect of what I propose today to those who work on the creation of their own models: to work from the objects and stories people do have in order, as theatre craftspeople, to manufacture the elements that are missing from their scenes and that are necessary to tell whoever is in front of us who we are. Knowing that the viewer's eyes become a prism that scatters light and a mirror that returns our own image to us.

By recovering the visual memory of a city and a time, of a historic moment and a political emotion, by creating a world of objects, images, speeches sounds and music, I was able to see the attentive, restless or moved gaze of the spectators, to receive the hugs of people who recognized themselves in those streets, who confirmed that they were friends, companions, contemporaries. With the train filming that universe in miniature and integrating the audience by projecting live the image of them surrounding the city, I realized that that display was becoming an original and effective tool to put in movement my personal experiences and to let them melt with other people's. It led me also to wonder if creating models could help others the same way it was helping me and to search for a way to make this creation procedure transmittable while extending it to other experiences of social breakdown in our societies. Both personal and collective.

This process also led me to the search for artists whose aesthetic work was traversed by the question of how we can, from within the arts, replicate complex experiences of our societies, providing metaphorical ways of dealing with the 'invisible wound'. What was initially an inspiring and stimulating review of multiple aesthetic languages, also found support in other fields of thought; some of the most inspiring works were: in the artists (including for example Ai Wei Wei, Lola Arias, Daniele Finzi Pasca, Maximo Corvalan Pinchera), in neuroscience (including for example Mellissa Walker, Bessel Van der Kolk (2014, Gabor Maté (2019), as well as among professionals in the field of research-creation around resilience (including for example Susana Pendzik (2008; Pendzik et al. 2016), Armand Volkas).

I was lucky enough to work with Frederic Vinot and Jean Michel Vives, co-directors of the Master's in Therapeutic Mediation through Art at the University of the Côte d'Azur in Nice. Both psychoanalysts and artists of great sensitivity. Their clinical look at the work that until then I had developed intuitively was very revealing for me. Vinot underpinned my reflection on the two-dimensional condition with which a traumatic experience is imprinted on us: as a flat image without nuances, invariable, which we perceive as stagnant and before which we are stagnant too. Our position and that of others, as well as the role of reality, seem frozen to us and define our position and reaction to experience. Images haunt us regularly without being able to have new perspectives. Until finally, we dissociate as a defense mechanism.

Building a model implies *standing up* a scene, giving it volume, thickness, and structure, thus moving from the two-dimensional mental image of the experience to the three-dimensional plastic creation and theatrical narration. Our story takes on layers and becomes part of a broader scene that contains it but where it does not exist exclusively, not allowing us to distinguish other scenes. This movement gives perspective to the person, creates a respiration between what has been lived and the story and gives a distance where even play and humor find their place.

Vives added a key notion: the traumatic experience takes us by surprise, bursts into our psychic apparatus producing an experience of psychic effraction, like a mirror that breaks into pieces and becomes impossible for us to reconstruct in its initial form. It is a nightmare experience, a loss of bearing. It is a passage from the dream space (which allows us to organize the loose ends of our experiences in a dreamlike space, giving meaning to the nightmare (where that order is transformed into meaningless chaos). The model allows to gather at the same time-space multiple fragments that in reality do not necessarily coexist. This metaphor procedure produced by the model would give the person a space to give a new shape to the experience, going from the nightmare (the trauma) to a dream (the model).

It was thus that the project began to unfold in new directions. In 2019. *Eighty Bullets in the Wing* was awarded two residencies by the French Institute of France and Argentina and by the French Centre for Contemporary Art Centquatre in Paris. With a team of Argentinian and French artists we developed the French version of *Eighty Bullets in the Wing*.

Presenting the model to a heterogeneous and multicultural audience confirmed for me an intuition that those passionate of theatre share: drama is universal. In conversation with colleagues and the French audience, people commented on their interest in the question of what the role of art could be in responding to traumatic experiences like the ones France was going through at that time – terrorist attacks in Paris and Nice, migration, and the Yellow Vests Protest, the series of populist weekly protests advocating first for economic justice and later encompassing a call for institutional political reform.

This led me to realize the vital need of working from the arts to propose pertinent questions to intimate and collective traumatic experiences but also the importance of facing horror not approaching frontally but indirectly, through an aesthetic transposition. The creation of narrative three-dimensional objects in a reduced scale, with the interaction between our lived experience with many other contemporary events happening at the same moment, could bring the possibility to disactivate the toxicity of a memory through metaphors, drifts, and humour.

III.

A new stage was beginning: I understood that it was worth using the performances of my Argentinian project to say, 'this is the story of my generation, this is my way of telling it and these are my questions; what are yours? What are the necessary questions for the French society? What is important to talk about in France today?' Both for native French people as well as for those who migrate there. This was the beginning of the *Intimate Model's Laboratory* (Table 29.1).

The Lab was conceived during a residency at the *Cité Internationale des Arts* in Paris to propose, to those who have had painful experiences, a space to meet. By exploring together with other people creative tools linked to the theatrical practice I belong to (movement theatre, objects, masks), they develop a metaphoric approach to the event, looking for new or renewed ways of 'reconstructing' the event by exploring and working with the way they/we embody their experience and the narration of it. Putting the emphasis not necessarily on what they know or have from the experience but rather working to elaborate the elements that are missing from their scenes and that are necessary in order to tell whoever is in front of them who we are, knowing that the viewer's eyes become a prism that scatters light and a mirror that returns our own image to us. Three principles guide the intense, five-day process where the participants move, make their models, and in groups reflect on Representation, Memory, and Social Bond.

Representation

The team uses the dynamics of movement theatre (expressive game work, object theatre) to seek out the universe conforming each intimate model and begin to work on the notion of transposition and diversion. This stage allows the participants to understand that a distance is necessary to be able not only to observe the experience, but also to be able to intervene aesthetically in it. This gap allows play, humour and fiction and opens a space to deactivate toxic or harmful elements that the experience brings to the person.

Memory

Composition and production of models. Each participant develops his modelling work according to his personal experience. Although each creation might be purely individual, the project seeks to create dialogues and exchanges where each model can be nourished by the work of the other participants. This stage focuses on exploring with the participants the different ways in which memories are formed in us (understanding memory as a phenomenon in permanent movement and transformation, modelled by the dynamic relationship between the person, the family, social, cultural context, the media, etc). In the same way, the group explores different artistic procedures to use memory as a source of creative resources.

Social Bond

Each person is involved in the conception of the model, research, recovery of existing files, or creation of missing or imaginary files, graphic design of the model, realization in volume (3D), definition of the points of view and narrators, creation of the story and individual or group presentation of the model to the public. A condensed and common time is dedicated to the interaction between participants for reflection, documentation, creation, and presentation of models to the public. I consider that every personal history is, in the same way, a social history. This is why the construction methodology of the models is based on integrating the lived experience into a space in which many other scenes are happening at the same time, many of them real, many others imaginary. This allows both the person who creates the model and the person who observes it to move together, identifying themselves or taking distance.

The week's work concludes by sharing the models with viewers. The participants are invited to present, although they always have the freedom to choose if they want to do it and in what way. At this point two aspects are important. Firstly, what a person presents when sharing her model is not the story he or she has lived (which is personal and belongs solely to him or her) but what he has been able to do with it. We cannot change the story but our relation to it, as Jean Paul Sartre, the French philosopher, says: 'We are what we do with what they have done of us'. In this re-writing we seek to playfully be in dialogue with the experience, generating, at the same time, the distance, and the closeness necessary to understand both individually and collectively the great suffering we go through, trying to pass from the wound to the scar, as well as trying to create dams so as not to repeat the social wounds that have been inflicted on our societies. Secondly, each narrator has the possibility of leaning on the group when presenting: the partners can be co-narrators, musicalize, manipulate, create atmospheres, and light the scenes.

IV.

Today, laboratories are being developed in collaboration with partners in Europe, Latin America, and North America. What was initially thought of as an experience to explore issues related to social trauma, and the way in which victims of these events and artists could treat it, has been expanding into encounters with different audiences and communities.

TABLE 29.1 Intimate Models Lab, process

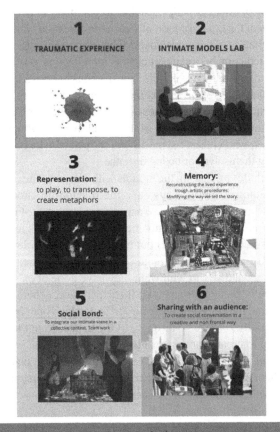

INTIMATE MODELS LAB

REPRESENTATION, MEMORY, SOCIAL BOND

"We are our stories printed in our bodies. Sometimes our story is like wings, sometimes it is a chain holding us down. Finding a way to narrate our story is an important tool to heal"...

1
TRAUMATIC EXPERIENCE

2
INTIMATE MODELS LAB

3
Representation:
to play, to transpose, to create metaphors

4
Memory:
Reconstructing the lived experience trough artistic procedures. Modifying the way we tell the story.

5
Social Bond:
To integrate our intimate scene in a collective context. Team work

6
Sharing with an audience:
To create social conversation in a creative and non frontal way

www.maquetasintimas.com

Since its creation in 2019, multiple laboratories have taken place in France, Switzerland, Belgium, Canada, the United States, Mexico, and Argentina. There have been many institutions, associations, companies, museums, universities, art centres, and art-science festivals that have expressed themselves through the models. Depending on the population to which the laboratory is directed, the proposal is defined and, in dialogue with the communities, the groups of participants are agreed and defined. Some of the most relevant collaborations have been with: The French Association of Victims of Terrorism (AfVT) (the participants were victims of the attacks suffered by French society in 2015 and 2016 of Charlie Hebdo); Bataclan, Nice; the French Institute; the Universities of Paris, Nice, Toulouse, Mexico, Concordia (Montreal); as well as artistic companies Miami New Drama, Cie Gare Central and Cie Bellova-Iacobelli (Belgium), among others.

By working with a wider range of themes, conflicts, identities and capacities within each community I've found a common denominator: the need not only to be able to 'externalize' and make visible the 'knot' (a personal, family or social experience that has not been able to be elaborated but kept stagnant inside a personal or social body) to oneself and others, moving the pain from the inside to the outside, but also to set in motion, to carry out a work of transposition. This means, the metaphorical procedure that allows for rewriting by re-positioning oneself in a place from where we can reinvent the way of narrating (to others and to ourselves) the lived experience.

The laboratories seek to create conversation in relation to issues such as migration and uprooting, gender, intrafamily, abuse, violence, environmental issues and, determined by the present times, the problems arising from the pandemic that we have all gone through, as well as the invisible wounds that COVID-19 has left in its wake. In the same way, the first laboratories focused on recovering past experiences in the question of how to modify our perception and interaction with them. Currently the laboratory seeks, through models, possible ways of facing the future, in which to develop new perspectives and useful tools to restore lost aspects of the social bond, as well as create new senses of community, common good and collective spaces.

References

Gershanik, P. (n.d.). *Intimate Models Teaser*. Available: Intimate models Teaser Video – https://youtu.be/2DJfR-ih0nU. [Accessed: 23 August 2023].

Institut Francais (2019) *Pablo Gershanik*. Available: www.institutfrancais.com/fr/magazine/portfolio/pablo-gershanik-artiste-argentin-en-residence-a-la-cite-internationale-des-arts. [Accessed: 23 August 2023].

Maté, G. (2019). *When the body says no*. Brunswick, Victoria: Scribe Publications.

Ochenta Balas sobre el Ala (*Eighty Bullets in the Wing*. (2016) Available: www.youtube.com/watch?v=hTD-QUAHuPI. [Accessed: 23 August 2023].

Pendzik, S. (2008). 'Dramatic resonances: A technique of intervention in drama therapy, supervision, and training', *The Arts in Psychotherapy*, 35(3).

Pendzik, S., Emunah, R. & Johnson, R. D. (2016). *The self in performance: Autobiographical, self-revelatory, and autoethnographic forms of Therapeutic Theatre.* New York: Palgrave Macmillan US.

Van der Kolk, B. A. (2014). *The body keeps the score: Brain, mind, and body in the healing of trauma.* New York: Viking.

30

HEALING PLACE

Creative Place-remaking for Reconstructing Community Identity

Katy Beinart

With increasing urbanization, cities show the scars of ongoing or recent conflict and violence or spatially reflect exclusion and segregation. Regeneration projects usually focus on economic reconstruction and physical rebuilding; while reconciliation projects usually focus on social and legal issues (Simpson, 1997; Winton, 2004). Creative place-*remaking* – a term I use intentionally in order to acknowledge the 'dynamic and dialectic relationship' between community and place (Anguelovski, 2014) – has the potential to address the 'healing' of places where conflict or violence has taken place, or is an ongoing problem.

This chapter investigates the possibilities for the role of artists and architects as 'healers' of community identity through an approach to place-remaking that foregrounds the community's relationship with 'place' and art and culture as tools for 'healing' place. If public space exists through social interaction (Massey and Rose, 2003), how can creative spatial interventions create these interactions, spark interdependencies between people and space, and build societal/community identities which 'heal' places? I offer criteria for 'healing' place that could be used by place-remaking practitioners, based on Gerd Junne's 'dimensions for an architecture of peace' (2006).

Section One – The Creative Place Re-making Framework

In defining where 'healing' place may be appropriate, we can identify different typologies of place. For instance, conflict to 'post'-conflict settings where a place has recently emerged from war or civil war/terrorism and a consequent slide from political to criminal violence with inherent impacts of mismanaged transitions from conflict (Simpson, 1997); to urban and urbanising settings where there are

DOI: 10.4324/9781003371533-35

ongoing issues of violence and insecurity, psychological hurt, material deprivation and symbolic disadvantage (Moser, 2004). There will be similarities between these settings across insecurity, loss of identity, exclusion for some groups, inter-community conflict (Du Plessis, 1999) as well as differences, for example issues of memorialization and the need for large-scale reconstruction, in the case of post-conflict settings. Exclusion and inequality can be seen as forms of 'structural violence', that is, 'violence built into the structure of society, showing as unequal power and consequently unequal life chances' (Galtung, 1969, p. 171). This has been shown to relate significantly to 'reactive violence', that is, violence in reaction to violence (Moser, 2004; Winton, 2004). While economic inequality is the primary measure of deprivation, inequality entrenched in the spatial structure of society can increase and contribute to the factors that cause reactive violence. Approaches to tackling urban violence vary widely and include criminal justice and public health interventions, conflict transformation approaches, urban renewal, environmental design, community security initiatives and other forms of social and spatial urbanism (Moser, 2004; Matzopoulos et al., 2020; Brown-Luthango, 2020). Post-conflict development is a huge and complex field, but key initiatives include peace-building, reconstruction and reconciliation (Junne, 2006b).

These places can hold 'spectral traces' (Till, 2009) as trauma memory – and creative practices can work ethically with place and memory to represent 'how these past presences occupy the realities of our lived worlds' (Till, 2010, p. 9). Visible transformations, through creative place-remaking activities, have a role to play in breaking down the barriers of exclusion and inequality, building new interactions and identities, and therefore reducing violence. In situations of urban violence and conflict, artists often look at the effects of the conflict both on the community and on the urban fabric and respond to this with site-specific works (Burnham, 1998). Artist Guillermo Gomez-Pena suggests that public artists are 'border crossers, cultural negotiators and community healers' (Lacy, 1995). These processes are more likely to be defined as 'public' or 'socially engaged' art but they have crosscutting impacts.

To understand the potential of these impacts I have defined a set of criteria, based on Gerd Junne (2006a) list of 'dimensions' for an 'architecture of peace' (Junne, 2006a, b). *Architecture of Peace* asks whether post-conflict physical reconstruction can make 'a broader contribution to peace, mutual understanding and recognition, to intensification of exchange, and to a common identification' (Junne, 2006b, n. p.), through providing a support system in a community – 'bonding' social capital (relationships within groups) can be a springboard to 'bridging' social capital (relationships between groups) (Junne, 2006b, n.p). Junne (2010, p. 20) therefore asks whether post-war reconstruction can be done in such a way that creates more peaceful relations among inhabitants, suggesting that the design of the physical environment can influence social relations. As Junne's criteria are architecturally based, I have adapted these to cover a broader creative place-remaking practice framework.

The criteria:

1. Does the work create transactions, new networks and interactions?
2. How is temporality considered? Does the work create a safe space for a specific period?
3. How is the memorialization and cultural context considered?
4. What is the decision-making process – in terms of participation, power and control?
5. How does the work build skills, increase social capital, or connect practitioners?
6. What new visual identity is created? What view of place does this promote? How does it foster a sense of belonging or identity?

Using these criteria, we can test questions such as:

1. How can temporary creative interventions remake place?
2. Can a link be established between creative interventions and improved social or community relations?
3. What sort of interventions can best be used to strengthen place identity?
4. What interventions can best preserve the specificity of places?
5. How do these interventions alter interactions and create new opportunities?

However, creative place-remaking practices need to engage with critical debate and not just advocate solutions. The following are possible pitfalls practitioners may encounter:

1 – A spatial/artistic intervention can have a 'healing' role as it can reinstate people's sense of identity and belonging (Hasic & Roberts, 1999). However, these interventions need to take place as part of holistic approach, not just as a 'sticking plaster'. Isolated improvements may just push problems elsewhere.
2 – Another danger is promoting 'miracle healing', or believing that because the external appearance is improved, social realities are as well: a nostalgia for the past without encountering present realities, where 'newcomers dwell imaginatively in that past they no longer live with' (Gingell, 2000).
3 – There is an increasing recognition of the role that art can play alongside architecture in transforming spaces of memory, but negotiating the complexities of memory in a post-conflict situation needs delicacy as processes such as memorialisation can re-open 'raw wounds', instead of healing.
4 – Art for healing can encounter different challenges. There is a danger in describing 'healing' attempts as art, precisely because audiences and participants expectations may then be raised. The difficulty is to establish effective criteria for measuring success. Projects evaluated as art or architecture face different criteria to those evaluated as development practice.

5 – The healing process is like a 'treatment period'. It may only be needed or appropriate as a temporary intervention and needs to be seen in context of the political environment of the time, the wider society, and a historical point of view. People can become entrenched in memory; spatial projects need to move them on, using a visual language which enables looking forward.

6 – Careful reading of the cultural context is necessary if true representation is to take place. This is made more difficult by the overlapping multiplicities of cultural contexts which exist in cities.

7 – The issue of who controls the process of (re)construction, of healing, in fact of memory, is an ongoing debate in public-art practice (Gingell, 2000; Miles, 1997). Artists have the potential to challenge this but must make a difficult choice about their complicity in projects. However, grassroots projects can be seen as too radical, which can make them difficult to sustain (Landry, 2000).

Section Two – Case Study: District Six and the Creative Remembering of Place

In 1948, in South Africa, the National Party, a white supremacist party, came to power and in 1950 that Government, in pursuit of its policy of apartheid, passed the 'Group Areas Act' which made it possible for them to legally force residents of racially mixed communities into new, racially segregated districts (Western, 1981; Bohlin, 1998).

District Six (D6), an area of some 3700 buildings on approximately 104 hectares of central Cape Town with a multi-ethnic community, thriving businesses, and a strong cultural scene, was one of the first areas to be fragmented and devastated. Whole streets were flattened, and 55,000–65,000 residents were ordered to move to hastily erected shacks in townships on the edge of the city, where they were dispersed, and where they encountered violence and insecurity (which remain to this day). D6 was not the only area subject to this treatment but became the most high profile because it was the largest and most publicly visible area close to the city centre. Some D6 residents, who had a strong identification with D6 as 'place of origin', had lost a symbol of their identity, one which had associations going back seven generations to the emancipation of enslaved peoples (Sauls, 2004). The 'sacred space' of the city centre had become a white place (Western, 1981).

The campaign to reinstate the land began soon after the first removals, culminating in a conference in 1989 with academics, activists and artists, which lead to the decision to form the District Six Museum Foundation (Minty, 2006). Its aims were to keep the memory of D6 alive and honour all those who had been affected by forced removals. After the fall of apartheid in 1991, the Foundation worked with former residents and artists to create a temporary exhibition in a number of sites around the city. One of these, in a former Methodist church in the old D6 site, became what is now the permanent District Six Museum as a volunteer force of former residents refused to let the exhibition close (Eager, 2005). Almost

30 years on, it is an award-winning community museum, a venue for conferences, seminars and book launches, and has expanded from its original building to include *The District Six Museum Homecoming Centre.*

The level of artistic, academic and community interest generated by the District Six Museum and other initiatives is reflective of the perceived symbolic importance of the site in South Africa today. D6 has alays been a place in which, and over which, struggles have been fought: 'fights which have ploughed deep furrows into the South African landscape, and from which have grown cultural and social practices... clearly the use to which District Six, in all its aspects, is to be put is of significance far beyond its size' (Mostert, 1992, cited in Layne in Meyer & Soudien, 1997). This idea of place as mirror or as sign has led to what some perceive as the 'mythologizing' of the true nature of the district. While former residents recall that 'all the religions were there. They all shared something' (Le Grange, 1996), these views present 'simplifications of or selections of more complex histories' (Eager, 2005). What is certain is that D6 had a wide cross section of race, class and religion, and became known for its cosmopolitan culture and as a hotbed of political activism.

Spatially, the former district was a successful example of 'Urban Place', whose streets combined domestic and public life and became a playground where children would interact, providing opportunities for interaction and encouraging mixing between the diverse occupants and users. The spatial approach was also crucial to the success of the museum: it appropriated a former site of resistance, previously used as a meeting place for banned anti-apartheid meetings, and it was positioned at a strategic interface between D6 and the city centre (Bohlin, 1998).

Since the 1970s, artists and community groups had been using cultural projects and dialogue to challenge the government and build resistance: 'the use of culture as a form of social activism was promoted heavily' (Minty, 2006). Spatial projects began to emerge; communities began building 'people's parks' in townships. The roots of culture as a tool of resistance could be traced further back to traditions created by former residents of D6 to find meaning and build community (Minty, 2006; Western, 1981). In 1997, the artist Kevin Brand curated the first *District Six Public Sculpture Festival*, bringing together artists, museum staff and former residents to work on over 90 artworks, using memories of the former district and sited on its empty land (Meyer & Soudien, 1997):

As a temporary project, it ... successfully engaged with the effect of Apartheid planning on communities and with the memory of people and place. In the process, it raised again the complex questions around the history of Cape Town and the effects of forced removals on all people in the city–Black and White.

(Minty, 2006)

The use of public art in the *Sculpture Festival* reflected a cultural activism. Roderick Sauls' piece, *Moettie My Vi'giette* (Don't forget me) (Figure 30.1) paid

FIGURE 30.1   Roderick Sauls, *Moettie my Vi'giettie* (2005) (image of), District Six
Sculpture Festival (1997). Image credit: The District Six Museum (2005).

tribute to the symbolism of the banning of the *Carnival* from District Six. Pieces of
the work, blown by the winds and creating sounds, 'added to the pathos' (Soudien
& Meyer 1997) experienced by former residents visiting the festival. By daring
to re-enter the site, the artists' works could interact with the environment of the
place, whereas inside a gallery they would have been static pieces, dislocated
from the temporal and environmental. However, Sauls (2004) has critiqued the
Sculpture Festival for its representation of race: as he notes, the majority of artists
were white, and were representing a popular version of the past, without any direct
experience of D6. Additionally, the lives of more privileged former residents were
not represented, creating an imbalance.

Returning to the criteria, and questions set out at the beginning of this chapter, how successful was the *D6 Sculpture Festival* and subsequent activities of the D6 Museum Foundation in using creative place-remaking to 'heal' place? How did these interventions and practices consider temporality, memory, cultural context, and visual identity? What were the processes of decision making and participation, and how were new interactions and skills created? Was a new or renewed sense of belonging or identity an outcome of the work?

Section Three – Practice Points

Practice Point One – Temporary Place: How Temporal Interventions Remake Places

The District Six Museum was the result of a collaboration between many people, giving it a formality and strength, which has allowed the D6 Museum Foundation to achieve not only its short-term goals but also its longer-term aim of enabling former residents to return to live in the district, through a restitution programme (D6 Museum website, 2023). The exhibits in the museum include a display of street signs rescued from the now destroyed streets, a huge map which was created by former residents with artists, photographs and displays showing what life was like in the district, a growing oral history archive and many texts of memories on display, giving visitors an insight into what it was like to live under apartheid, and how its legacies continue to affect people's lives. The ongoing education programme continues to share these legacies and memories with both local and international audiences.

The challenge facing the District Six Foundation is how to use this strong sense of collection and exploration of memory, and the practice and process of the *Sculpture Festival*, in the context of the permanent site, and the 'return' of the former residents. A *Memorial Park*, to 'contribute to the process of healing' (Le Grange, 2003), designed by architect Lucien Le Grange through activities with former residents has been endorsed by the museum and is part of ongoing work to develop the area (District Six LSDFBA Report, 2022; Malan & Soudien, 2002). The difficulty facing the designers, planners and artists of the future D6 area is the ongoing contestations over place and memory in post-apartheid South Africa. As Maddell and Murray (2019), quoting Bach (2013), note, heritage landscapes are often both imperfect and contentious and hold tensions. In the case of D6, these are between two potential approaches: of conservation, memorialization, retrieval and recovery *or* of development (considering the site as a *tabula rasa*) (Malan & Soudien, 2002). Perhaps what the ongoing temporary, unfinished space of the *Memorial Park* offers is an agonistic space where these tensions and contestations can be explored, including continuing debates over the representation of race and identity in South Africa.

Practice Point Two – Symbolic Healing: Collaborative Creativity to Reclaim Relations in Place

The deserted site of the former district can be seen both negatively, as a giant wound, and positively, as a symbol of apartheid's ultimate futility and failure, and a symbol of healing. The site has been called 'salted earth': a reference to the emotion carried in the land, and the protests which made it impossible to sell (Bohlin, 1998; Bedford & Murinik, 1997). However, by returning and remembering, the former residents can begin the process of healing, through creative work with the spectral traces of trauma.

The *Sculpture Festival* healed in several ways. It provided spaces within which to grieve and find solace, places for rest and contemplation and it reclaimed the 'salted earth'. Andrew Porters swings brought children's laughter back, symbolizing the triumph of human spirit. By creating new interactions, a new identity for the place was being born. The artists were in a sense recreating the interactions that apartheid destroyed. Through participation, a sense of protection over the site was instigated. D6 Museum's participatory approach and interactive exhibitions have made the imagined and real, past and present D6 communities visible. While engagement is limited in one sense to those who 'belong' to the former territory, it is also inclusive: it offers a 'forum for healing' for all who suffered under apartheid. Vicarious memories present the possibility for transmission of feelings of belonging, so others identify with a certain group (Bohlin, 1998). These vicarious memories extend beyond the borders of the city to become a symbol of the possibility of equality and justice, and of healing, for the nation and beyond.

The District Six Museum and *Sculpture Festival* had a specific healing role in working with former residents' memories and providing a safe place for these to collect and be exhibited through artworks and exhibitions. However, while the potential of this approach offers healing through interaction, using culture as social activism can lead to work being critiqued therefore losing status (Burnham, 1998). Ongoing debates over the value of socially engaged practices mean that practitioners doing the work have to make choices over what is important to them. In the case of District Six, artists were able to develop networks which offered 'an alternative to mainstream art-making' in Cape Town (Hayes-Roberts, 2020, p. 126).

Practice Point Three – Strengthening Place: Re-appropriating Space to (Re)build Identity

How have projects like this had a 'healing' impact through building place identity? District Six Museum directly rebuilt a cultural heritage which re-makes community and place identity. However, projects must remain socially active in order to remain spatially present. Therefore, a further challenge for the museum is that of how to ensure sustainability while not losing their connection with the community (Eager, 2005). To sustain their spatial presence, projects must establish funding and support, which adds new types of responsibility and brings claims of 'selling out' and 'professionalising' and risking their connection to the community who helped establish them in the first place. Therefore 'mainstreaming', while it can create more opportunities and a longer-term benefit for the community, carries with it challenges of negotiating how this is done, and who is being represented. On the other hand, if they remain at grassroots level, they can face problems in getting recognition and sustaining funding (Landry, 2000). We tend to want to see creative spatial interventions as permanent, but perhaps instead should see them as the aesthetic of a particular moment. These interventions can only be seen as part of a process of healing, not as a permanent solution to problems – they can create a safe space for a 'treatment period'.

Practice Point Four – Specifying Place: Preserving the Specificity of Place

The site was particularly evocative for all involved and gave those creating and inhabiting the space a unique means to reinterpret the site, while preserving its authenticity, especially in the post-apartheid Cape Town context of culture and art as a form of social activism (Minty, 2006). This gave the Foundation a legitimate reason to use visual and spatial creative tools to emphasize the former identity of the place. On the other hand, it also gives those creating new permanent public art works and architecture for the site the challenge of developing a visual language which has relevance for future generations. Therefore, the use of memorialization and cultural context in healing place needs to be carefully balanced with allowing space for future specificities of place to emerge, acknowledging that places are not static.

Practice Point Five – A Place for Transactions: New Interactions, Opportunities, and Networks

The value of these projects, in working outside art and architecture conventions, is that they can take independent stances, and be provocative, opening up debate. Through attracting different users, and visitors, projects can bring investment into an area, increasing its value and building community worth. Through making visual statements, the projects create possibilities for new interactions and for community members to spread their horizons. Spatial and artistic interventions often make visible debates over place, highlighting place, which can translate to social and economic benefits. Those who maintain involvement with the project develop new skills, increasing social capital, and becoming part of a 'community of practice' (Wenger, 1998). These projects often work in an underground economy, exchanging 'gifts' of time, presence, skills, and participation, which have no commodity value (Beardsley, 1995). This avoids the dilemma of artists 'becoming handmaidens to service economy' (King, 1996).

Conclusion: The Medicine Cabinet – Implications for Theory and Practice

Returning to Junne's 'architecture for peace' which suggests that design has a part to play in the creation of healing social relations in post-conflict settings, this chapter has sought to think through how Junne's criteria, adapted to creative place-remaking, offer ideas for 'healing' place through the rebuilding of interactions and identities. I have defined key criteria which must be considered, including: how practices and projects consider temporality, cultural contexts, and memory; what decision-making processes are used and how new interactions, transactions and networks are created; and what new visual and social identities are produced. Exploring the case study of the D6 *Sculpture Festival* and further activities, I have addressed questions of how criteria can be successfully applied to strengthen identity, create new interactions, make symbolic forms of healing and preserve specificities of place whilst allowing new forms and identities to emerge. However, the chapter and case study have also suggested some reasons that creative placemaking may be less successful, for instance how they sustain the process of rebuilding identities, and also how spaces for tensions to continue to be worked out are made in ways that are safe whilst also allowing difference. Other potential reasons discussed for failure of interventions include lack of relevance to a place, failure to carefully examine the criteria discussed above. It is vital to understand the cultural context(s), and memory and acknowledge the positionality of who is being represented, by who.

There are other contexts where it will not be possible to even consider or match the criteria. Where basic infrastructure is dangerous, it would be difficult to create a safe place. Where there is a strong resentment of 'outsiders', a 'stranger in the streets' may be unwelcome. Experienced practitioners using visual and spatial methods may need to take on different roles, as observers, facilitators, or documenters, listening to locally-generated imaginations about how to improve the city (Robinson, 2006). We can also speculate the most likely arenas for success, for instance where there is strong community motivation to explore memory and identity; or in urban settings where a community feels spatially excluded and ignored. Such interventions can also break down divisions of definition and encourage debates as part of post-colonial urban theory (Robinson, 2006) transcending global divides.

Tactics used in countries in transition from violence could inform debates in the developed world, potentially transforming debates on participation and urban development in urban theory (Blundell-Jones et al.; 2005; Borden et al., 2001). In adapting Junne's work on architecture for peace into creative place re-making, through work on healing place I offer a more holistic and cross-disciplinary approach to tackling urban violence, linking environmental justice action to design and artistic practices, and understanding the need for what Anguelovski (2014) calls the 'physical and psychological dimensions of community rebuilding and place remaking' to be met.

Acknowledgements

My thanks to Gerd Junne for sharing his work. Thanks to District Six Museum for permission to use the image featured in this article.

References

Anguelovski, I. (2014). *Neighborhood as Refuge: Community Reconstruction, Place Remaking, and Environmental Justice in the City.* Cambridge, Mass.: MIT Press.

Bach, J. (2013). 'Memory landscapes and the labor of the negative in Berlin', *International Journal of Politics, Culture, and Society*, 26(1), pp. 31–40.

Beardsley, J. (1995) *Gardens of Revelation: Environments by Visionary Artists.* London; New York: Abbeville Press.

Bedford, E. & Murinik, T. (1997). 'Re-membering that place: Public projects in District Six' in Soudien, C. & Meyer, R. (eds.) (1997) *The District Six public sculpture project.* Cape Town: The District Six Museum Foundation.

Blundell-Jones, P., Till, J., & Petrescu, D. (eds.). (2005). *Architecture and Participation.* London: Spon Press.

Bohlin, A. (1998). 'The Politics of Locality: Memories of District Six in Cape Town' in Lovell, N. (ed.) *Locality and Belonging.* London: Routledge.

Borden, I., Kerr, J., Rendell, J., & Pivaro, A. (eds.) (2001). *The Unknown City.* Cambridge: MIT.

Brown-Luthango, M. (2020). 'Neo-liberalism (s), socio-spatial transformation and violence reduction in Cape Town—lessons from Medellin', *Cogent Social Sciences* 6: 1827524.

Burnham, L. F. (1998). 'The Artist as Citizen' in Burnham, L. F. & Durland, S.(eds.) *The Citizen Artist: 20 years of art in the public arena-an anthology from High Performance Magazine 1978–1998.* New York: Critical Press.

District Six Local Spatial Development Framework Baseline and Analysis Report (2022). City of Cape Town.

District Six Museum website (2023). Available: www.districtsix.co.za/. [Accessed: 23 March 2023].

Du Plessis, C. (1999). 'The Links between Crime Prevention and Sustainable Development', *Open House International*, 24(1).

Eager, C. (2005). Looking to the Past, Thinking towards the future: The District Six Museum and the Redevelopment of 'Community' in Cape Town. MA Development Studies. *International Development Centre*, Oxford University.

Gingell. (2000). 'The Barry Job: Art, sentiment and commercialism' in Bennett, S. & Butler, J. (eds.), in *Advances in Art and Urban Futures Volume 1: Locality, Regeneration and Divers(c)ities.* Bristol: Intellect.

Hasic, T. & Roberts, A. (1999). 'Opportunities for sustaining human settlements in a post-conflict zone: The case of Bosnia and Herzegovina', *Open House International*, 24(1).

Hayes-Roberts, H. E. (2020). *Frameworks of Representation: A Design History of the District Six History of Cape Town.* PhD. University of the Western Cape. Available: http://etd.uwc.ac.za/xmlui/handle/11394/7419. [Accessed: 25 March 2023].

Junne, G. (2006a). 'Architecture.doc'. Word file sent as e-mail attachment, received 20 August 2006.

Junne, G. (2006b). 'Indesem-Delft-9 Juni 2005'. PowerPoint file sent as e-mail attachment, received 20 August 2006.

Junne, G. (2010). 'Designing Peace', *The Broker*, 20/21, July 2010, pp. 30–33.

King, A. D. (1996). *Re-Presenting the City*. Basingstoke: Macmillan.

Lacy, Suzanne (ed.). (1995). *Mapping the Terrain: New Genre Public Art*. Seattle: Bay Press.

Landry, C. (2000). *The Creative City: A Toolkit for Urban Innovators*. London: Earthscan.

Le Grange, L. (1996). 'The Urbanism of District Six', *The last days of District Six: photographs by Jan Greshoff* (exhibition catalogue). Cape Town: The District Six Museum Foundation.

Le Grange, L. (2003). *District Six Heritage Impact Assessment*.

Le Grange, L. (n.d.) Architects website: www.lucienlegrangearchitects.com/district-six-memorial-park. [Accessed: 23 March 2023].

Madell, C. & Martin, M. (2019). 'Contested collective memory in the segregated city of Cape Town', Ristic, Mirjana & Frank, Sybille (eds.) *Urban Heritage in Divided Cities: Contested Pasts*. London: Routledge.

Malan A. & Soudien C. (2002). 'Managing heritage in District Six, Cape Town: conflicts past and present' in Johnson, William Gray, Beck, Colleen M; Schofield, A. J (eds.) *Matériel culture: the archaeology of twentieth-century conflict*. London: Routledge.

Massey, D. & Rose, G. (2003). *Personal views: Public art research project*. Milton Keynes: The Open University.

Matzopoulos, Richard; Blocha, Kim; Lloyd, Sam Lloyd; Berens, Chris; Bowmane, Brett; Mayers, Jonny Myers; Thompson, Mary Lou. (2020) 'Urban upgrading and levels of interpersonal violence in Cape Town, South Africa: The violence prevention through urban upgrading programme'. *Social Science & Medicine*, 255, 112978.

Meyer, R. & Soudien, C. (eds.) (1997). *The District Six public sculpture project*. Cape Town: The District Six Museum Foundation.

Miles, M. (1997). *Art, Space and the City: Public Art and Urban Futures*. London: Routledge.

Minty, Z. (2006) 'Post-Apartheid Public Art in Cape Town: Symbolic Reparations and Public Space', *Urban Studies*, 43(2).

Moser, C. (2004). 'Urban Violence and Insecurity: An introductory roadmap', *Environment and Urbanization*, 16(2), October 2004.

Mostert, N. (1992). 'Frontiers: the Epic of South Atheca's Creation and the Tragedy of the Xhosa People' cited in Layne, V. (1997) 'Whom it may, or may not, concern, but to whom this appeal is directed anyway' in Meyer, R. & Soudien, C. (eds.) *The District Six public sculpture project*. Cape Town: The District Six Museum Foundation.

Robinson, J. (2006). 'Inventions and Interventions: Transforming Cities-An Introduction', *Urban Studies*, 43(2).

Sauls, R. K. (2004). *Identity: A study of representation with reference to District Six*. MA Fine Art. University of Cape Town. Available: https://open.uct.ac.za/handle/11427/8009. [Accessed: 25 March 2023].

Simpson, G. (1997). 'Reconstruction and reconciliation: emerging from transition', *Development in practice*, 7(4).

Till, K. E., Ed. 2010. *Mapping Spectral Traces*. Edited volume and exhibitions catalogue. Virginia Tech University and Blacksburg, VA USA.

Till, K. E., & Jonker, J. (2009). 'Mapping and excavating spectral traces in post-apartheid Cape Town', *Memory Studies*, 2 (2009).

Wenger, E. (1998). *Communities of practice: learning, meaning, and identity.* Cambridge: Cambridge University Press.

Western, J. (1981). *Outcast Cape Town.* London: George Allen & Unwin.

Winton, A. (2004). 'Urban Violence: A Guide to the literature', *Environment and Urbanization,* 16(2).

31

A RECONCILIATION FRAMEWORK FOR STORYTELLING

A Trauma-Informed Placemaking Approach

Katie Boone, Wilfred Keeble, Rita Sinorita Fierro and Sharon Attipoe-Dorcoo

Introduction

Throughout the years, we (Wilfred and Katie) have been exploring and developing our questions together around what reconciliation looks like here in the United States, specifically what it means here in Mahkato (Mankato), Minnesota. The Reconciliation Sense-Making Framework was born from these explorations. It started with an idea for a truth, healing and reconciliation game, formed through kinship between Wilfred Keeble and Katie Boone. The game is designed for story sharing and co-inquiring, learning from the truth of what unfolded in Mankato. This place is the site of the largest mass execution on United States soil, and the game works to elicit new questions through new understandings from the intercultural spaces where the oral histories can be shared. We have found that it is through the questions that we find our way to the 'how' we do this work.

In this chapter, we offer up how the Reconciliation Sense-Making Framework can be used for trauma-informed placemaking through a case study from Mahkato, and then two applied examples of use in an organizational and personal context.

Reconciliation Sense-Making Framework

Frameworks help us to reflect on our thinking as we approach our life and work. The *Reconciliation Sense-Making Framework* (Table 31.1) for trauma-informed placemaking was developed in 2012, through the question of 'What does collective healing look like?' First Nations people have explored questions like this for centuries; Katie's journey into this inquiry first began on a hill in Mankato, with a small intercultural group of participants led by a Dakota elder, engaging participants through storytelling.

DOI: 10.4324/9781003371533-36

TABLE 31.1 *Reconciliation Sense-making Framework*, Wilfred Keeble & Katie Boone (2019)

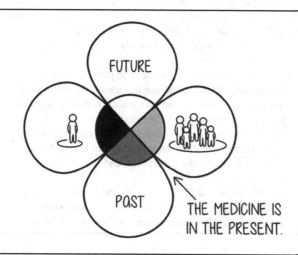

The framework has evolved with a shared sense of clarity and purpose through years of praxis. We are excited to offer this as a contribution to trauma-informed placemaking, situated in our own cultural context, relations, and understanding. We've learned that trauma-informed placemaking requires an awareness of multiple dimensions of understanding. There are functional and tangible aspects of understanding, like resources and planning, but to do this work well, it requires more than that. This work is not mechanistic, it requires great care and long-term relationships, this form of placemaking cultivates fields of caring and belonging. The outer dimensions of the framework are interdependent and interconnected to the center, which forms a Medicine Wheel and represents the present. Black Elk, an Ogalala Lakota visionary and healer said that 'the world always works in circles'. If you would like to learn more about the Medicine Wheel, we encourage you to read *Black Elk Speaks* (Elk & Neihardt, 2014) for deeper learning and insights. We now invite you to co-inquire and learn as we share what this framework has taught us so far (below, and Table 31.2).

Understanding the Framework: Mahkato Case Study

Individual – Deepening Our Roots

Katie: My feet are on stolen land, at a place that holds the deep wounds of broken promises in treaties that were never honored. My heart aches for a space where justice and reunification can be felt, seen, and heard. My mind can't comprehend the harm that has been caused by the people I descend from.

TABLE 31.2 Five dimensions of *Reconciliation Sense-Making Framework*

Individual: Deepening Roots	Understanding ourselves as individuals: who we come from, how we got here helps us discover our places of belonging.
Collective: Nurturing Intercultural Spaces	Decolonizing our minds and hearts in the relationships we nurture; this helps us learn to show up in new ways in intercultural spaces.
The Past: Looking Back to Look Forward	We need to understand the stories and people who are connected to the land that our feet touch, the journeys of our ancestors, and the paths their feet created through the journey that brought us to this place.
The Future: Developing as We Move Forward	When we honor, listen, see, feel, and learn together we develop the capabilities and capacities to learn what it means to regenerate from place, and see place as a participant in our work.
The Present: Cultivating the Soil	The medicine is in the present. Truth and healing are possible when we deeply listen and shift our inner and outer awareness of what needs healing within and among us.

We, in Minnesota, are using the *Reconciliation Sense-Making Framework* to heal our legacy of chronic trauma and historical, unresolved grief across generations (Brave Heart & DeBruyn, 1998). Minnesota holds a complex and traumatic history of colonization and displacement of Dakota people from their homelands. The loss of land breaks sacred connections to the plants, animals and places that hold sacred relationships that are deeply connected to place (Brave Heart & DeBruyn, 1998). The Dakota people have a saying, *Mitakuye Oyasin*, which means, 'all my relations'. This saying recognizes the relationships to all living things, from the plants and trees, the birds, fish and four-legged, from the ants, and to the stars – we are all related and connected to life around us, working in and with nature. This framework has engaged participants in a sense-making process connected to sacred sites, where participants learn from the story of place, while gaining new understandings of the cultural, social, historical, and ecological history of each site and location as we go.

Our process centers the voices of Dakota elders. Participants learn from elders' stories, oral histories are passed on at each site and location. After each site, participants are given Post-It notes and asked to reflect on the questions that are coming up for them. Participants place Post-It notes onto the framework where their question(s) lands within the framework. The framework supports intercultural learning and the development of new practices to bring about the truth of what unfolded here. Stories are passed down in the cultural tradition of storytelling, surfacing the unwritten history of this place. Healing and reconciliation doesn't happen until the truth is shared, heard and honored.

Wilfred: American history is not true at all. It was made up to please the writer in how they view such events and accounts. With the history of Minnesota, you need to ask the original people who lived here. As Dakota people, our belief system makes it possible to remember and honor our ancestors and relatives. It happens daily throughout the *Oceti Sakowin* (Seven Council Fires). When we were placed on reservations, ceremonies were outlawed, we couldn't practice our belief system. We were punished if caught practicing; ailed, starved, made to perform hard labor, tortured, or worse, sent to the Hiawatha Insane Asylum in Canton, South Dakota never to return. Yet Grandma and Grandpa practiced what they could. We remember them for their sacrifices. Many more paid the ultimate sacrifice, we honor and remember them. The Dakota 38 + 2 *Akicita* (*Akicita* is a Dakota word for soldier or warrior) made the ultimate sacrifice, we remember and honor them with an annual ride. The *Dakota 38 +2 Wokiksuye Horse Ride* takes place in December, with riders traveling by horseback over 330 miles, starting from the Lower Brule Reservation in South Dakota to arrive in Mankato, Minnesota, on the anniversary of the day and time the 38 Dakota warriors were hanged (the largest one-day execution in American history, during the Dakota War of 1962). In remembering and honoring the Dakota 38 + 2, we used the horses to help because of their ability to heal.

🎥 80 min Activity: Offering an intercultural learning experience to ground you into the stories being shared with readers of this chapter, we invite you to watch the Dakota 38 + 2 HD Full Length Documentary on YouTube.

Collective – Nurturing Intercultural Spaces

Katie: I come from a place that remembers the aboriginal caretakers, whose people long for reconnection to the ancestors that make up the ground I am standing on. I show up to listen and be of service to the work. At the time of this writing, it has taken 23 years to begin to understand this work. My body is covered in whiteness, my blood carries the memories of my ancestors, and my mind and heart are continuously striving towards a decolonized state – understanding of my place, role, and responsibility. I do my best to help people understand what this work requires of them. Nurturing intercultural spaces requires developing capabilities in people who come from settler-lineages. It requires a commitment to staying in dialogue through the discomfort, to stay grounded in the dysregulation that happens when truth is shared and hard questions are asked with no answers. These inner capabilities help us to stay present, curious, and compassionate in ways that honor the whole truth. We get to the whole truth by deeply listening to the stories and voices of the communities whose histories have not been taught. This deep listening starts to shift us into an authentic, relational 'we' space. In the 'we' space, we no longer center our comfort and privilege, no longer see an 'us and them'. We recognize that the work requires us to stand shoulder to shoulder, seeing all of what is here in the present, shifting our gaze towards a reunified and restored whole – centering a shared purpose that begins to help us find our way, collectively moving towards shared understanding of what is needed to move forward.

Wilfred: When the *Dakota 38 + 2 Ride* came to Crow Creek Agency, I followed the horse, he is what got me into this in 2005. The *Dakota 38 + 2 Ride* is a prayer ride, it provided a whole new understanding for me and brought me into this work. The land tells us that intercultural exchanges were happening long before Europeans arrived. Altars were built for prayer and offerings for all tribal nations, these altars still exist all across Turtle Island. All along the *Dakota 38 + 2 Ride*, there are memories of historical exchanges that show up in the present-day behaviors of people from the settler-lineage, reminding us of what it was like for our relatives as they were marched through towns and exiled from our homelands. The past is alive today with landmarks that help us remember the sacredness of spaces and places where our people took care of each other and where our ancestors and relatives reside.

The Past – Looking Back to Look Forward

Katie: We cannot move towards a more equitable, just and regenerative future if we do not understand where and who we come from. By looking back, we begin to understand our present and become more clear about what it is we need to do in order for future generations to thrive.

Wilfred: Wilfred asks, 'I guess the big question here would be how did we get in this situation to begin with?' A good starting point is the *Doctrine of Discovery*.

History about the *Doctrine of Discovery*: sovereignty is key, it can easily be lost if it is not put at the center of what we are exploring in this work. The *Doctrine of Discovery* was written in May of 1493 by Pope Alexander the VI, this informed the process of discovery in Christians' dominion over the land, and the Indigenous people who inhabited it (Castanha, 2015). The era of discovery removed the sovereignty of non-Christian people, making it so that the inhabited land was instead 'discovered' through 'Christian invasion, conquest, and possession' (Newcomb, 1992).

The stories Wilfred's ancestors passed down, shared it, was a process of 'classifying us Indigenous people as enemies without knowing them'. Westward expansion was fueled by this judgment, making colonialism the norm. These practices were carried out through the tactics and strategies of the Treaties that were signed. Wilfred shares that 'Native people held their treaty agreements, we didn't break our word given to uphold the treaties signed by our ancestors, treaties were broken by the US government'.

The land our (Wilfred and Katie's) work is connected to in Mahkato falls under the Treaty of Traverse des Sioux, signed in 1851. The Crow Creek Agency where Wilfred is from is one of the reservations that resulted from the Dakota women and children who were exiled and taken to Fort Thompson. Wilfred shares that here, 'they endured colonialism, torture, rape and dehumanization'. Wilfred acknowledges that 'to forgive this is hard – truth telling, to forgive anything that happened to Indigenous people is hard'.

The Future – Developing as We Move Forward

Katie: We come together to seek the truth by listening to the stories that were never shared and to feel our way forward on a journey that unfolds one conversation at a time. One of the greatest joys in this work was seeing the faces of Dakota elders and leaders from the *Oceti Sakowin* when they learned white-settler descendants were asking 'how' questions as a result of utilizing the *Reconciliation Sense-Making Framework*. We have learned that the 'how' question is shared across the intercultural spaces, 'how' is the opening that we've found in the reconciliation, restoration, and reunification work here in this place we all call home. 'How' opens up a space of shared not-knowing, inquiry and exploration. We follow the questions exploring the 'how' – here we begin to see what's possible, where we begin to find community. Through the 'how' more voices come together. The land remembers us, it holds and heals us as we gently shape and form the path as we journey together and move forward.

Wilfred: There's a sense of reconnection, of trying to reconnect. There is no right or wrong way to do this, it is how you understand it and express it. Every band is just finding these things out again, we have different ways of saying things, expressing things, each band has their own unique ways of doing it. When our language was attacked after the Westward Expansion, it made it difficult for us to reconnect. We are just now realizing these things again. I don't know how long it is going to take us, but we are in that process.

The Present – Cultivating the Soil

Katie: In order to cultivate the soil, I must work on my own healing first. My family's history was not shared or passed on. My earliest memory of discovering that there is something here for me to learn was when I was very young, at my Grandmother's table, paging through a family scrapbook where I came across a newspaper clipping telling the story of our family and the connection to the history of the Dakota people who were here. I asked about the article and the book was closed, taken away, never to be seen again. When I was 21 years old, I had a mental breakdown. I was scared because I didn't understand what was happening to me. A very clear question came to me throughout this time, 'Who is the Chief?' – having no context or understanding of what this question meant, I asked my father if we have any Native American blood in our veins or on our ancestors' hands. Puzzled by the question, he shared that we have history with our family that is connected to what happened here with the Dakota people, this marked the beginning of my healing and reconciliation journey. Decades later, the scrapbook newspaper article surfaced again, given to me from my grandmother. Remembering this, it led me on my healing journey, reconnecting with place and the original caretakers who helped me heal into my whole self and integrate my history into a larger context. Cultivating the soil enabled me to find my role in this work. For me, trauma-informed placemaking requires understanding reconciliation, 'learning by doing' and showing up to unlearn and relearn my relationship to this place and the Dakota people – decolonizing my mind and heart.

Wilfred: We come from the stars. There are creation stories that remind us that originally the Lakota, Nakota, and Dakota people were one tribe, one band. Our creation story is about how Turtle Island came to be. We have people returning our artifacts back to us that have been found in North Carolina, Canada, as well as South America. This confirms what we have been saying all along. We traveled this continent numerous times, only carrying with us what we needed. Tipi poles and materials like stones for ceremony and tools, we would leave behind for the next band or group coming through. Across the continent, every 15 to 20 miles there are campsites scattered across the land. We shared a lot of things with each other, our belief in the creator is a shared belief across all nations of Turtle Island. Our artifacts are still being dug up. The land holds a lot to be remembered, reminding us of the sacredness of place. We remember these things on the *Dakota 38 + 2 Ride*, and the many rides that take place to remember and honor our ancestors. On these horse rides, we are a Dakota nation on the move across our homelands once again.

Understanding The Reconciliation Sense-Making Framework – Applications

Home for Good Coalition: Organizational Application

I (Rita) used the *Reconciliation Sense-Making Framework* for self-reflective practice during a grassroots initiative: to get unstuck and find the flow.

I co-founded the Home for Good Coalition as a translocal place-based grassroots initiative for national large-systems transformation. Rooted in 30 years of study of systemic racism and the ways in which family-policing system tears financially impoverished families apart, the coalition intended to bring organizations together for systems transformation. We wanted families at the center of our work. Over the course of the 10 years, it seemed hard to: 1) find organizations willing to work across multiple systems (welfare, child welfare, criminal justice, juvenile justice, housing, etc.); 2) place families impacted at the center of the work; 3) retain African American women on our team; 4) organize supporters.

Individual: Deepening Roots

I was born in the United States from Italian parents who moved us back to Italy at the age of 10. In my 20s, my activism was rooted in my own wounds. I felt more in sync with African Americans, because I felt more embraced, loved, and mentored – as a forever outsider – than I had in white spaces. My women of color mentors warned me against pushing versus listening for how things could grow organically.

They also warned me against how whiteness taught constant activity to escape my own trauma.

Collective: Nurturing Intercultural Spaces

My PhD dissertation on Black mothers who lost their children to foster care in Philadelphia grew my desire for community organizing. I co-founded Home for Good Coalition to see transformation on a larger scale. Over 10 years, we learned how to organize supporters and families and how to place family experiences at the center of our decision making in a trauma-informed way.

The Past: Looking Back to Look Forward

Rewind to me age 10, girl born in the Bronx, moved to Southern Italy. I could feel, but not explain, the generations of political, cultural, and economic exploitation. Years later that feeling helped me resonate with African Americans and give voice to what I felt in Italy. During COVID, my ancestral Land called me. My mentors invited me to give in to the call and allow activism to shift towards joy. What if the best thing I could do for the movement was to go back? I couldn't create a future by escaping the past. I started digging into my family history to discover what ancestral healing was itching for resolution. I uncovered secrets: degrading patterns of servitude.

The Future: Developing as We Move Forward

I was stunned when Kimmerer said in Braiding Sweetgrass (Kimmerer, 2022), if white people want to decolonize, we must become 'indigenous to place', choose a plot of land we can devote our lives to, and do it. My long-time activism against systemic racism is always on my heart and soul, but that plot of land, of me, is in Italy. Writing the book *Digging Up the Seeds of White Supremacy* (2022) meant harmonizing my past and my future. Shortly thereafter, I began to plan a place-based initiative that would revive my town's community through organic farming.

The Present: Cultivating the Soil

A year after these reflections, every time Home for Good Coalition tried to jump to the next level, there seemed to be a roadblock. In February, we decided it was time to stop pushing. We paused for three months. When we reconvened, most of us were ready to say goodbye, but something magical happened. We found a way to yield a big result – a trained cohort of parents and youth – with minimum effort: supporting an existing partnership. In only two hours, we had a timeline, a flyer, and a plan. We were back in flow. It's important to pause the work sometimes to ask bigger questions. Are our souls still inspired? Is this the right time? Are we the right people? Cultivating the soil isn't about executing an old plan. It's about taking time to discover a new path.

Identity and Immigration: Personal Application

Trauma-informed placemaking work assumes that there is a connection to place in the work. Therefore, one has to start off by defining what place means in the various contexts of navigation. I (Sharon) express the use of the *Reconciliation Sense-Making Framework* from the context of my identity and connection to place with the continent of Africa. Africa, my homeland, is 'one cultural river with numerous tributaries articulated by their specific responses to history and the environment' (Asante et al., 1985).

Collective

Our connections to place and time allow us to stay connected to our ancestry. As the river flows across the continent to connect each of us in the diaspora, I deepen my roots by remembering my own ancestor Corporal Attipoe who was killed unjustly by the British for demanding promises made to service veterans in World War Two, and his death resulting in the Ghana riots of 1948 (*BBC News*, n.d.).

Past

Traveling across the Atlantic across different time zones to make my home in the United States did not only come with cultural differences, but identity trauma related to encountering racism as I went from the child of my parents, granddaughter of my amazing grandmother, to having an identity defined by the color of my skin. I became a Black woman in the United States.

Individual

So, when I do my work in the United States, the connection of place and healing for me is deepening my roots both within my genetic bloodline to those who fought against injustices both on my home continent in Africa and beyond. Ancestors including Yaa Asantewaa, Kwame Nkrumah, Thomas Sankara, Nelson Mandela, James Baldwin, Audre Lorde, W. E. B. DuBois, John Lewis, and many more. Why is this important, because when we identify ourselves from our inner beings, we are able to cast our visions wide into the expanse of the world with grounding in our ancestry, our identity, and what binds us spiritually and culturally. I am a mom of three kids, two of them Black boys; a foreign-born individual in the United States who wonders if I am better off going back to my motherland from the fear of the future of my Black babies in this country. A home where my everyday work involves providing evidence that will hopefully inform equitable policies, but my advanced degree does not clothe the color of my skin, and the equitable future I dream of can involve highs and lows of trying to scream of a better humanity where racial inequities are not ignored.

Future

In developing capabilities in our trauma-informed placemaking practice we need to reinforce respect. Culturally derived and culturally oriented forms of service can heighten community engagement and participation. Community is an existential aspect of humanity with various ways of being and indigenous wisdom. The innate power of this connection as a means of creating intercultural spaces in order to be present in our practice, and to begin to recognize and harness this gift rather than leaning solely into formal/institutionalized ways of learning or being. Elder Malidoma Somé teaches us that 'This spark of the ancestral flame, which I have brought to the land of the stranger, is now burning brightly' (Somé, 2016). So, as I nurture intercultural spaces in my work, I am reminded of the 'I' of how I show up in my identity, and how that inner work of healing fuels the flame of mind and heart to connect with the stories of others as I build community within my work collectively.

Present

As an individual of Ghanaian origin, the essence of the word *sankofa* a Twi word meaning 'to go back and retrieve' (African Burial Ground), shows a deep connection to memory, learning and place. A grounding to place in trauma healing establishes a need for setting an intention to recognizing the past, naming the present, and working through the trauma without avoidance towards a liberated future for all peoples.

Conclusion

As we all weave our stories together to share how storytelling is impactful in trauma-informed placemaking practices, our goal is that readers will embrace their own stories of place, to deepen our roots, nurture intercultural spaces, co-create futures, cultivate our soil, and tap into our own hearts and the power of place as a participant in this work. Our narratives are informed by the lenses through which we view the world. As we craft these stories, we face the challenge of learning only from our view and not considering our blind spots. If we do not create a thread of our identity to others, and nurture our belonging to place throughout time and the healing of our traumas, we risk missing out on the gifts that these experiences yield to us.

Joe Whitehawk, a Dakota elder, said that the longest road we travel is the 12-inch journey from our mind to our heart, what is known as the *Red Road*. We hope that this weaving of voices, stories, places and experiences helps you begin the journey. We pray that more people listen and learn from the original caretakers of the places we love and call home. Over the course of this writing, there have been Dakota and Lakota elders who have passed that have nurtured the learnings reflected in the stories that are shared here. We honor and remember Dave Brave Heart, his legacy of reconciliation in Mahkato will live on for future generations. We honor and remember Jim Miller, the dreamer who started the *Dakota 38 + 2 Ride*, begun in 2005 and coming to a close in 2022, contributing to 17 years of healing and remembering of what unfolded here. We send prayers for them, their families and communities as Dave and Jim have journeyed into the spirit world and leave behind a legacy of healing and reconciliation. May we rise to the roles and responsibility their teachings call us to be of service to, may we reunify through truth-telling to enable the healing that is needed as we move into a more equitable and just future reconnecting all people and their places. May we care for each other and the land as our mother, and be of service to past, present, and future generations. May our ancestors stand at our backs and provide us with gentle guidance and wisdom as we journey together, with hearts, hands, and minds working as one, to continue to tend to this work in good ways.

References

African Burial Ground. National Monument. Available: http://npshistory.com/publications/afbg/index.htm. [Accessed: 17 August 2023].

Asante, M. K., Welsh-Asante, K., & Asante, K. W. (eds.). (1985). *African culture: The rhythms of unity,* 81. Westport: Praeger.

BBC News (n.d.). *Ghana Veterans and the 1948 Accra Riots.* Available: www.bbc.co.uk/programmes/p01t10s9. [Accessed: 17 August 2023].

Brave Heart, M. Y. H., & DeBruyn, L. M. (1998). 'The American Indian holocaust: Healing historical unresolved grief' in *American Indian and Alaska Native Mental Health Research*, 8(2), 60–82.

Castanha, T. (2015). 'The doctrine of Discovery: The legacy and continuing impact of Christian 'Discovery' on American Indian populations', *American Indian Culture and Research Journal*, 39(3).

Elk, B., & Neihardt, J. G. (2014). *Black Elk Speaks.* Lincoln: University of Nebraska Press.

Fierro, R. S. (2022). *Digging Up the Seeds of white Supremacy.* Philadelphia: Collective Power Media.

Kimmerer, R. W. (2022). *Braiding Sweetgrass.* Vancouver: Langara College.

Malidoma Somé. (2016). *An Introduction into a Dagara-Inspired way of walking on the Earth.* Available: https://malidoma.com/main/. [Accessed: 17 August 2023].

Newcomb, S. T. (1992). 'The evidence of Christian nationalism in Federal Indian law: the doctrine of discovery, Johnson v. McIntosh, and plenary power', *New York University Review of Law & Social Change*, 20(2), 303–342.

32

ALLOWING A CONVERSATION TO GO NOWHERE TO GET SOMEWHERE

Intra-personal Spatial Care and Placemaking

Sally Labern, Sophie Hope and Rebecca Gordon

Preamble

This chapter is about beginning differently, holding space, and wasting time. It acknowledges that placemaking is not always about physical space/place but can be thought of as *intra*-personal spatial care: a space being opened up for relationality in which radical care is practised. Historically, so-called public spaces of engagement have seen disenfranchisement and violence against people based on intersectional identities. Based on the ideas of Karen Barad, intra-personal spatial care works against such spatial violence (Barad, 2006, p. 151). The placemaking process becomes porous and shaped by the entanglement of a community of people. That intra-personal space is a place of resistance: not necessarily a remaking, but a wrestling with, and recognition of, the resistances to spatial violence by those gone before. As Lauren Berlant asserts, it is not about 'being purposive but inhabiting agency differently' (Berlant, 2011, p. 116). It is a humbling process of ally-ship, of lateral solidarity and a place of trust and co-dependency that can become only through time and entanglement.

The main body of this chapter is a conversation between three critical thinkers (the authors): a social-practice artist, a practice-based researcher, and an independent scholar. To prepare, the three looked to *Cruel Optimism* and ideas of 'lateral agency' and 'unheroic agency' (Berlant, 2011). The conversation then took its cues from phrases and keywords on the cards of the game, *Cards on the Table* (COTT) (Bas et al., 2016) (Figure 32.1), created by five cultural workers (including co-author, Sophie Hope) as a tool to critique the process of collaboration (Hope, 2022a). *Cards on the Table* is a literal card game that helps people think and talk critically about a specific project that they are all working on as a group.

DOI: 10.4324/9781003371533-37

FIGURE 32.1   Ania Bas, Sophie Hope, Sian Hunter Dodsworth and Henry Mulhall (2016), *Cards on the Table* in play. Image credit: COTT/Leontien Allemeersch.

The project in this case is a troubling of (trauma-informed) placemaking around the positionality of artists, with the game being played by the co-authors over Zoom. It was recorded and transcribed using AI software Otter.ai, then edited to present curated and collective reflections.

As white, educated women from the Global North, our positionality is that of relative privilege and agency. Our entanglement with each other must not fail to be viewed through this intersectional lens. The structure of the chapter as a dialogical fragment was inspired by the triangular exchange of interdisciplinary scholars in *Ferenczi Dialogues* (Soreanu et al., 2023). It is with a dialogic impetus and collaborative working we come to the troubling of placemaking.

The Conversation

Prompt: 'Allowing Conversations to Not Go Anywhere'

Conversation

Sophie: Our prompt, *Allowing conversations to not go anywhere*. I find this so important. Inevitably, you feel like you need to know where you're heading before you set off. I'm thinking of the *1984 Dinners* project (Hope, 2022b) which, through different iterations, was about bringing people together to reflect on difficult histories. The dinners in Singapore were formed of a group of women theatre practitioners, revisiting that period in 1984; the dinner in Johannesburg was about the anti-apartheid struggle from different factions of the movement. ... So, there's different contexts of trauma, violence or destruction, and how and why people continued to try and change the world and change society. All of that necessitates an ethics of care and responsibility around that, with the fluidity of memory and creating new stories out of current memories of past events.

The *Dinners* have a structure and a menu of questions, and they have a framework. It's not a public conversation. It's curated in that sense. But within the structure you allow for things to go off in different directions. And you have to trust what people bring to that. The whole point is that you set something up for things to go in directions that you were not anticipating. One of the issues that I have is where that goes next, and who continues to work with that material.

Rebecca: Where do those conversations go? How do they continue? ... On a practical level, what happens... is it recorded?

Sophie: It is audio recorded and there's a website with a growing audio archive. I have edited most of them, and thematized some of the extracts. It's not a funded project. Maybe that helps. Sometimes the funding makes you go, actually, 'where is this going?'

Sally: Maybe that's what gives quite an open, forceful drive, the fact that there is no external institutional body that is controlling what these outcomes could be. I love the idea of '*ad hoc* trust', because I think it has to be *ad hoc*. Also, the not knowing where you're going. If I knew where I was going, I wouldn't actually want to go. And the only way you often don't know where you're going is if you have that kind of cultural equity, where you actually have to relinquish some of the thoughts that you might have as you go along about where you might want it to go. You have to drop that power.

Reflection

The positionality of the artist, as bridge between institution and community, enables the opening up of spaces where different experiences, knowledges, and vulnerabilities can be present. For that to be possible, the building of trust is essential. The artist's inevitable position of power in the dynamic is often challenged in the process of opening space for different voices to be aired and relationships of trust to develop. That takes time and intentionality. It may mean introducing co-curated frameworks – rules of the game – to provide safe parameters and expectations along the way for all those involved. It may also mean holding that space while conversations are allowed to go nowhere; where questions, confusions, tensions, creative wanderings are free to unfold. This conceptual holding is an under-appreciated skill gained by social practice artists over time, through interactions, mistakes, and learnings: bringing people together and seeing what emerges.

Prompt: 'You Always Need the Driving Force'

Conversation

Sally: I am not convinced that one can ever hold an untroubled relationship with this prompt, *You always need the driving force*. But being trauma aware, I think We can know a bit more about how to perhaps navigate it. We as artists, are often the driving force with the things that we do or want to make happen. And partly, that's because people's lives are so full up with what they've got to juggle – not that mine isn't as well – but we hold that emotional or purposive space for others, and it feels like it comes with the territory a bit. I think also when slowing something down and moving away from being that project driver and making spaces and different structures for people's voices to be heard in different ways – and for things that we never even thought we were going to hear, catalyses new ideas – then we start to see that a shared driving force is actually creating multiple driving forces. But actually, perhaps the word 'driving' this force, this pushing through, is not what we're wanting to do as artists. It's actually creating a much slower, more relaxed pace. But by doing that, we therefore have to recognise that slow violence that has shaped everything that we do.

Sophie: We use this term 'holding a space' don't we? Maybe there's a slow violence within even that idea of 'holding.' It's interesting in terms of that driving force being quite a military type of language compared to this more psychology-based domestic idea of holding, which comes from child psychology (Winnicott, 1953, 1960). Driving force and holding a space are both loaded in terms of potential trauma.

Rebecca: At a workshop recently, I heard someone – who had worked as a volunteer for the Samaritans – saying that your role as a Samaritan, you don't necessarily have training, you're literally just the walls, a safe space, for that person for the duration of the call. So that brings up for me this picture of containment, of someone out of control, needing to be contained. If that is dealt with, either in a tokenistic way – so, insubstantial, or not with due diligence – or on the flip side, if that space holder ends up being too involved, so much of that chaos is transferred to that person. Is there another way for artists to curate these spaces of sharing, acknowledging that nine times out of ten trauma experiences will be brought to the table. Are there other safe ways to hold back, rather than it all being on a person, namely the artist, in this context?

Sally: Maybe our part to play in that as artists is actually to hold that uncomfortable place I don't find being uncomfortable, uncomfortable, but I know other people would. So that, in effect, this 'holding a space open' means a project can be porous with a community and, therefore, there's more opportunities for things to come in that have never been agreed at the beginning, and begin to take some form and shape that hold a central place... Through an *ad hoc* trust, a mutuality can perhaps throw up some very different ways of doing things together.

Reflection

By intentionally opening up spaces for entanglement – physically, socially, creatively, emotionally – these collective spaces must be demarcated by care. It is a purposeful going nowhere together; a dedicated 'holding space' for slow discovery and reflection; a slow walk towards somewhere else. Or perhaps even allowing for walking in circles. At points it may involve a slippage into chaos: confusion, conflict, turmoil. This takes the holding space to a new level: ringed debating chamber or therapeutic platform that must be grounded by peacemaking. It requires of its indwellers respect, reciprocity, openness, and care. These spaces of care may only be set apart momentarily in the midst of ongoing spatial violence. Commitment to making those spaces is a form of resistance and resilience-building (Chamberlain, 2020). While artists and community makers may be the ones to initiate the opening of those spaces, they must not occupy them but become conduits – witnesses to those stories brought and holders of 'critical hope' (Labern, as practice; In/Visible Fields 2018-), to 'care with' (Tronto, 2013) rather than propose a 'cure of'. There must also be the understanding that each person entering that space, and the experiences they bring, are not a priori isolated entities, but are becoming through this process of intra-action. (Barad, 2006)

Prompt: 'All Parties Have to Feel Their Voices are Being Heard'

Conversation

Sophie: For *Cards on the Table*, for example, all parties have to feel their voices are being heard. Imposing a timing structure on the game is a little tiny step towards equalising – although that is never really entirely possible – the voices in the room around the table. If everyone's got two minutes each to speak, it means that the dominant voices give way to less dominant voices. Although it is awkward keeping the timings. This idea of *feeling* like your voice is being heard, is different to your voice *being* heard. Similarly, having time to speak does not necessarily mean that you feel able to say what you want to say. Again, the politics and ethics of that, in projects I've worked on… there's a violence in putting into the 'aims and objectives' of a project that people must feel their voices are being heard. I mean, that's such bullshit because… what does it mean to actually have your voice heard? That realisation that no-one's listening, there are no pathways for listening, really listening, to devote to different voices.

Rebecca: I was really struck by what you said: what are the pathways to really listening? How often do we in life, let alone in a research project, or in a creative exchange, actually curate those pathways for listening? And that's something we've thought about, isn't it, that deep listening. What is that deep listening? How would you do that? And then I'm reminded of Indigenous storywork (Gordon, 2023).

When you mentioned, Sophie, about what are the expectations – every project has expectations, whether we know them or not. ... There's often ritual that surrounds the telling of stories. And that in itself is a kind of structure, rules of the game, as it were, that allow for, or prompt, that deep listening. And what do we have that is similar to that in a trauma-aware practice, actually?

Sally: I was just thinking about that, whilst working with a particular resident on a social housing estate, and realising that the only way she and I can work together, and therefore we can work with others, is for us to do that over the period when she makes the family meal, because that time is built into her day as 'essential'. And if we can share some of the labour of that meal-making process, we can do that deep listening together, and she can be part of leading the project. Without it, the structure of that listening, so much stress could be caused for her to juggle family and work responsibilities where she says, 'this week I want to meet outside the family, but tomorrow I have to be... ah, we'll have to do it over the making of dinner!' You know, 'can you take direction? Can you take direction over what I need you to do, Sally?' Yes, of course. And it is so gender and class orientated! So, it's also about looking at how funds could be found and made available for that person's time to be paid for so that being involved in a project could ease some of the burden within the family, perhaps also what that does – potentially – to deep listening across the family. And yes, it's structural. So, it's how to manipulate funds? But you're also manipulating a situation, to involve somebody in shaping what is in fact an experiment?!

Reflection

What does it mean for your voice to be heard? What does it mean to really hear someone else's voice? Gender, class, and race are the defining characteristics that allow for projection and reception of voice. Is it possible to create equality of space so that all voices hold the same respect, volume, care, and traction? This is what the horizontal model of mutual aid seeks to action through consensus decision making and unity of purpose (Spade, 2020). Its outworkings however require the communication of clear expectations and structure. The conversation above highlights some of the rules and frameworks that have been designed to govern games like COTT and intra-personal times of working and listening. There are strategies of operation: e.g., working to predetermined units of time; giving everyone the same temporal space; giving everyone the same time to listen. Yet assigning uniform structure is not necessarily equitable. Support must be given based on need: there is no point giving everyone in the group money or assistance with childcare when not everyone has caring responsibilities. The need and nature of support will be specific to each individual. In the context of social practice art and placemaking, flexibility, unknowability, and specificity of need must be written into funding applications, recognising that not everyone's voice is equal, not everyone's time is their own; to value intersectional voices from the project's outset through validatory and equitable remuneration and support, in order to shape and respond throughout, rather than seek such inputs on a tokenistic and unsupported basis.

Prompt: 'We Don't Want It to Be a Waste of Time for People'

Conversation

Sophie: I think more time-wasting is needed. How do we create space for wasting time together? Because obviously, we're in a hyper-productive world in which we have to prove our productivity all the time. Let's subvert it. I don't know. It goes back to those things, about conversations not going anywhere. I think that we need to keep questioning this idea of what it means to hold space, because that is contentious and difficult. But how does that act as a counter to the tyranny of productivity... to the idea of overcoming trauma, of becoming better, of getting over it, so you can contribute to the fricking economy and go get a job! [all laugh] Is there something about the idea of unwrapping that a bit with wasting time?

Sally: ... to trouble the space so that, actually, more people can come in with ideas no-one's thought about. Talking about emotional labour, and that idea again of the 'driving force', if I think it's going to create that thing of the driving force, then actually, we won't do anything together. Because I can't be the driving force; I don't want to be the driving force. So, it's intentionally about trying to create some sort of porous space... allowing people to come in on different terms, without even laying down what the terms are. Something is brewing, coming. You don't know what it's going to be, the shape of it. Time has to be wasted for that to emerge.

Sophie: And in a way it's not about wasting time because... we're just rethinking how to use our time. It's not wasting time preparing dinner or eating food together. I guess it's sort of rethinking 'how do we want to spend time together' and it's just too hard to... keep cracking open those times and spaces for that.

Reflection

In Western neo-liberal society, time is wasted when nothing is produced: when no money is made, no transactions completed, no products produced, no interactions documented. Time continues to be framed by 'short-termism', dominated by 'temporal stresses' (external 'perspective-shortening forces hidden in plain sight') and 'temporal habits' ('internal processes and understandings within our own psychology and that of others') (Fisher, 2023, pp. 3–9). Taking not only a longer-term view, but also a person-centred view (i.e., intra-spatial view), time can be allowed to extend through intentional (perceived) stasis. Stasis could be viewed as inactivity. However, used as a suffix – e.g., haemostasis (stopping a flow of blood) – it means a slowing down. Slowing down, in that sense, is wholly productive, not wasteful. To slow, to catch breath, to be unhurried – all deliberate actions to create space and time. Ultimately to bring life. In the context of social practice art, the slowing down of intra-spatial connections opens up new perspectives and prospects.

Prompt: 'There's Never Such a Thing as an Open Brief'

Conversation

Sophie: There's never such a thing as an open brief in my understanding, it's just whether agendas and expectations are not always spoken or transparent.

Rebecca: Something that troubles me in the conversations that I have with artists, practitioners, and some funders as well, is this issue of having to be specific; if they are commissioning something, or trying to find a brief, there are constraints of having a brief, of having that inflexibility with a funder. I'm coming from thinking about the trauma, say, second-hand trauma, the burnout of social practice artists who have to operate within that funding structure, but how do you safely and with any kind of 'honesty'... how do you interact, and create something together with people when you are handed down this inhospitable structure of... I was gonna say 'being creative', but actually, the structure of having to produce something, to be productive, where you're unable to nurture that *ad hoc* trust that only comes through time and shared experiences with people. That can't be manufactured.

Sally: Being political does help, it means that we can hold more... we can translate more of that uncertainty, or to be frank, the certainty that a funder wants, into the uncertainty needed for cultural equity to be shaped by lived experiences.

Sophie: We can play multiple games at the same time. Which, let's be honest, that's what most people are doing most of the time, operating on different levels... I think we probably share that with other sectors as well, whether it's in social work sectors, or health sectors, care sectors... you're doing a job, but, actually, your mind is elsewhere, or your goal is elsewhere. It's kind of absurd... how you live with that absurdity.

Rebecca: Talking of the absurd, I just finished a wee book by philosopher Tim Morton, *All Art is Ecological* (2021). He talks about an uncanny gap between 'little me' and 'me as a member of a species'. I'll read a bit:

> *I find that I am, and I am not a human insofar as I did, and did not contribute to global warming, depending on what scale you think I'm on. So, these scales don't have a smooth transition point between being one human and being part of a total population of humans, suddenly we find ourselves on one scale or another. It's that paradox again, it seems absurd.*
>
> *(Morton, 2021, p. 24)*

How do we sit in that space? The vagueness and absurdity make it a comic tragedy (ibid., p. 27). In placemaking, the question becomes do I, 'little me', make a difference? And also, who is invited into that place – whose space is it? It seems that there is a crucial role for art and cultural practice. To linger, to slow down, to question. How is the uncanny, tragi-comic, vague space to be held and boundaried? By whom; for whom? And by what means?

Sophie: In this way, this is also an opportunity to think about the practice in a non-celebratory way. That's why I get a bit annoyed about the placemaking label because of it being like, 'aren't we all amazing? This place is great, invest!' rather than actually going, 'we're awful, this place is full of all sorts of experiences'. What we're trying to do through these various practices is to at least try to listen and to share and to give each other an excuse to have some time together to understand something about each other and where we're coming from, where we're going. The violence, the trauma discussion, allows us to wallow in that a bit because it's so risky... I think there's a tyranny of celebratory, impact-type narratives that assume positive, non-violent, non-harmful results. But we can sit with this and go, 'no, life is shit for most people', and how can we not sugarcoat this, but also have some fun. Laugh in the face of all of this nonsense. People would still keep going. We still keep going, if you can, if you want to, because you often have to, but also, because you want to. It's not just 'where is this going?' as we talked about, but 'how?' I guess that's a big question, isn't it? How do you keep going?

Sally: It's not comfortable to make those decisions. But in a way we do... the artist may be the one to do the absurd thing, performing dissention through actions and the making of objects without permission. Something, somehow, has to be done, in that moment, to join that body of resistance, one of a radical imagination (Haiven & Khasnabish, 2014). It isn't either about attracting attention to these 'acts in place' in the way of it being *pioneering*. No, it's more a recognition of a body of work that we all hold. When we're working in this way, as artists, activists and researchers, we recognise that it's an essential resistance to collectively received traumas. These traumas are the many spatial injustices that are experienced unevenly across human lives, across an earth that is broken by human induced violence, as fostered by the capitalist system and the multiple traumas of extraction, appropriation and accumulation that shape our public places.

Reflection

Through experience and creativity, holding ambiguity and uncertainty, artists can offer new possibilities and methods for testing and collaborating across the ecosystem of funders, organisations, communities and practitioners. Offering spaces and relationships of trust, denouncing emphases on deliverables, new resistances can join past resistances through making space for the absurd – the tensions of humour and sadness, agency and voicelessness. In order for new, valuable voices to be heard, those holding the public purse must understand the value of space to go nowhere, of building relationships, to 'laugh in the face of this nonsense'. These traumas may never heal or disappear completely. But a collective response, in parallel to the quiet witnessing of individuals' traumas, recognising the value of the individual as empathetic 'witness' in survival and healing, brings the hope of making new narratives. It means allowing new parts of ourselves to grow into the socialised world as we think about self, place, and time differently, whereby this continual shuffle can be radically relational. It opens up richer possibilities for radical creativity to come out of catastrophe (Soreanu et al., 2023, p. 145). Ferenczi suggests that fragments of trauma and catastrophe can be seen as parts of a 'scene', a mosaic of experiences and therefore places, whereby the trauma of the fragmented self, the catastrophe of fragmented societies, can grow new healthier parts that can feel, try, hesitate, play, resist, survive, experiment anew (Soreanu et al., 2023). This enfolds what Haraway talks about as 'tentacular knowledge' (Haraway, 2016, p. 32). These new parts can grow out of fissures, fractures and splits, offering the potential for new, almost visual parts of ourselves, to shape emergent new physical/non-physical places.

Conclusion

By creating a space of conversation, interacting collaboratively in co-authorship, this chapter has sought to trouble, then re-trouble, the notion of trauma-aware/-informed placemaking and the role of the artist in that process. It proposes the need for intentional spaces to be opened up and held, by artists and appropriate funding, in which time, conversations, relationships, have the freedom to go nowhere. Yet that seeming nowhere is a path to somewhere. The format of a conversation led by prompts from the *Cards on the Table* card game has enabled us to take a route through our experiences. The interspersed reflections trigger more questions, feelings and references. As with the practices we are learning from, the conversation itself is a space to hold unknowns, and this conversation is already leading to many more. The dialogue may be cyclical, nonlinear, unmeasurable. But this is where social art practice acts as a process of intra-personal spatial care that holds the radical plasticity of love and critical hope that we need.

Sally Labern, Sophie Hope and Rebecca Gordon are members of Social Art Network UK (SAN), an open network that supports artists' publishing, symposia, alliances and new tools to support playful radical imaginaries. Before, during and beyond COVID the authors' contributions to SAN's collective resources committed to furthering cultural democracy, social and spatial justices through collaborative and intersectional artist-led practices co-produced equitably with communities. References below come out of this critical location to multiple forms of social arts practice and its capacious and lucid challenges.

References

Barad, K. (2006). 'Agential Realism: How Material-Discursive Practices Matter', *Meeting the Universe Halfway: Quantum Physics and the Entanglement of Matter and Meaning.* Durham: Duke University Press.

Bas, A., Hope, S., Hunter Dodsworth, S., Mallet, S., & Mulhall, H. (2016). *Cards On the Table.* Available: www.cardsonthetable.org/copy-of-about. [Accessed: 25 August 2023].

Berlant, L. (2011). *Cruel Optimism.* Durham: Duke University Press.

Chamberlain, L. (2020). 'From Self-care to Collective Care: Institutionalising self-care to build organisational resilience and advance sustainable human rights work', *International Journal on Human Rights,* 17(30).

Fisher, R. (2023). *The Long View: Why We Need to Transform How the World Sees Time.* London: Wildfire.

Gordon, R. (2023). 'Indigenous Storywork as an Ethical Guide for Caring with Social Practice Artists', *Prioritizing People in Ethical Decision-Making and Caring for Cultural Heritage Collections,* Nina Owczarek (ed.). London: Routledge.

Haiven, M. & Khasnabish, A. (2014). *The Radical Imagination: Social movement research in the age of austerity.* London: Zed Books.

Haraway, D. J. (2016). *Staying with the Trouble: Making Kin in the Chthulucene.* London: Duke University Press.

Hope, S. (2022a). 'Affective experiments: card games, blind dates and dinner parties', *Affective Experimentation Anthology,* Britta Timm Knudsen, Mads Krogh & Carsten Stage (eds.). London: Palgrave Macmillan.

Hope, S. (2022b). 'We thought we were going to change the world! Socially engaged art as cruel optimism', *The Failures of Public Art and Participation,* Cameron Cartiere and Anthony Schrag (ed.). Milton Park: Taylor & Francis.

Morton, T. (2021). *All Art is Ecological.* London: Penguin.

Otter.ai. Available: https://otter.ai/. [Accessed: 31 August 2023].

Soreanu, R., Staberg, J. & Willner, J. (2023). *Ferenczi Dialogues: On Trauma and Catastrophe.* Leuven: Leuven University Press.

Spade, D. (2020). *Mutual Aid: Building Solidarity During This Crisis (and the next).* London: Verso.

Tronto, J. C. (2013). *Caring Democracy: Markets, Equality, and Justice.* New York: New York University Press.

Winnicott, D. (1953). 'Transitional objects and transitional phenomena: A study of the first not-me possession', *International Journal of Psycho-Analysis,* 34.

Winnicott, D. (1960). 'The theory of the parent-child relationship', *International Journal of Psycho-Analysis,* 41.

Closing Remarks
Being Accountable as Placemakers

PLACEMAKING AND THE MANIPUR CONFLICT

Urmi Buragohain

Introduction to the PlaceMaking Foundation and Myself

I am Urmi Buragohain, a placemaking practitioner from the state of Assam in Northeast India, with a background in architecture and urban planning. After completing my postgraduate studies in India, I, along with my husband, migrated to Australia in 2003. Having lived and worked for almost two decades overseas, we decided to return to our roots.

With the intent to continue the newfound zeal for placemaking that I discovered while working in Melbourne, I registered my own non-profit social enterprise, PlaceMaking Foundation, in Imphal, Manipur, in 2018. My intent was to apply design thinking to come up with alternative answers to the question, 'How can we create a society in which everyone has a chance to live a healthier and more dignified life?' I sought to innovate through the eyes of the end user, resulting in deeper insights about their unmet needs.

Over the past two years, PlaceMaking Foundation has been working with a government agency in Manipur, providing expert advisory services and undertaking pilot placemaking initiatives in the capital city. PlaceMaking Foundation is also spearheading the placemaking movement in Northeast India, with the formal launch of the regional hub earlier this year, and fostering collaborations amongst creative placemaking practitioners, thought-leaders, and change-makers.

The Manipur Context

Many of you may not have heard of Manipur. It's a hilly north-east Indian state that sits east of Bangladesh and borders Myanmar and consists of a valley surrounded by mountain ranges. It is home to an estimated 3.2 million people belonging to 39

DOI: 10.4324/9781003371533-39

ethnic communities following different faiths, including Hinduism, Christianity, and Islam, as well as Indigenous religious traditions. Manipur's context can be understood in the geo-political context of the North Eastern Region (NER) of India which includes eight states other than Manipur – Arunachal Pradesh, Assam, Meghalaya, Sikkim, Tripura, Mizoram, and Nagaland. The NER has the dubious reputation for including some of Asia's most militarised and politically volatile societies. It also has the highest number of indigenous peoples in India, characterised by self-determination movements that have taken the form of armed struggle against the Indian state (Kakati, 2021). Consequent counter-insurgency measures have included the imposition of broad-based powers for the military and paramilitary groups, sometimes leading to the abuse of power. On top of this, the ongoing instability in the neighbouring country of Myanmar spills over to Manipur, which shares a porous 400 km boundary with the country. As a result of the longstanding tumultuous situation, the state has become the site of rampant gun-running and narco- and human-trafficking (Nepram & Schuchert, 2023).

In recent times, Manipur has been in the news because of an ethnic conflict that started in May 2023. To date, over 180 people have been killed and 400 wounded in violence. More than 60,000 have been forced from their homes as the army, paramilitary forces, and police struggle to quell violence. While the violence in Manipur is some of the worst witnessed in the state in decades, it is not an unfamiliar occurrence in India's North East, where the identities of different ethnic communities have repeatedly been weaponised to serve the interests of a powerful few who control the guns, drugs, and politics (ibid.).

My partner (who is from Manipur) and I are children of conflict. Born in the mid-'70s and growing up in the '80s and '90s meant we were no stranger to conflict – bombings, shootings, kidnappings, curfews, black-outs, violent agitations, you name it. Although not directly exposed to a catastrophic event, we were indirectly exposed to traumatic experiences almost incessantly – adult conversations, graphic news items in print and visual media, heavy military presence in public places, untimely death of people we knew to drugs and violence, and being frisked everywhere we went. 'I never realised how I had normalised the presence of guns in day-to-day life until I thought about it while I was in Delhi after the conflict broke out', said a friend. But this exposure to traumatic events is not new. It's just that our previous generations learnt to suppress their memories and feelings as a coping strategy, and that code of silence has been passed down to us.

We are considered to be the lucky ones who got the opportunity to move out from the region and subsequently overseas, seeking a better life. During our journeys, we found ourselves revelling in how 'safe' places are and the freedom it afforded us, both in abstract and physical terms. When we decided to move back to Manipur and North East India a few years back, the region was going through a period of relative stability that started around the mid-2000s. We, along with many others, were naïvely optimistic and hopeful of doing something meaningful on our return. Despite the pandemic, things had started looking up until the fateful night on 3 May

2023. When I mentioned to a friend that the sound of gunshots, tear-gas shelling and military aircraft passing overhead was a new experience for me, he responded with a wry smile, 'welcome to Manipur, which is now a member of a select group of conflict-affected regions.'

My Questions to Placemaking, *vis-a-vis* the Manipur Context

The barbarity and senselessness of the events that ensued since the fateful day has deeply shaken my belief in humanity. What is most appalling is that it's the women and children belonging to poorer sections of society who are caught in the crossfire, irrespective of which community they belong to. Perhaps being the mother of an autistic child has made me more sensitive to the psychological damage that is being caused to the minds of a young generation that has never experienced anything like this before.

The role of women has been prominent in this crisis, albeit a conflicting one as reported by numerous media outlets. They have been both the protectors and instigators of violence. Where on one hand they are putting up a formidable resistance to state forces, they are also being subjected to sexual violence. The sickening event of stripping and parading of two women belonging to a particular community by a mob has grabbed headlines recently, sparking an outrage that reverberated across the globe. More worryingly, cultural barriers to expressing trauma – such as shame, family honour, stigma – particularly prevalent amongst women could potentially be hiding many more such heinous crimes.

The impact on children belonging to displaced families also presents a disturbing scenario. According to government data, more than 14,545 displaced students in five districts worst affected in the ongoing ethnic clashes have taken shelter in relief camps. More than 3000 are below five years of age (Zaman, 2023). With many schools converted to relief camps, classes have resumed in a staggered way with batches of children coming in turns every few days, since rooms are housing displaced people. With patchy counselling services being provided by the state, the teachers have the added task of addressing their trauma. Faced with an uncertain future, children are at a heightened risk of slipping into depression and/or being drawn into the cycle of violence and drug abuse. Already, reports are filtering out as to how children are being 'recruited' and 'trained' in using guns to defend their communities.

In response to the situation, my instinct is to jump right in and try to help with relief work that many individuals and organisations are carrying out. But I started asking myself – what happens after the immediate physical needs are met? What about the hidden psychological needs? I started delving deeper into understanding the underlying causes that led to the current situation. What I unearthed, and am still processing, is a complex picture that continues to evolve with each passing day. It involves intergenerational trauma and trust deficit, geo-political issues, historical inequity, socio-economic disparity, human rights violations, flawed policies, polarising ideologies, mass displacement of people, systemic failure of state machinery and institutions and narco-arms-human trafficking amongst others. The problems seem so big that solutions at the moment are impossible to comprehend. The 'enormity problem', as coined by *ABC News* correspondent Bill Blakemore, is leading to feelings of hopelessness and despair where a future horizon does not seem to exist (Mollica, 2021).

But standing by and doing nothing is also not an option. If placemaking is about bringing people together and getting them to talk to each other, surely our duty-of-care extends to finding enabling pathways for traumatised people to heal and one day get talking to each other again? In my quest for an answer, I started researching and reaching out to people in my placemaking network. This led me to discover many principles, concepts and approaches that have evolved in recent decades that operate at the nexus between placemaking, human psychology, mental health and trauma. Some of these ideas include trauma-informed design/placemaking, ethical placemaking, urban trauma, traumascapes, therapeutic landscapes, psychosocial crisis, to name a few. A common thread running through all these is the acknowledgement that the knowledge and practices of these disciplines need to be imbued with a deeper understanding of the effect of trauma on cities and their contingent realities. This further crystallised three questions in my mind:

1. WHEN is the right time for a placemaking intervention post a conflict situation, considering prolonged stress caused by trauma could potentially turn toxic? What conditions predetermine the critical moment when intervention is appropriate?
2. HOW do we repair places after physical trauma in a way that also profoundly heals its inhabitants and helps them reconcile with their environment which they perceive as hostile?
3. WHAT placemaking practices could help younger generations and women exposed to trauma in overcoming the mental barriers and get them on the path towards healing from the psychological damage caused by these events?

What Do These Questions Mean for Placemaking?

It's been almost nine months since the crisis started. The incidents of violence are still playing out in terms of sporadic gunfights, killings, arson, looting, protests in areas where the real and imagined territories intersect. Entire landscapes have now experienced place-associated trauma that would take a long time to heal – and perhaps may even outlive this generation. The law-and-order situation is still volatile and the poorly resourced state forces are struggling to contain the situation. The centralised approach to relief and response measures is inadequate and gets increasingly patchy as you move away from the capital city. Children in relief camps in the valley are getting just two meals comprising of rice and dal (lentil soup) and going hungry to schools in the morning – but it seems they are the fortunate ones.

Everyone I spoke to expressed hopelessness, frustration and anger at a situation not of their own making. Everyone wants accountability and leadership at this time of reckoning, but there seems to be no political will to find a speedy solution to the crisis. In the guise of security measures, democratic rights are being insidiously eroded like the free movement of people, internet access, and free speech.

In the current situation, where the affected people are struggling to meet their basic needs, mental health doesn't seem to be a priority, which in turn means the need for a psychological healing process has taken a back seat. Although experts would advise on place-based healing as a way forward, there are multiple barriers to be overcome on the ground to make this happen, including lack of awareness, lack of access, and cultural barriers. Such an approach would also need to be carefully phased over a period of time to be effective, as rushing to superficial recovery instead of healing deeply may leave people and places more vulnerable to future crises (Pitter, 2021). Subsequently, a responsive trauma-based placemaking approach that continues to show concerned attention over time as conditions evolve (Eckenwiler, 2016) could potentially address historic inequities and be an enabler for equitable participation in the processes for shaping places. This approach would require a concerted effort by practitioners from diverse areas of expertise, including mental health, humanitarian aid, public policy, community work, placemaking, architecture, urban planning, etc. But such a holistic approach to placemaking is still at a nascent stage in terms of practical, on-ground application and would likely require a sophisticated level of coordination within an established institutional setup. This seems like a big challenge for a developing country such as India, and particularly in an unstable region such as North East India.

In conclusion, although my newfound knowledge of this area of placemaking has given me some insights as to what the ideal way forward might be in the context of Manipur, I still feel there is a significant gap in knowledge in terms of practical measures that could be undertaken on the ground. So, I take this opportunity to call out to the global placemaking community to give particular attention to furthering this evolving practice of trauma-based placemaking and help develop a roadmap for its practical application in a world that's experiencing humanitarian crises at an unprecedented level.

References

Eckenwiler, L. (2016), 'Defining Ethical Placemaking for Place-Based Interventions', *Am J Public Health*, November 2016; 106(11): 1944–46.

Kakati, B. (2021). 'Conflict and development in Northeast India: Stories from Assam', *Transnational Institute* [online]. 16 May 2021. Available: www.tni.org/en/article/confl ict-and-development-in-northeast-india. [Accessed: 26 August 2023].

Mollica, R. F. (2021). 'Moving beyond the enormity problem: Tackling the global refugee crisis', *Psychiatric Times* [online], 38(12). Available: www.psychiatrictimes.com/view/moving-beyond-the-enormity-problem-tackling-the-global-refugee-crisis. [Accessed: 26 August 2023].

Nepram, B. & Schuchert, B. W. (2023). 'Understanding India's Manipur Conflict and Its Geopolitical Implications: Violent ethnic clashes in northeast India have a long local history – but the effects could reverberate across South Asia', United States Institute of Peace [online], 2 June 2023. Available: www.usip.org/publications/2023/06/unders tanding-indias-manipur-conflict-and-its-geopolitical-implications. [Accessed: 26 August 2023].

Pitter, J. (2021). 'We need to heal traumatized urban landscapes–and people–after COVID', Policy Options Politiques [online], 9 August 2021. Available: https://policyoptions.irpp. org/magazines/august-2021/we-need-to-heal-traumatized-urban-landscapes-and-peo ple-after-covid/. [Accessed: 26 August 2023].

Zaman, R. (2023). '"One generation is going to lose out": Manipur strife is taking a toll on children's education', Scroll.in [online], 15 August 2023. Available: https://amp.scroll. in/article/1054072/one-generation-is-going-to-lose-out-manipur-strife-is-taking-a-toll-on-childrens-education. [Accessed: 26 August 2023].

INDEX

Note: Locators in *italics* refer to figures and those in **bold** to tables. Author names are indexed using initials only for forenames; artists and historical figures are indexed using full names.

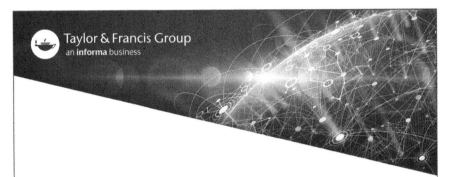

Printed in the United States
by Baker & Taylor Publisher Services

Printed in the United States
by Baker & Taylor Publisher Services